International Automobile Scientific Forum (IASF-2017)

Intelligent Transport Systems

IOP Conference Series: Materials Science and Engineering
Volume 315

Moscow, Russia
18 – 19 October 2017

ISBN: 978-1-5108-5891-6
ISSN: 1757-8981

Printed from e-media with permission by:

Curran Associates, Inc.
57 Morehouse Lane
Red Hook, NY 12571

proceedings.com

Some format issues inherent in the e-media version may also appear in this print version.

Printed by Curran Associates, Inc. (2018)

For permission requests, please contact the Institute of Physics
at the address below.

Institute of Physics
Dirac House, Temple Back
Bristol BS1 6BE UK

Phone: 44 1 17 929 7481
Fax: 44 1 17 920 0979

techtracking@iop.org

Additional copies of this publication are available from:

Curran Associates, Inc.
57 Morehouse Lane
Red Hook, NY 12571 USA
Phone: 845-758-0400
Fax: 845-758-2633
Email: curran@proceedings.com
Web: www.proceedings.com

Table of contents

Volume 315

International Automobile Scientific Forum (IASF-2017) Intelligent Transport Systems

18–19 October 2017, Moscow, Russian Federation

Accepted papers received: 7 February 2018
Published online: 22 February 2018

Preface

Papers

IOP Conf. Series: Materials Science and Engineering **315** (2018) 011001 doi:10.1088/1757-899X/315/1/011001

International Automobile Scientific Forum (IASF-2017) "Intelligent Transport Systems"

Prospects of the creation, production and operation of autonomous (driverless) vehicles and related infrastructure became the main topic of IASF-2017, which took place at FSUE "NAMI". More than 350 people participated in the Forum, including representatives of the Ministry of Industry and Trade of the Russian Federation, Ministry of Transport of the Russian Federation, KAMAZ PTC, PJSC AVTOVAZ, GAZ GROUP, GLONASS NP, Yandex, FEV, VOLVO, Bosch, AVL, National Instruments, Cognitive Technologies, Mechanical Engineering Research Institute of the Russian Academy of Sciences, Joint Institute for High Temperatures of the Russian Academy of Sciences, Joint Institute of Mechanical Engineering of the National Academy of Sciences of Belarus, Yandex, Bauman Moscow State Technical University, Moscow Automobile and Road Construction State Technical University (MADI), Moscow Polytechnic University, South Ural State University and other institutions. Among the participants, there were young scientists and engineers developing technologies in the field of intelligent transport systems (ITS).

Within IASF-2017, ITS development trends, creation and improvement of regulatory and legal framework in this field, development of processing equipment and ITS testing procedures, risks related to the implementation of such systems in everyday life were discussed.

Alexander Morozov, Deputy Minister of Industry and Trade of the Russian Federation, has invited the audience to participate in the activities of AUTONET National Technological Initiative, which supports projects aimed at development of the related components, programs and systems.

There was also an exhibition of R&D achievements in the field of autonomous vehicles and their components organized within the Forum. FSUE "NAMI" presented some projects, which are already well known to the public – SHATL driverless electric bus and electric driverless vehicle – as well as electronic control systems and vehicle components. KAMAZ together with FSUE "NAMI" presented the "Driverless KAMAZ" Project. GAZ GROUP presented a driverless bus based on a light electric platform and a new-generation frame bus, which completely meets the requirements of the "Accessible Environment" program. MADI showed the "Open Automobile Platform for Autonomous Vehicle Technology Testing" project.

On the second day of the Forum, October 19, a joint session of young professionals, students and postgraduates' section and "Vehicle Intelligent Systems" section was held.

All postgraduates and students who presented their papers within the session were granted IASF-2017 Certificates of Participation.

IOP Conf. Series: Materials Science and Engineering **315** (2018) 011002 doi:10.1088/1757-899X/315/1/011002

Peer review statement

All papers published in this volume of *IOP Conference Series: Materials Science and Engineering* have been peer reviewed through processes administered by the proceedings Editors. Reviews were conducted by expert referees to the professional and scientific standards expected of a proceedings journal published by IOP Publishing.

IOP Conf. Series: Materials Science and Engineering **315** (2018) 012001 doi:10.1088/1757-899X/315/1/012001

Prospects of development of land driverless trucks

S Bakhmutov[1], A Saykin[2], D Endachev[3], V Evgrafov[4], A Shagurin[5], I Kulikov[6] and K Fedoseev[7]

[1] Deputy CEO for Science, FSUE "NAMI", Moscow, Russia
[2] Special Vehicles Center, FSUE "NAMI", Moscow, Russia
[3] Information and Intelligent Systems Center, FSUE "NAMI", Moscow, Russia
[4] Intelligent Systems, Information and Intelligent Systems Center, FSUE "NAMI", Moscow, Russia
[5] Electronics, Information and Intelligent Systems Center, FSUE "NAMI", Moscow, Russia
[6] Power Units Center, FSUE "NAMI", Moscow, Russia
[7] Design Department, Style Center, FSUE "NAMI", Moscow, Russia

E-mail: s.bakhmutov@nami.ru

Abstract. The paper is dedicated to the problem of creation and efficient use of driverless trucks. It also contains general analysis of foreign experience and development trends. The concept of FSUE "NAMI" for development of a platoon of driverless trucks for cargo transportation between specialized terminals on public roads is provided. A road train with virtual couplings where the lead vehicle has a driver and the rest are driverless is proposed as a base for the driverless platoon. The platoon is formed by such road trains, which ensures extended functional features, including the possibility to form platoons of road trains from different logistics centers and with different destinations, quick reforming of the platoon in the logistics centers, possibility to split the platoon when driving on public roads, higher precision of control over the platoon during driving, etc. It is also shown that in order to improve the efficiency of cargo transportation, driverless cargo platforms (trucks) shall be designed without a cabin. The paper also considers the ways to reduce the price of driverless trucks by creating a special road infrastructure, which will allow transfer of the most expensive subsystems of driverless vehicles to such infrastructure. Upon that, the following subsystems will remain within the driverless chassis: navigation, communication, chassis control and ADAS. It is noted that creation of such infrastructure will require high investments, therefore, gradual implementation of such systems is seen as rational. It is shown that the most appropriate area for these purposes is transregional cargo transportation. It is noted that the economic reasons for implementation of the proposed concept are as follows: significant reduction of the number of drivers; significant simplification of the driverless platform control system and reduction of the price of subsystems; no need for the driver's cabin within the driverless platform; fuel economy when driving within a platoon. Power unit options for driverless cargo platforms are investigated and the reasons for selection of a particular power unit are given.

1. Introduction

Creation of driverless vehicles (DLV) and driverless vehicle control systems (DLVCS) is a priority task in global automotive industry. Such types of vehicles can be rather various and different. Foreign automotive manufacturers are developing DLVs including passenger cars, buses, in-plant or in-factory vehicles, trucks, farm vehicles, special-purpose vehicles, etc. The most intensive development activities are underway in the USA, Japan, Germany and Sweden in such corporations as General Motors, Ford, Toyota, Volkswagen, Audi, BMW, Volvo, etc. According to experts' forecasts, DLV integration would enhance efficiency, decrease cargo-and-passenger transportation costs, ensure traffic safety, reduce a quantity of motor vehicle collisions (MVC), cut down on operational fuel consumption, reduce travel time and minimize emissions of hazardous substances and greenhouse gases into the atmosphere.

Currently, the level of development of foreign DLVs ensures their automatic driving without human intervention on the highways with a good road infrastructure, road marking and without intersections. The DLV is controlled by the driver when driving within the city with intersections and traffic lights and also when driving on roads in winter under icy and snowy conditions. The main type of information generated in the DLVCS is a surrounding situation and vehicle positioning data. Depending on configurations of computer vision systems, the digital, infra-red, laser and ultrasonic scanning of the territory and barriers is implemented in the DLVCS. The data received from computer vision system sensors are holistically analyzed by their priority. Following the analysis results, the control unit transmits commands to actuators, which perform autonomous driving of the DLV according to the developed algorithms and software.

It should be noted that activities related to driverless trucks are especially intensive overseas. Using driverless trucks within the truck platoon (truck convoy) is seen as promising and in prospect is expected to have the following results:

- road safety improvement, as this would lead to minimization of the negative impact of human factors being statistically the main reason of almost 80% of MVC;
- fuel saving up to 20%;
- increase in transportation efficiency by 1.3-1.4 times;
- provision of comfortable working conditions for drivers inside the driven trucks;
- minimization of harmful effects on the maintenance staff and environment;
- reduced need in a large pool of professional drivers with high remuneration;
- possibility to integrate driverless transport systems in technological processes of enterprises including, first of all, large transport and logistic centers, ports, etc., providing their continuous and round-the-clock functioning.

The examples of driverless trucks being developed by the leading global companies are presented in figure 1.

The Volvo driverless truck platoon control system - Safe Road Trains for the Environment (SARTRE) - makes it possible for several vehicles to move on the road within the organized platoon (convoy) - figure 1i [5]. The vehicles follow the lead vehicle which is the truck driven by a professional driver. The vehicles line up to keep the distance of 6 m from each other and fully repeat the movement of the lead truck. In the course of movement, each vehicle can leave a convoy at any time. Volvo's best practices and deliverables in the field of active safety systems, including adaptive cruise control, are used for the SARTRE system creation. At present, such convoy is at the stage of operational tests.

The principles of development of driverless trucks are practically the same as that of passenger DLVs [8-14]. For fuel-saving purposes, the Japanese state organization - New Energy and Industrial Technology Development Organization (NEDO) - developed the platoon of four trucks, which move at the distance of 4 m from each other. When driving in the platoon, aerodynamic air resistance for the following vehicles decreases, thus, fuel consumption by driverless trucks is reduced - figure 1g.

The problem of truck performance and capacity increase is of high-priority. In general, development of driverless trucks is conducted based on modern classical designs of trucks. At the same time, with creation of perspective driverless trucks, the focus on classical configuration solutions cannot

IOP Conf. Series: Materials Science and Engineering **315** (2018) 012001 doi:10.1088/1757-899X/315/1/012001

significantly increase labour efficiency in the transport sector (by 1.34 – 1.54 times as specified in "Transport Strategy of the Russian Federation until 2030").

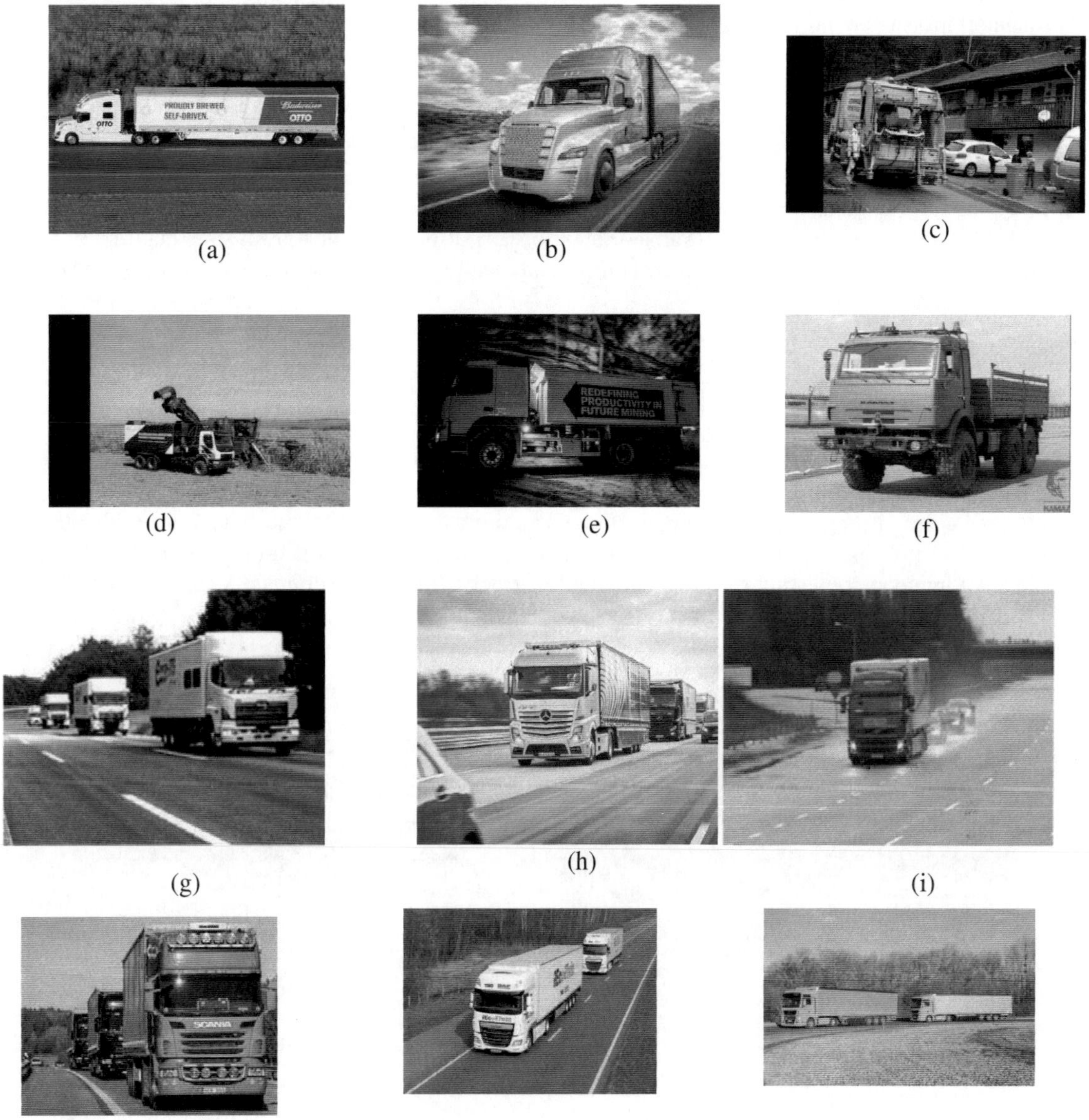

(a) (b) (c)

(d) (e) (f)

(g) (h) (i)

(j) (k) (i)

Figure 1. Driverless trucks of the leading global manufacturers: a) Otto driverless truck; b) Inspiration driverless truck (Inspiration Truck) by Freightliner; c) Volvo-Renova driverless garbage truck; d) BMW DLV with Connected Drive Connect; e) Volvo FMX underground driverless truck; e) driverless truck based on KAMAZ-5350; g) Japanese driverless trucks; h) Mercedes-Benz Actros semi-autonomous vehicle platoon/convoy; i) Volvo DLV platoon/convoy (with Safe Road Trains for the Environment); j) Scania driverless truck platoon; k) DAF driverless truck platoon; l) MAN driverless truck platoon.

2. Concept of creation of driverless truck platoon

Prospects of logistic transportations are based on transport and infrastructure technologies development. At present, foreign and domestic transport enterprises introduce modern logistic technologies of cargo transportation and handling: telecommunication systems for cargo transportation maintenance, terminal systems for cargo transportation, door-to-door (house-house) transportations, etc.

According to various estimates [15], the transportation costs amount to 20 - 70% of total logistic expenses, at that, the share of transport costs in the price of goods is different and depends on the product type: 2-3% - for electronics, 5-6% - for food, 7-12% - for vehicles and equipment, 40-60% - for raw products, 80-85% for mineral and construction products. According to experts' calculations, integration of information systems could increase the average vehicle speed by 10-30 km/h.

From the perspective of cargo transportation organization, there is a promising step to combine the DLV and truck platoon concepts. As a result, we are dealing with the two types of transport systems and infrastructure for them as follows:

- "road train" with a virtual automatic coupling;
- truck platoon (incl. virtual links);
- intelligent logistic center.

The main difference between the road train with virtual automatic couplings and the truck platoon with virtual links consists in the number of vehicles included. It is proposed to introduce the following quantitative indicators: for the road train with virtual couplings – from 2 to 5 vehicles, for the truck platoon – 2 and more road trains with virtual couplings. At present, the road train with only three mechanical links is already a vehicle which is complicated in control and development (Fig. 2b), and the truck platoon consisting of a large number of trucks (Fig. 2a) requires a large number of drivers, given the fact that they need to substitute each other in the course of driving.

(a) (b)

Figure 2. Modern truck platoon and road train: a) truck platoon; b) three-link road train with mechanical couplings.

In order to ensure modularity and unification, it is suggested that a perspective truck platoon must consist of several road trains with virtual links (virtual automatic couplings) inside of and between them. At that, there is a possibility of platoon uncoupling in the following cases:

- driving across intersections, railroad crossing (road trains (trucks) must drive along one after another);
- maneuvering (also when driving in the city and among other vehicles).

Thus, the road train with virtual couplings is a key item and module of the driverless truck platoon. Accordingly, a different amount of driverless road trains with virtual couplings may be used to form the driverless truck platoon. They may be taken from various logistic centers and intended for different destinations. Truck platoon formation is necessary to organize a joint movement of driverless road trains along highway sections that are comparably long and common for all vehicles.

Let's consider the composition and purpose of such a road train. It consists of several DLVs with virtual automatic couplings, at that:

- the lead vehicle has a driver;
- the driven vehicles are driverless (following behind the lead vehicle, maneuvering under tight space conditions, turning movements/U-turns by separate driverless vehicles, etc.);
- driverless truck platforms are to ensure two main driving modes: automatic driving within the road train and maneuvering, which is controlled by operator.

Reformation of logistic centers in terms of their role and purpose (which will make it possible to exercise effective management of transportations) is required for new categories of vehicles.

- driverless road train and truck platoon formation / reformation;
- vehicle maintenance;
- change of drivers and operators of lead vehicles.

For ensuring the effective formation and reformation of driverless road trains and truck platoons, separate logistic centers must exchange information that will make it possible to focus on general cargo traffics by accepting and sending the driverless vehicles. It is proposed to change the roles of professional drivers. The road train and truck platoon are controlled by a driver and an operator whose task is to monitor correct functioning of all systems providing autonomous driving.

Besides, it is necessary to change the nature of activities fulfilled by drivers and operators accompanying them, ensuring fulfilment of functions in the local territory. During the working shift, they move a truck platoon from one point to another at a distance of several hundred kilometers. After that, the driver and operator are replaced by other colleagues who continue to move the truck platoon further up. The previous ones get a possibility to have a rest and start their way back with another truck platoon. On return, they finish their shift and have a possibility to have a rest at home before picking up a shift again.

Thus, the truck platoon consisting of driverless vehicles (Fig. 3) is designed for cargo transportation between specialized terminals, at that, the driverless vehicles must be driven on public roads. The main requirement defining the driverless vehicle image is that the driverless vehicle platoon must be a part of the traffic, i.e. move along depending on other vehicles movement, traffic lights, road signs, etc.

Figure 3. NAMI's prototype road train with the driver-operated lead truck and driverless links with virtual couplings.

The truck movement organization systems built that way will ensure the movement of driverless truck platoons and road trains with virtual automatic couplings. Virtual couplings between the driverless trucks will ensure movement of a large number of vehicles, lane-changing, turning movements, U-turns, driving across intersections, driving together with other vehicles on the road, setting and fulfillment of various tasks related to cargo transportation. The driverless vehicles integrated in one platoon can follow the lead vehicle, and at the same time, they have capabilities for some types of autonomous driving: change the distance when necessary, pass a sharp turn, U-turn, get to the unloading / loading point, etc. All of these will allow to reduce the transportation costs and increase efficiency thereof.

When developing perspective driverless trucks, it is reasonable to consider the design of the driven driverless truck without a driver's cabin, with the traction module and motion control system (MCS) arranged within chassis dimensions. In such configuration, the length of the cargo platform of the driverless truck can be extended almost to the front border of the frame due to the driver's cabin absence. The maximum extension of the cargo platform of the driverless truck is also possible when using the electric drive. Implementation of the electric driverless truck design without a driver's cabin makes it possible to extend the cargo platform on average by 20 - 30%.

Today, the price of the DLV control system can reach up to $100,000, though the system reliability is far from perfect. Striving for higher reliability will cause further increase in price, which is one of the constraints for mass introduction of DLVs. Increasing the DLV production volumes will allow significant price decrease, but will not solve the problem. This constraint can be overcome only by creating a specialized road infrastructure, which shall ensure the following functions of DLV in minimum configuration:

- DLV identification (information subsystem);
- road situation monitoring (computer vision subsystem);
- real-time transmission of data on road situation to DLV (communication subsystem).

Introduction of such road infrastructure will allow to transfer the most expensive DLV subsystems to the infrastructure, while the driverless chassis will feature only the following functions:

- Navigation and orientation / positioning (navigation subsystem);
- Acquisition of information on the situation from the road infrastructure (communication subsystem);
- Calculation of the trajectory and routing based on the information received from the infrastructure (chassis control subsystem);
- Collision prevention (ADAS).

Such approach will allow reduction of the price of the DLV control system down to $2,000-$3,000 (even before mass implementation), while the stationary system reliability will be higher than that of a moving object.

It is obvious that creation of such infrastructure will require high investments and long time for debugging and refinement; therefore, it is reasonable to introduce and refine such technologies gradually. The most appropriate area for these purposes is transregional cargo transportation. First, the price of the control system will be minor compared to the price of the truck, and second, such automation will have a significant economic effect due to the transportation cost reduction.

The economic reasons for introduction of the proposed concept are as follows:

- significant reduction of the number of "accompanying" drivers;
- transfer of the main infrastructure subsystems (computer vision, operating (control) center, communication system) to the lead truck vehicle;
- significant simplification of the driverless platform control system and reduction of the price of subsystems (down to $2,000-$3,000);
- no need for the driver's cabin within the driverless platform (including life-support subsystems);
- fuel economy when driving within a platoon (up to 10-20%);
- easier maneuvering (in remote control mode) and, as a result, relaxation of requirements for logistic terminals.

It is proposed to implement the cargo transportation automation concept stage by stage. At the initial stage, before the DLV road infrastructure deployment, creation of a "mobile infrastructure" for the driverless cargo platforms is required. The mobile infrastructure shall be installed onto a special truck, which shall be the lead vehicle of a road train with virtual couplings. The driverless cargo platforms shall strictly follow the lead vehicle automatically, keeping the minimum safe distance. The lead vehicle, which is controlled by the driver, shall feature main subsystems of the mobile infrastructure and an operating (control) center of the road train with virtual couplings.

3. Rationale for choosing the power unit for driven driverless trucks

Since the lead links of the road trains feature virtual couplings, it is necessary to solve the issue of power supply of these links when driving within the platoon. In the short term, three solutions can be considered: diesel ICE, combined power unit (plug-in hybrid) and electric platform (Fig. 4).

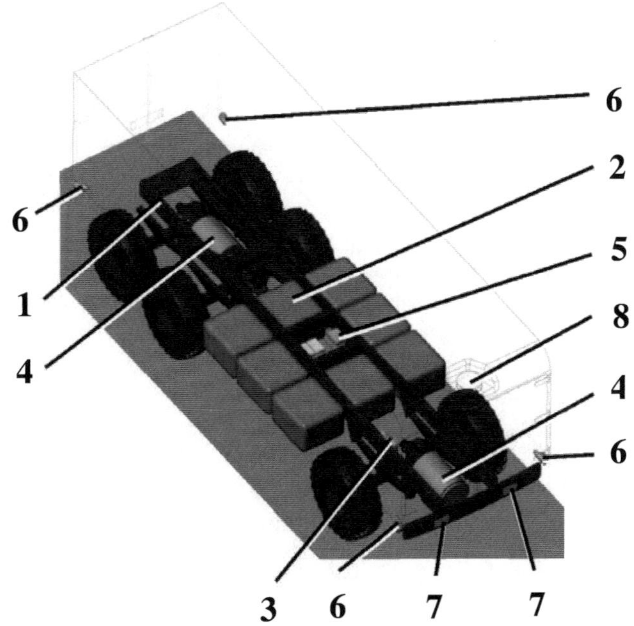

Figure 4. Driven electric driverless truck: 1 – chassis; 2 – batteries; 3 – steering gear electric drive; 4 – electric machines; 5 – electronic motion control unit; 6 – video cameras; 7 – radars; 8 – lidar.

Running 400…700 km per shift is typical for regional transportations. Currently, there are no affordable and available electric power accumulation technologies, which would make it possible to use electric vehicles for regional cargo transportation in terms of economic feasibility (Table 1).

Table 1. Calculation of 500 km run of the road train driven link (in electric mode, on a single charge, at the speed of 85 km/h, with the fully loaded weight of 35 t).

Required power capacity of the battery	620 kWh
Battery weight	3,100 kg
Battery cost	$315,000
Battery cost with 30 kWh Range Extender: (battery power capacity – 440 kW, battery weight – 2,200 kg)	$220,000
Battery cost with 50 kW Range Extender: (battery power capacity – 320 kW, battery weight – 1,600 kg)	$160,000

Besides, charging of the battery making such runs possible will demand either a long time or high current loading networks, reducing charge efficiency and battery service life. The lifespan of accumulators will not provide for using them without replacement during the entire operating life of the truck, which equals to the run of up to one million kilometers and more. Therefore, the alternative mid-term solutions might be the Combined Power Units (CPU), which have a rechargeable battery making it possible to drive several dozen kilometers in electric mode under certain conditions. Usage of a diesel engine without an energy recuperation system will not provide the required level of energy efficiency.

Heavy-load vehicles and road trains are characterized by a large weight, relatively low resistance to a steady movement within the low and medium speed range and high diesel efficiency in a wide operational range. In this regard, a considerable fuel saving is possible for these types of vehicles due to recovery or recuperation of vehicle kinetic and potential energy. The regenerative braking system of category "B" according to the classification of UNECE Regulation No. 13 shall be used to realize the full recovery potential. This means that the recovery system functioning is fully coordinated with service brake system and active safety systems functioning. A significant effect of energy recovery may be expected when driving in the mode of "acceleration / deceleration" cycles. Such mode is typical for cities or suburbs which heavy-load vehicles can pass only in transit. When driving on highways, the recovery energy may be accumulated as a result of speed decrease in turning movements or other service decelerations and also when driving downhill. As for heavy-load vehicles, it is possible to accumulate a significant amount of energy after several decelerations. When driving across the city or in suburbs, especially in traffic jams, it is possible to switch off the ICE for a certain period of time. Another variant is to consume the recovery energy for additional power take-off. First of all, this refers to refrigerator trucks. According to publications, road trains with the weight of around 32 t consume up to 20% of fuel for operation of an additional diesel engine actuating the refrigeration unit supply generator. Using certain generating unit parameters known from publications and technical materials, it is possible to calculate that 15% of average fuel consumption by the road train without a refrigeration unit (about 35 l / 100 km) correspond to about 5 kg of additional fuel consumption per 100 km. With the average speed of 60 km/h and energy recovery of 150 … 200 kW (typical capacity of a traction electric machine for hybrid vehicles of this type), the vehicle must drive 6.5…8.5% of the route in the recovery mode in order to save 15% of fuel. The required capacities may be provided either by 0.25 … 0.35 m/s^2 service decelerations (for example, speed decrease when turning) or by 1.5…2-degree downhill driving or by combination thereof.

References

[1] Information on http://www.google.com (accessed: 31.07.2017)
[2] Information on http://robocv.ru (accessed: 03.08.2017)
[3] http://www.volkswagenag.com/content/vwcorp/content/en/innovation/driver_assistance/Temporary_Auto_Pilot.html (accessed: 04.08.2017)
[4] http://www.bmw.com/com/en/insights/technology/connecteddrive/2013/index.html (accessed: 10.08.2017)
[5] Information on http://www.volvocars.com/uk/top/about/news-events/pages/default.aspx (accessed: 21.08.2017)
[6] Information on http://www.cadillac.com (accessed: 20.08.2017)
[7] Information on http://matlab.ru/products/matlab (accessed 23.08.2017)
[8] Saykin A, Bakhmutov S, Terenchenko A, Endachev D, Karpukhin K and Zarubkin V 2014 Tendency of creation of "driverless" vehicles abroad *Bioscience Biotechnology Research Asia* **11** pp 241-46
[9] Saykin A M, Buznikov S E and Karpukhin K E 2016 The analysis of technical vision problems typical for driverless vehicles *Research journal of pharmaceutical, biological and chemical sciences* **7** (4) pp 2053-59
[10] Information on http://www.nhtsa.gov (accessed: 30.08.2017)
[11] http://spectrum.ieee.org/cars-that-think/transportation/self-driving/otto-robotruck-hauls-budweiser-while-backup-driver-sits-in-the-back (accessed: 25.10.2016)
[12] https://www.eutruckplatooning.com/News/495554.aspx?t=Press+release%3a+Dutch+business+community+welcomes+truck+platoons+
[13] http://media.daimler.com/marsMediaSite/en/instance/ko/Start.xhtml?oid=4836258
[14] http://jalopnik.com/heres-the-first-look-at-googles-self-driving-semi-truck-1796465587
[15] Sergeyev V, Fedorenko A and Gerami V December 2012 Role of logistics in development of the Russian Federation transport industry: in the context of amending the Transport Strategy of the Russian Federation until 2030 *Sci. and Analytical J. "Logistics and Supply Chain Management"* **6** (53)

IOP Conf. Series: Materials Science and Engineering **315** (2018) 012002 doi:10.1088/1757-899X/315/1/012002

Development, modeling and research of the system of automatic control and equalization of the charge state of a battery pack of a hybrid engine of a vehicle

S Bakhmutov[1], Y Sizov[2] and M Kim[2]

[1] FSUE "NAMI", Moscow, Russia
[2] Moscow Polytech University, Moscow, Russia

E-mail: sizovua@yandex.ru

Abstract. The article is devoted to the topical problem of developing effective means of monitoring and leveling the charge state of batteries in a power unit of hybrid and electric cars. A system for automatic control and equalization of the charge state of a battery pack of a combined power plant, the originality of which is protected by the Russian Federation patent, is developed and described. A distinctive feature of the device is the possibility of using it both in conditions of charging (power consumption) and in operating conditions (energy recovery). The device is characterized by high reliability, simplicity of the circuit-making solution, low self-consumption and low cost. To test the efficiency of the proposed device, its computer simulation and experimental research were carried out. As a result of multi factorial experiment, a regression equation has been obtained which makes it possible to judge the high efficiency of detecting the degree of inhomogeneity of controlled batteries with respect to the parameters of an equivalent replacement circuit: voltage, internal resistance and capacitance in the magnitude of the obtained coefficients of influence of each of these factors, and also take into account the effects of their pair interactions.

1. Introduction

In recent years, a fairly large number of works by devoted to the determination of the electrical parameters of elements in a battery has appeared. However, existing devices are disadvantageous: the design solution does not allow monitoring battery blocks with voltages above hundreds of volts, the charging currents are small in magnitude and charging of batteries is carried out only in the pulsed mode. The device for its implementation requires many additional components for each element of battery, there is no measurement of parameters during operation, automatic mode of operation and other shortcomings are not provided.

2. Purpose and Methods

The purpose of this work is:

1. Development of a system for automatic monitoring and equalization of the charge level of the battery pack of the hybrid power unit of the car according to the parameters of the internal resistance R, battery capacity C, voltage U, providing:

1.1. the ability to work both under conditions of energy output and under conditions of electric consumption;
1.2. automatic determination of the parameters of each battery in the composition of the AB and their alignment;
1.3. relative prostate of the circuit solution with a slight complication of the equipment with an increase in the number of monitored batteries in the unit.
2. Computer simulation of the developed scheme, experimental research and analysis of the results obtained.

To implement the tasks, an original automatic control system [1] was developed, the structural diagram of which is presented in Figure 1. The system works as follows. The signals from the microcontroller are fed to the address switching input of the dual demultiplexer with the number of outputs n. In each demultiplexer, the n accumulators of the battery are sequentially connected through n + 1 single-position keys with a small internal resistance to the common wires A and B. Pros of the accumulators with odd numbers are connected to the wire A and the pluses of the accumulators with even numbers are to the wire B. On the wires A and B occurs an alternation of positive and negative voltages, which are transmitted to the information processing unit.

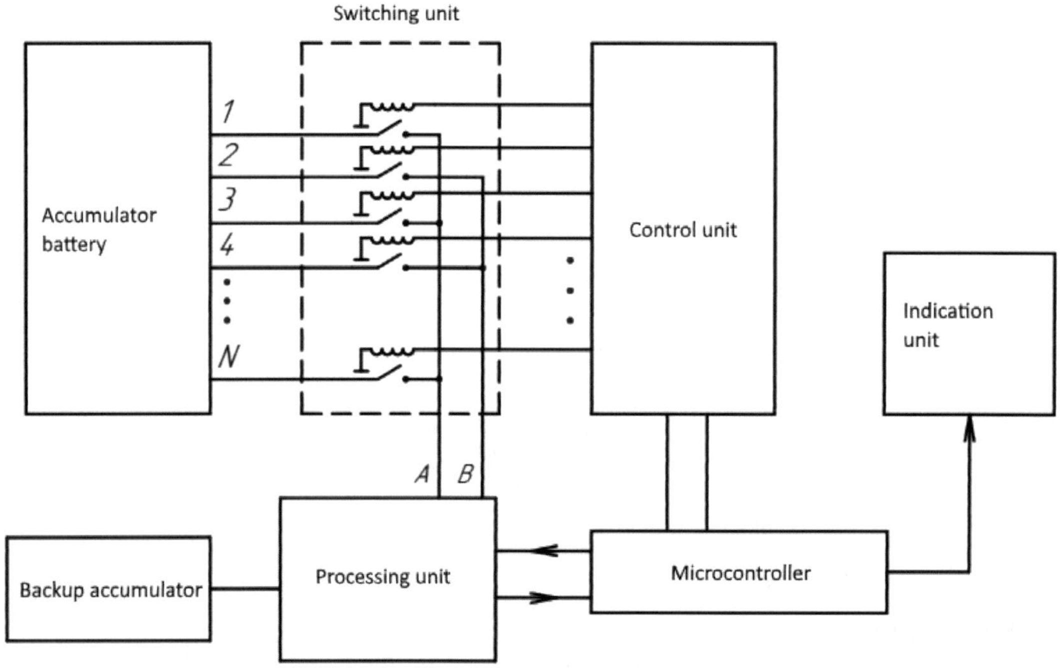

Figure 1. Structural diagram of the control and balancing system.

If the batteries have the same degree of charge, the voltage at the output of the information processing unit will be close to zero. If at least one of the battery packs is overcharged or undercharged, then the output will be a positive or negative mismatch pulse, with an amplitude proportional to the magnitude of the imbalance.

If the amplitude of the mismatch pulse exceeds the output of the information processing unit of the specified boundary voltage, the microcontroller sets the process of interrogation of the battery cells, fixing on the problem battery and indicates the serial number of the problem battery and the type of malfunction (recharge or undercharging) on the indicator display.

Simultaneously, the microcontroller generates a signal of the duration of the Tz recharging time. During that period the problematic battery is connected through the information processing unit to the backup battery cell and the charging state of the backup cell and the problem battery cell are equalized. Since problem accumulators can be either recharged or undercharged, the backup battery cell will maintain a charge level close to the arithmetic mean of all battery cells. At the end of the Tz period, the microcontroller will continue to poll the battery packs with the intermittent multiplexers of the control unit until another problem battery is encountered, etc. Thus, the system can work both in traffic conditions of the vehicle and when replenishing power from the charger.

A distinctive feature of the device is its ability to use it both in conditions of charging (power consumption) and in operating conditions (energy recovery), while the device is characterized by simplicity of circuit design, low power consumption, low cost and potentially high reliability.

If it is necessary to increase the number of elements in the battery, for example, up to 100 or more, the complexity of the electrical circuit is only due to a proportional increase in the number of keys and the number of outputs of the dual demultiplexers, which slightly increases the cost of the device.

3. Theoretical and experimental research, evaluation of results

To determine the operability of the developed electrical circuit, when measuring deviations of the battery parameters from the nominal values in the Multisim environment, a theoretical model of the system block was constructed, including eight controlled batteries, each of which was represented by the equivalent replacement circuit shown in Figure 2.

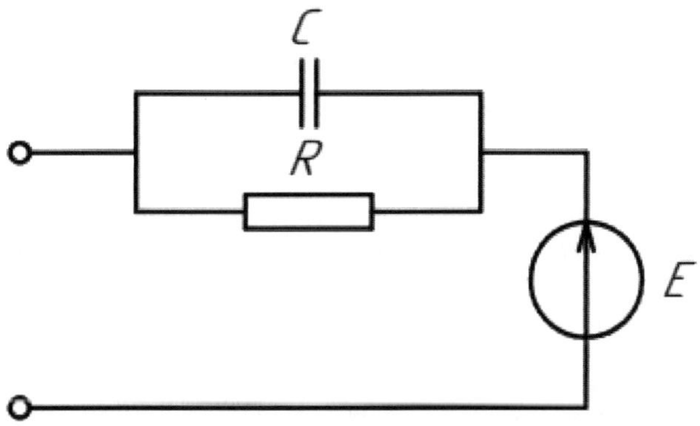

Figure 2. Equivalent battery replacement circuit. R-internal resistance; C - total capacity; E is the internal e.m.f.

In simulation, the input signal was generated by a virtual pulse generator, and the analysis of the output data was performed by a virtual oscilloscope. Figure 3 shows the time waveform of the output voltage of the signal processing unit at a battery sampling frequency of 100 Hz. The upper graph shows how the output voltage changes when the internal resistance of the fourth problem is deflected at 10% of the rated value. As it can be seen from the presented oscillogram, the useful signal is significantly allocated at the level of interference. The lower graph shows the time variation of the voltage on the batteries relative to the ground when they are sequentially interrogated, which shows that the fourth battery is problematic. Thus, the developed theoretical virtual model allows to carry out experimental research of the developed device with rather high accuracy.

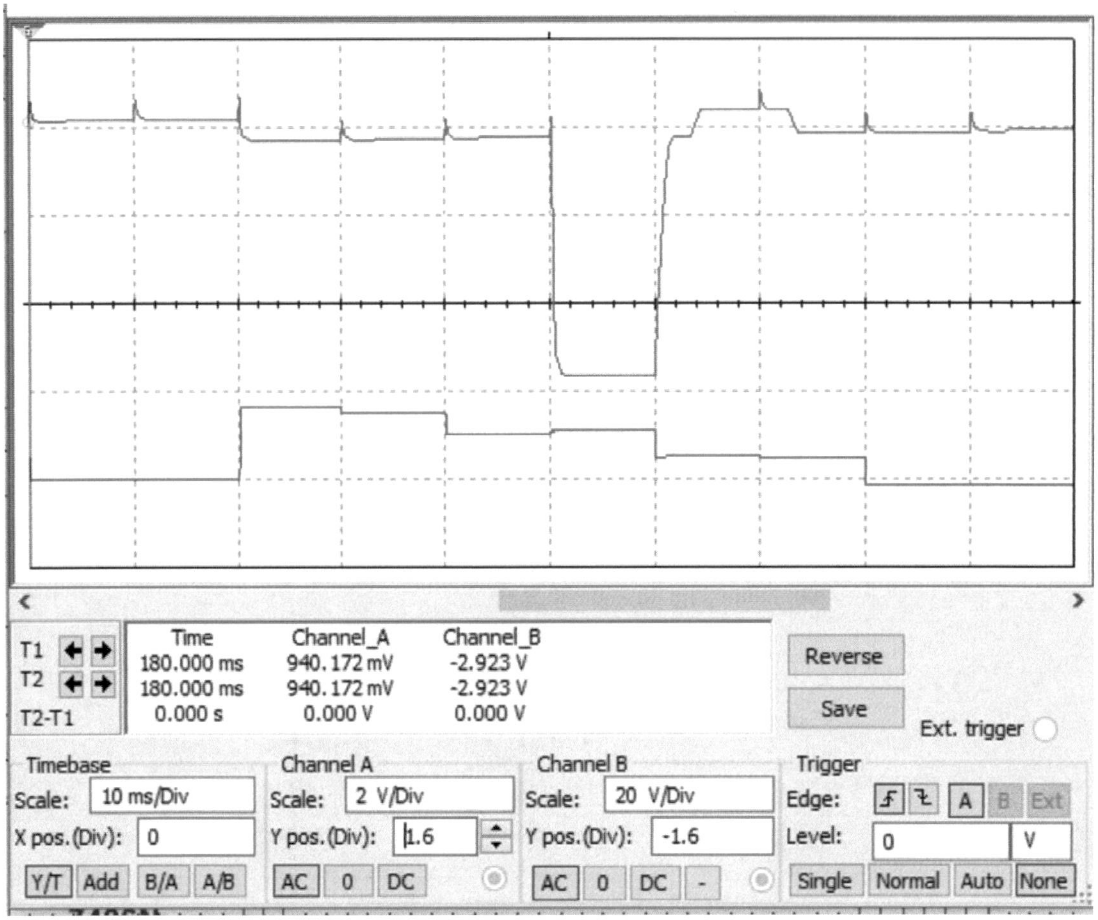

Figure 3. Oscillogram of the output voltage when a problematic battery is detected.

With a simple one-factor measurement of the degree of influence of individual parameters on the output voltage value due to their misalignment, it is impossible to estimate the real situation when all the influencing factors change simultaneously. It may be necessary to consider paired and triple effects and second-order effects. Therefore, as a method of experimental multifactor study, in determining the coefficients of influence of the three parameters of accumulators (U, R, C) on the virtual model, the B-plan [2,3,4] is chosen, which allows to reveal the second-order dependences. The results of the experiment are summarized in Table 1.

Table 1. The results of the experiment.

	Normalized values of factors			The natural values of factors			Output values				Average output value
	x_1	x_2	x_3	X_1	X_2	X_3	y_1	y_2	y_3	y_4	$\bar{y}$
1	1	1	1	5	0.03	12	0.38	0.378	0.382	0.381	0.380
2	-1	1	1	3	0.03	12	3.315	3.312	3.315	3.309	3.313
3	1	-1	1	5	0.01	12	0.809	0.812	0.811	0.812	0.811
4	1	1	-1	5	0.03	4	0.825	0.823	0.819	0.823	0.823
5	-1	-1	-1	3	0.01	4	2.04	2.043	2.039	2.04	2.041
6	1	-1	-1	5	0.01	4	0.808	0.808	0.81	0.809	0.809
7	-1	1	-1	3	0.03	4	3.318	3.315	3.318	3.32	3.318

	Normalized values of factors			The natural values of factors			Output values				Average output value
	x_1	x_2	x_3	X_1	X_2	X_3	y_1	y_2	y_3	y_4	$\bar{y}$
8	-1	-1	1	3	0.01	12	2.028	2.031	2.027	2.03	2.029
9	1	0	0	5	0.02	8	0.818	0.82	0.818	0.818	0.819
10	0	1	0	4	0.03	8	0.861	0.862	0.859	0.862	0.861
11	0	0	1	4	0.02	12	0.869	0.872	0.87	0.869	0.870
12	-1	0	0	3	0.02	8	2.683	2.679	2.685	2.68	2.682
13	0	-1	0	4	0.01	8	0.851	0.849	0.855	0.851	0.852
14	0	0	-1	4	0.02	4	0.858	0.862	0.859	0.86	0.860

The output value of Y in the experiments is the output voltage of the signal processing unit, measured in volts. To increase the accuracy of the measurements, the experiments were carried out at four-fold duplication at each point of the experiment, so the arithmetic mean values of the output value were written in the right column of the table. After checking the adequacy by the Fisher criterion, removing the insignificant coefficients by the Student's criterion and bringing it to the natural values of the influencing factors of the regression equation takes the form:

$$\hat{y} = -0{,}075 + 0{,}962x_1 - 0{,}824x_2 + 0{,}154x_3 + 0{,}109x_1^2 + 0{,}465x_2^2 - 0{,}232x_3^2 - 0{,}38x_1x_2 - 0{,}323x_1x_3 + 0{,}402x_2x_3$$

Where:
$\hat{y}$ - output voltage predicted by the regression equation;
x_1- voltage U;
x_2- resistance R;
x_3- capacitance C.

As can be seen from the regression equation obtained, only the effects of single factors (U, R) are significant, but also their paired effects. Also, the coefficients of linear coefficients at U and R were most significant, and the influence of the C-value is practically insignificant, and the coefficients for quadratic effects turned out to be of little significance. From what has been said, it follows that the regression equation has a practically linear form. Other advantages of the obtained regression equation include the ability to predict the magnitude of possible imbalance in the operating conditions from a known variation in battery battery parameters and to construct a more adequate model of a real battery [4].

4. Conclusion

Thus, the device allows:

- to carry out continuous monitoring of the battery charge level of the battery by the main parameters of voltage, internal resistance and capacity, indicating the serial number of the problem battery and the nature of the malfunction (undercharging or recharging);
- provide the ability to work both in conditions of energy output and in conditions of electric consumption;
- to provide relative simplicity of the circuit solution with insignificant complication of the equipment with increasing number of controlled batteries in the unit;
- provide selective recharging and alignment of problematic batteries from the backup battery with a high current for a certain period of time.

References

[1] Bakhmutov S, Sizov Y, Revonchenkov A, Chernokozov V, Gordeyev A and Kim M 2017 *The device for automatic control and equalization of the charge state of a battery pack of a combined power plant* (Russian Federation 2626378)

[2] Sizov Y, Kharitonov V and Chekmaryova Y 2014 Calculation and modeling of electric parameters of accumulators on transient processes *Bulletin of the Moscow State University Series "Engineering and Technology"* No. 4

[3] Kharitonov V, Gordeyev A and Lozhkina Y 2013 Modeling and research of electrical parameters of series-connected batteries *Bulletin of the Moscow State University Series "Engineering and Technology"* No. 4

[4] Bakhmutov S, Sizov Y and Kim M 2015 Development of a mathematical model and study of a traction battery of a hybrid power plant *Proc. NAMI* p 262

Creation of operation algorithms for combined operation of anti-lock braking system (ABS) and electric machine included in the combined power plant

S V Bakhmutov[1], **V G Ivanov**[2], **K E Karpukhin**[3] **and A A Umnitsyn**[4]

[1] Deputy CEO for Science, FSUE "NAMI", Moscow, Russia
[2] Research Professor, TU Ilmenau, Ilmenau, Germany
[3] Head of Combined Power Units Department, FSUE "NAMI", Moscow, Russia
[4] Design Engineer of Combined Power Units Department, FSUE "NAMI", Moscow, Russian Federation

E-mail: artem.umnicin@nami.ru

Abstract. The paper considers the Anti-lock Braking System (ABS) operation algorithm, which enables the implementation of hybrid braking, i.e. the braking process combining friction brake mechanisms and e-machine (electric machine), which operates in the energy recovery mode. The provided materials focus only on the rectilinear motion of the vehicle. That the ABS task consists in the maintenance of the target wheel slip ratio, which depends on the tyre-road adhesion coefficient. The tyre-road adhesion coefficient was defined based on the vehicle deceleration. In the course of calculated studies, the following operation algorithm of hybrid braking was determined. At adhesion coefficient ≤0.1, driving axle braking occurs only due to the e-machine operating in the energy recovery mode. In other cases, depending on adhesion coefficient, the e-machine provides the brake torque, which changes from 35 to 100% of the maximum available brake torque. Virtual tests showed that values of the wheel slip ratio are close to the required ones. Thus, this algorithm makes it possible to implement hybrid braking by means of the two sources creating the brake torque.

1. Introduction

In the modern automotive industry, sufficient attention is given to the study of the deceleration process of the hybrid vehicle using regenerative braking with combined operation of the anti-lock braking system and electric machine included in the combined power plant [1, 2, 3]. The justification to that is as follows:

- Growth of a share of electric vehicles and vehicles equipped with the combined power plants in the total number of sold vehicles. As a consequence, study and search for the optimum algorithms of the electric machine operation within the vehicle (V) transmission become more and more important for ensuring minimization of hazardous emissions and fuel consumption. [4].
- The ratio between the braking power and total power consumption of the vehicle varies from 34% to 82% depending on a driving cycle [5]. Thus, regenerative braking can be considered as one of the important factors, ensuring extension of distance on a single charge and reduction of

hazardous emissions from the internal combustion engine (ICE) as a result of reduction of its operation time.

- The analysis of the foreign works concerned with the studies of this subject has shown their significant growth [6-11].

The majority of works dedicated to the hybrid braking processes, i.e. braking at the same time by the means of friction brakes and regenerative braking using an electric machine, deal with the issues connected to the maximization of power gained from the electric machine during the service braking with a small wheel slip [6, 9-11]. At the same time, the attention given to the combined operation of electric machines and anti-lock braking systems is not enough.

Therefore, development of the anti-lock braking system (ABS), which will enable combined braking with the use of friction brakes and regenerative braking by one or several electric machines and will at the same time meet requirements of Annex 13 to UNECE Regulation No. 13, is currently a long-term and crucial task, solving which will allow improvement of fuel efficiency and ecological compatibility of the vehicles equipped with the combined power plants, and extension of their run distance on a single charge as well.

2. Development of operation algorithms for abs system with hybrid braking ability

For the development of algorithms of hybrid braking in question, N2 category electric vehicle has been chosen as a study subject, and its necessary specifications are shown in table 1.

Table 1. Specifications of N2 category vehicle.

Fully loaded weight	4200 kg
Weight on front axle	1653 kg
Weight on rear axle	2547 kg
Unladen weight	4060 kg
Weight on front axle	1555 kg
Weight on rear axle	2505 kg
Height of the center of mass	
unladen condition	998 mm
fully loaded weight	1001 mm
Tires	185/75R16C
Electric machine	Siemens 1PV5135-4WS14

The electric machine characteristics are shown in figure 1, [12].

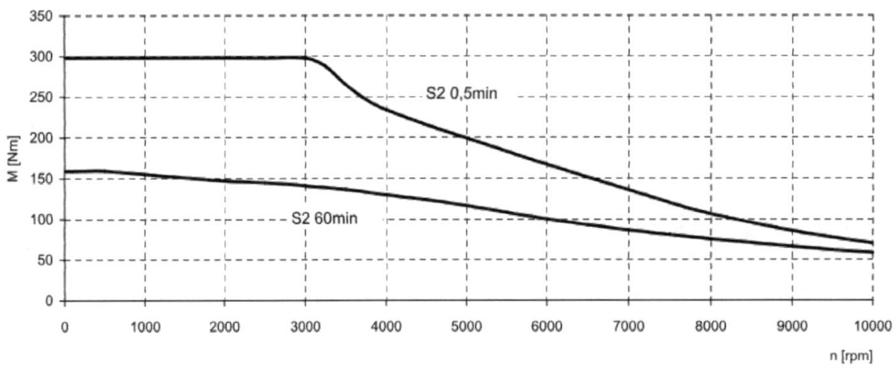

Figure 1. Specifications of the Siemens 1PV5135-4WS14 electric machine.

In order to define maximum longitudinal reaction of traction wheels, we shall examine forces acting on the vehicle at a linear motion (neglecting aerodynamic resistance forces), figure 2.

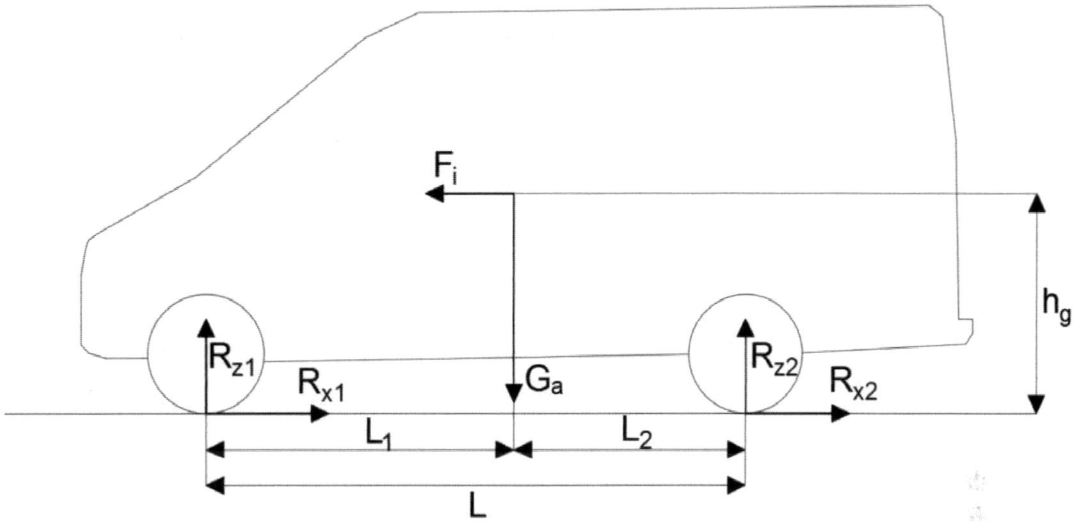

Figure 2. Distribution of forces during braking.

The maximum force acting on the rear axle wheels:

$$F_{x2} = R_{z2} \cdot \varphi_x, \tag{1}$$

where F_{x2} is axial or longitudinal (braking) force on the wheel, R_{z2} is vertical reaction of the rear wheels, andφ_x is the tire-road friction coefficient.
By setting the sum of moments (torques) relative to the front wheel contact equal to zero, we shall find the vertical reaction of the rear axle wheels:

$$R_{z2} = \frac{G_a}{L} \cdot (L_1 - \varphi_x \cdot h_g), \tag{2}$$

where G_a is the vehicle weight, L is the vehicle wheelbase, L_1 is the distance from the center of the front wheel to the center of mass, h_g is the height of the center of mass. By inserting (2) in (1), we shall obtain:

$$F_{x2} = \frac{G_a}{L} \cdot (L_1 - \varphi_x \cdot h_g) \cdot \varphi_x \tag{3}$$

The axial force on the vehicle traction wheels, which the electric machine operating in the energy recovery mode is capable to create:

$$F_{xel} = \frac{T_{el} \cdot i_r \cdot i_{fd}}{r_d}, \tag{4}$$

where F_{xel} is the axial brake force transferred to the wheel from an electric machine, i_r is the reduction ratio, i_{fd} is the final (axle) drive ratio, and r_d is the effective radius. By equating the (1) expression to the (4) expression, we shall obtain:

$$\frac{G_a \cdot L_1 \cdot \varphi_x}{L} - \varphi_x^2 \cdot h_g = \frac{T_{el} \cdot i_r \cdot i_{fd}}{r_d} \tag{5}$$

Taking into account that dependence between the electric machine braking torque and its speed (rpm) is decreasing (figure 1), we shall set $T_{el} = T_{elmax}$. By solving the equation for φ_x, we shall obtain that the maximum tire-road friction coefficient, which allows the electric machine to ensure enough braking torque without causing wheel lock-up within the whole variation range of the motor shaft rotation speed, is $\varphi_x = 0.1$. At the same time, we need to consider part load modes of the electric machine: since, on the one hand, the braking torque at high rpm of the electric machine shaft is not enough to ensure slip control without using friction brakes, and on the other hand, at low rpm of the electric machine shaft the braking torque excess can appear, which shall cause wheel lock-up. We shall accept the following hybrid operation strategy:

- for $\varphi_x \leq 0.1$, braking of the driving axle is only carried out with the use of the braking torque created by the electric machine. The wheel slip coefficient is adjusted by changing the braking torque created by the electric machine operating in the energy recovery mode;
- for $0.1 < \varphi_x \leq 0.2$, the electric machine shall work with 35% of the maximum torque. The wheel slip coefficient is adjusted by changing the braking torque created by the friction brakes.
- for $0.1 < \varphi_x \leq 0.3$, the electric machine shall work with 50% of the maximum torque. The wheel slip coefficient is adjusted by changing the braking torque created by the friction brakes.
- for $\varphi_x > 0.3$, the electric machine and friction brakes shall jointly create braking torque and the electric machine shall operate at full capacity, and the wheel slip coefficient is adjusted by the friction brakes.

3. Definition of transfer function of abs hydraulic unit

In order to define the transfer function of the ABS hydraulic unit, an experiment has been conducted using the bench equipment of the Ilmenau Technical University, Germany. The experiment objective was to define characteristics of a commercial ABS hydraulic unit by sending a step signal to the module input and to construct a transfer function.

The bench represents a fragment of the vehicle braking system (figure 3), including the following components: brake pedal with control robot (1), vacuum booster (2), master brake cylinder (3), ABS hydraulic unit (4), four brakes and two braking discs (5). Since the bench imitates only the hydraulic module operation, the brakes are installed pairwise on two fixed brake discs. The bench is equipped with pressure sensors installed at all hydraulic inlet and outlets of the braking module.

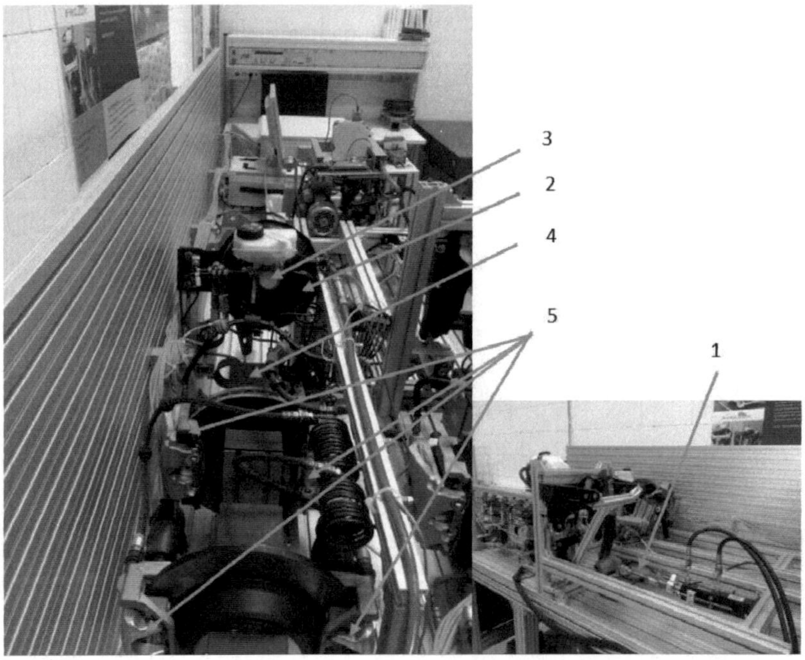

Figure 3. Test bench for ABS study.

The brake pedal is driven by the robotized drive connected to the real-time computer and controlled by Matlab State Flow. The valves and hydraulic module pump are controlled using the algorithms created in the Simulink environment. Information obtained from the pressure sensors is recorded by the computer.

Figure 4 shows pressure change in the brake line of the front right cylinder. Blue color shows change of the target pressure, black shows the real pressure measured on the front right wheel, and red shows the pressure measured using the transfer function.

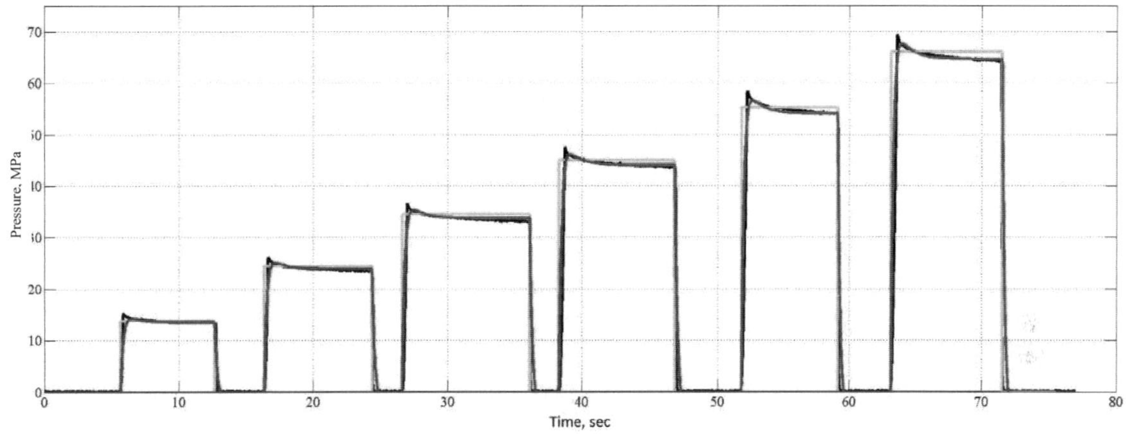

Figure 4. Pressure change in ABS hydraulic unit.

It may be noted that the transfer function describes real change of the pressure with high precision. That particular function is applied in the further work on the algorithm of the ABS module operation.

4. Development of ABS operation algorithms

One of the main purposes of the ABS system is to maintain necessary wheel slip coefficient. In this study, maintenance of the set slip is ensured by using a PID controller with an integral windup filter. As it is known, this controller is a control loop feedback mechanism. It produces the control signal using the sum of three components, which are proportional, integral and derivative components.

The control program, the diagram of which is shown in figure 5, adjusts the braking pressure in the ABS hydraulic module and controls the electric machine operation during braking. Input parameters for the system are pressure measured in the master brake cylinder, slip coefficient of each wheel, vehicle speed, its deceleration and rotation rate of the electric machine shaft.

In this system, search for the optimum value of the slip coefficient depending on the road condition is implemented. The tire-road friction coefficient is defined as a function of the maximum longitudinal deceleration of the vehicle. The decision to shut down the rear brakes is made based on the current tire-road friction coefficient.

The brake cylinder pressure is adjusted in the brake control units. The pressure is adjusted by the PID controller equipped with an integral windup filter. The feedback is performed by monitoring of the values of the wheels' slip coefficient. Control error is calculated as the difference between the target and current wheel slip coefficient.

The rotation and braking torque at the electric machine shaft are controlled by the electric machine control unit. Depending on the current tire-road friction coefficient, decision on the electric machine operation mode is made as follows: full load, part load or torque control using PI controller. In the latter case, the feedback and control error are set in the same way as for the brake control units.

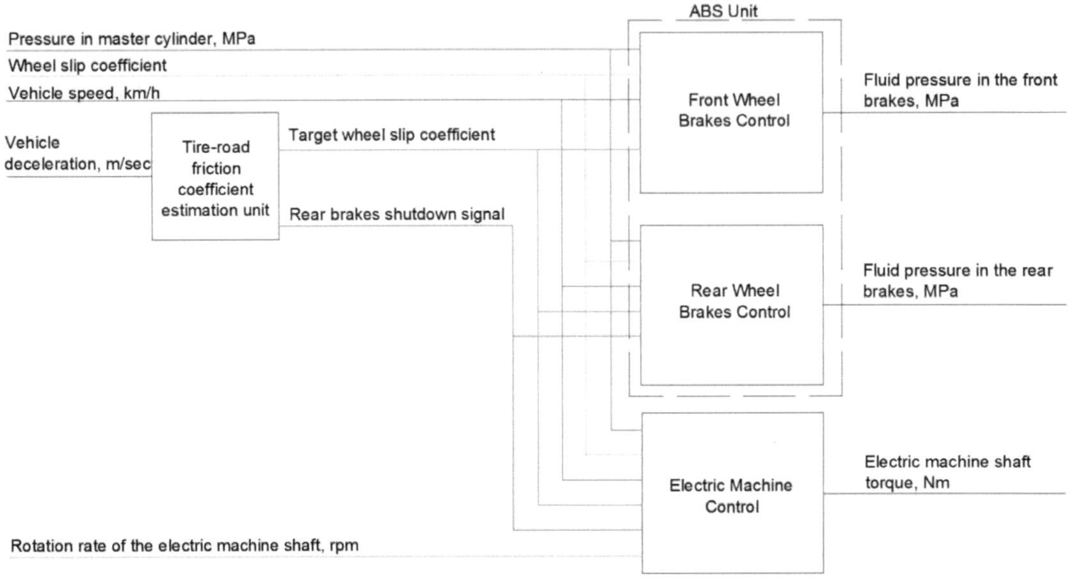

Figure 5. ABS Diagram.

The following virtual testing plan has been accepted for the system testing:

- Initial driving speed - 98 km/h, accelerator pedal travel - 100%.
- After reaching 100 km/h: accelerator pedal travel - 0%, brake pedal travel - 100%.
- Braking until complete stop.

The number of virtual tests were carried out on the road pavement with the tire-road friction coefficient varying from 0.8 to 0.1.

The system operation results for the front left and rear left wheels are shown in figure 6-9. Red shows the values of the wheel slip coefficient, blue shows the pressure change, and green shows the torque at the electric machine shaft.

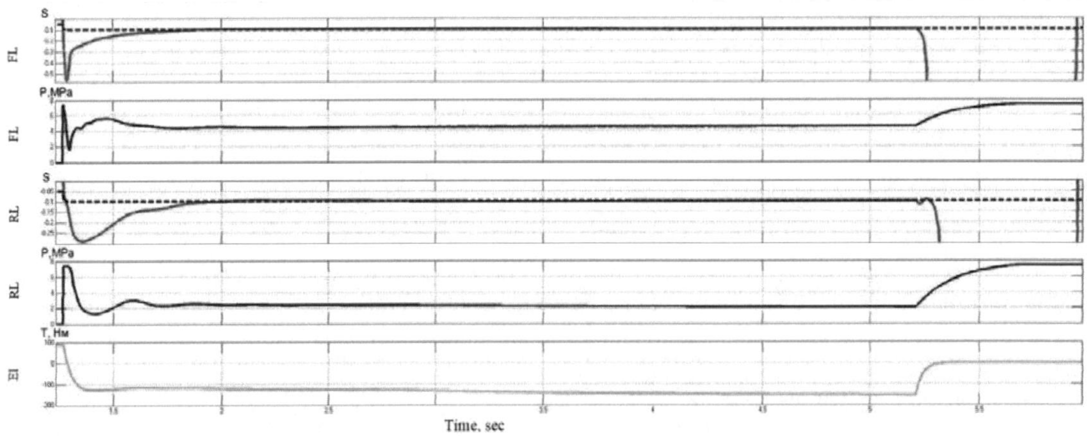

Figure 6. Slip coefficient and braking torque of the vehicle wheels when driving on the road with the friction coefficient of φ_x=0.8.

IOP Conf. Series: Materials Science and Engineering **315** (2018) 012003 doi:10.1088/1757-899X/315/1/012003

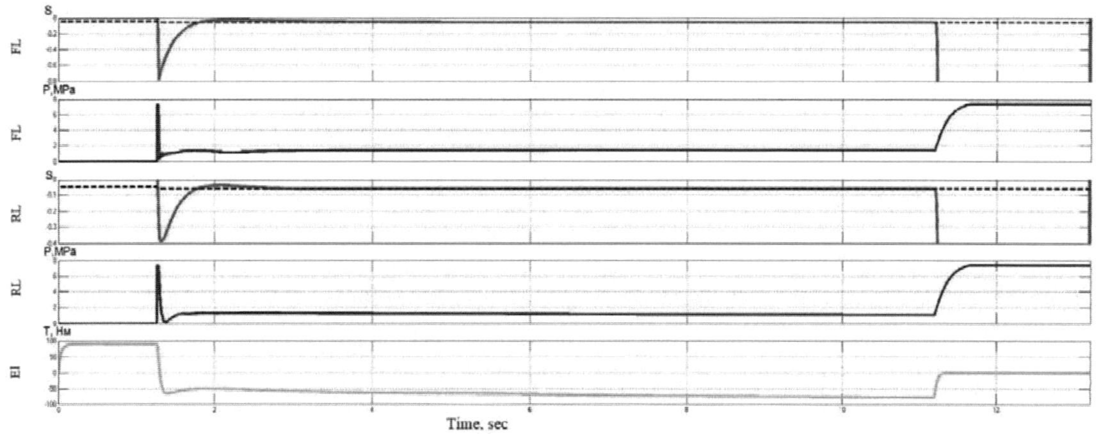

Figure 7. Slip coefficient and braking torque of the vehicle wheels when driving on the road with the friction coefficient of φ_x=0.3.

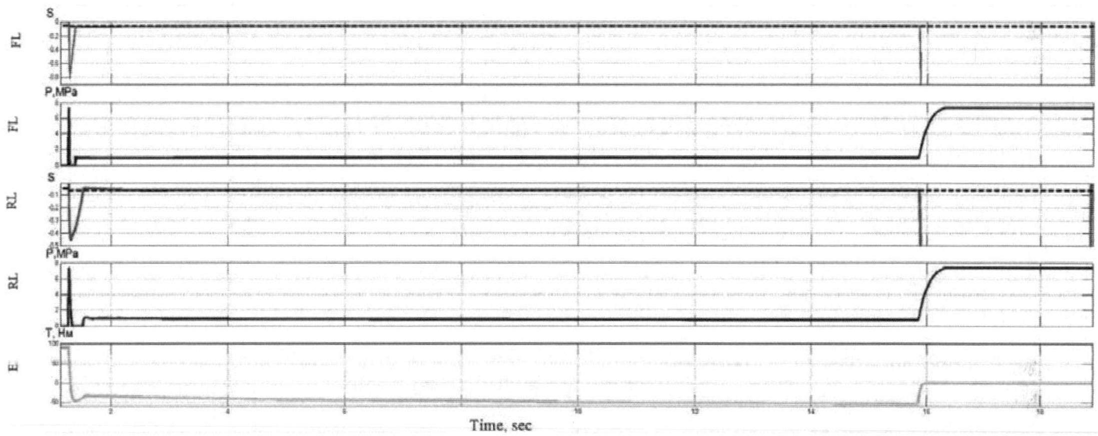

Figure 8. Slip coefficient and braking torque of the vehicle wheels during moving on the road with friction coefficient amounted to φ_x=0.2.

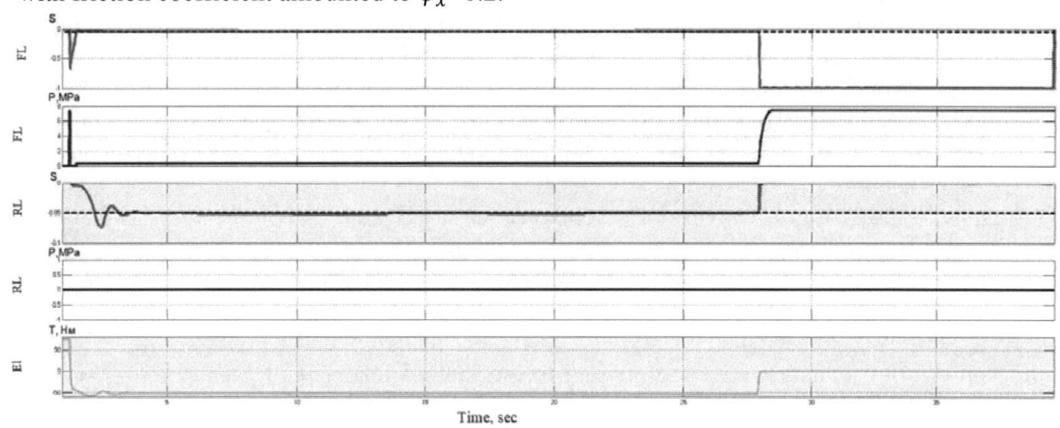

Figure 9. Slip coefficient and braking torque of the vehicle wheels when driving on the road with the friction coefficient of φ_x=0.1.

The virtual testing results showed that the system allows to combine braking using two different sources of the braking torque - friction brakes and electric machine operating in the energy recovery mode. At the same time, the wheel slip coefficient values are close to the target one.

At this stage, the wheel slip is set in an idealised way, and that leads to the smooth shape of the curve characterizing the wheel slip value. Further, it is planned to implement an energy storage device model and more detailed wheel model, and refuse from defining the pavement condition based on longitudinal deceleration of the vehicle.

5. Conclusion

This work allows us to make the following conclusions:

- For the system with hybrid braking, 3 operation options can be defined:
    a. Braking of vehicle driving axle using only electric machine.
    b. Braking of vehicle driving axle using friction brakes and electric machine operating with partial braking torque available when operating in energy recovery mode.
    c. Braking of vehicle driving axle using friction brakes and electric machine operating with full braking torque available when operating in energy recovery mode.

- The braking torque of the electric machine operating in the energy recovery mode can reach from 40% to 100% of the braking torque on the driving axle.
- The preset target value of the slip coefficient of the driving axle wheels is reached in 1.1 sec. when using only friction brakes and in 1.3 sec. when using hybrid braking. Thus, performance of the ABS when braking using the friction brake and the electric machine operating in the energy recovery mode is comparable to the "classical" ABS performance.
- The used approach for the tire-road friction coefficient definition based on the vehicle deceleration is low-efficient due to the high inertia of the vehicle. It is reasonable to study the mathematical model more deeply and to define the tire-road friction coefficient based on the dynamic parameters of the wheel motion.
- In order to get quantitative evaluation of the energy, which can be recovered by using the developed system, it is necessary to study the electric machine in more details and implement the model of the energy storage device.

References

[1] Savitski D, Ivanov V, Augsburg K, Pütz T and Barber P 2016 The new paradigm of an anti-lock braking system for a full electric vehicle: Experimental investigation and benchmarking *Proceedings of the Institution of Mechanical Engineers, Part D: Journal of Automobile Engineering* **230** pp 1364-77

[2] Kulikov I, Shorin A, Bakhmutov S, Terenchenko A and Karpukhin K 2016 A Method of Powertrain's Components Sizing for a Range Extended Electric Vehicle *SAE Technical Paper* 2016-01-8096

[3] Ivanov V, Augsburg K, Savitski D, Schreiber V, Els S and Dhaens M 2017 Collaborative engineering of integrated chassis control for ground vehicle: Case study of lifelong learning technologies in automotive mechatronics *IEEE International Conference on Mechatronics (ICM)* pp 449-54

[4] Panchishny V I, Luksho V A, Kozlov A V, Terenchenko A S and Karpukhinn K E 2015 The problem of toxicity of gas engines and their solutions *Biosciences Biotechnology Research Asia* **12** pp 217-24

[5] Ehsani M, Gao Y and Emadi A 2009 *Modern Electric, Hybrid Electric, and Fuel Cell Vehicles* pp 109-16

[6] Gao Y, Chen L, and Ehsani M 1999 Investigation of the Effectiveness of Regenerative Braking for EV and HEV *SAE Technical Paper* 1999-01-2910

[7] Cikanek S R and Bailey K E 2002 Regenerative braking system for a hybrid electric vehicle

IOP Conf. Series: Materials Science and Engineering **315** (2018) 012003 doi:10.1088/1757-899X/315/1/012003

American Control Conference pp 3129-35

[8] Ahn J K, Jung K H, Kim D H, Jin H B, Kim H S and Hwang S H 2009 Analysis of a regenrative braking system for hybrid electric vehicles using an electro-mechanical brake *International Journal of Automotive Technology* **10,2** pp 229-34

[9] Wang F and Zhuo B 2008 Regenerative braking strategy for hybrid electric vehicles based on regenerative torque optimization control *Proceedings of the Institution of Mechanical Engineers Part D: Journal of Automobile Engineering* pp 499–513

[10] Chen Y and Wang J 2012 Fast and global optimal energy-efficient control allocation with applications to over-actuated electric ground vehicles *IEEE Transactions on Control Systems Technology* **20** pp 1202–11

[11] Zijian Zhang Weimin Li Guoqing Xu Liang Zheng 2010 Regenerative Braking for Electric Vehicle based on Fuzzy Logic Control Strategy *2010 2nd International Conference on Mechanical and Electronics Engineering* (Kyoto) pp V1-319-V1-323

[12] http://media3.ev-tv.me/Azure300VDC-400Acurve.pdf

[13] Singh K B and Taheri S 2015 Estimation of tire–road friction coefficient and its application in chassis control systems *Systems Science & Control Engineering* **3** pp 39-61

Modeling of the motion of automobile elastic wheel in real-time for creation of wheeled vehicles motion control electronic systems

E V Balakina, N M Zotov and A P Fedin

Volgograd State Technical University, Volgograd, Russia

E-mail: fahrgestell2011@yandex.ru

Abstract. Modeling of the motion of the elastic wheel of the vehicle in real-time is used in the tasks of constructing different models in the creation of wheeled vehicles motion control electronic systems, in the creation of automobile stand-simulators etc. The accuracy and the reliability of simulation of the parameters of the wheel motion in real-time when rolling with a slip within the given road conditions are determined not only by the choice of the model, but also by the inaccuracy and instability of the numerical calculation. It is established that the inaccuracy and instability of the calculation depend on the size of the step of integration and the numerical method being used. The analysis of these inaccuracy and instability when wheel rolling with a slip was made and recommendations for reducing them were developed.

It is established that the total allowable range of steps of integration is 0.001..0.005 s; the strongest instability is manifested in the calculation of the angular and linear accelerations of the wheel; the weakest instability is manifested in the calculation of the translational velocity of the wheel and moving of the center of the wheel; the instability is less at large values of slip angle and on more slippery surfaces. A new method of the average acceleration is suggested, which allows to significantly reduce (up to 100%) the manifesting of instability of the solution in the calculation of all parameters of motion of the elastic wheel for different braking conditions and for the entire range of steps of integration.

The results of research can be applied to the selection of control algorithms in vehicles motion control electronic systems and in the testing stand-simulators

1. Purpose

The analysis of the inaccuracy and the instability of the results of numerical calculation of motion parameters in real-time of elastic wheel rolling with a slip and development of recommendations for reducing them.

2. Introduction and relevance

Modeling of the motion of the elastic wheel of the vehicle in real-time is used in the tasks of constructing different models in the creation of wheeled vehicles motion control electronic systems, in the creation of automobile stand-simulators etc. [1-13]. During the motion of the vehicle wheel the lateral force is almost always present [9, 11]. Therefore, the models of his motion must take into account the phenomenon of slip, by which the motion of the elastic wheel is accompanied in the presence of lateral force. The accuracy and the reliability of simulation of the parameters of the wheel motion in real-time when rolling with a slip within the given road conditions are determined not only by the choice of the model, but also by the inaccuracy and instability of the numerical calculation. It is established that the inaccuracy and instability of the calculation depend on the size of the step of integration and the numerical method being used [11]. The authors of the article found that the inaccuracy and instability of the calculation depend on the size of the step of integration and the numerical method being used.

3. Problem statement

To achieve the purpose of research it is necessary to determine the model of the motion of a wheel, numerical methods and steps of integration.

Lateral force can be defined using the slip angle, calculated by the deformation theory [1, 9]:

$$P_y \approx \frac{\mathrm{tg}(\delta)\cdot C_{ty}\cdot (l_c/2)}{1 \pm \sin(\alpha)}; \tag{1}$$

where P_y – the lateral force, δ – the slip angle; C_{ty} – the coefficient of the tyre lateral stiffness, l_c – the length of a tyre contact patch with the road, α – the wheel camber angle.

The sign in the denominator is defined by the correspondence between the tilt of the wheel and lateral force directions.

The value of the lateral force is related to the value of proportion of the static friction coefficient in the longitudinal direction [1, 9]:

$$f_{stx} = \begin{cases} \sqrt{f_{st}^2 - \left(\dfrac{P_y}{P_z}\right)^2}; \text{ if } f_{st}^2 - \left(\dfrac{P_y}{P_z}\right)^2 \geq 0; \\ 0; \text{ if } f_{st}^2 - \left(\dfrac{P_y}{P_z}\right)^2 < 0. \end{cases} \tag{2}$$

where P_z – normal wheel load, f_{st} – static friction coefficient.

$$\begin{cases} M_{bi} = \begin{cases} f_{st}\cdot P_z \cdot R\cdot (t_i/T) \text{ if } M_{bi} \leq f_{st}\cdot P_z\cdot R\cdot (t_i/T); \\ f_{st}\cdot P_z\cdot R \text{ if } M_{bi} > f_{st}\cdot P_z\cdot R\cdot (t_i/T); \end{cases} \\ \dot{\omega}_i = \dfrac{1}{J}\cdot\left(-M_{bi} + P_z\cdot R\cdot \varphi_{x\ i-1}\right); \\ \omega_i = \omega_{i-1} + \dot{\omega}_i \cdot dt; \\ \dot{v}_i = -\varphi_{x\ i-1}\cdot g; \\ v_i = v_{i-1} + \dot{v}_i\cdot dt; \\ s_{xi} = 1 - \dfrac{\omega_i\cdot R}{v_i}; \\ CUSF_i = f(s_{xi}); \\ \varphi_{xi} = f_{sl}\cdot s_{xi} + CUSF_i \cdot f_{st\,x}(1 - s_{xi}); \\ L_{xi} = L_{xi-1} + v_i\cdot dt; \\ L_{yi} = L_{xi-1}\cdot tg(\delta). \end{cases} \tag{3}$$

where i – the step number; t – the current value of time; M_b – the braking moment applied to the wheel; R – the dynamic radius of the wheel; T – the time of braking moment change; $\dot{\omega}$, ω – angular acceleration and wheel velocity; $\dot{v}$, v – linear acceleration and velocity of the wheel axis; J – the inertia moment of the wheel; g – acceleration of free fall; $CUSF$ – the coefficient of use of static friction in the contact patch; φ_x – the friction coefficient between wheel and road in the longitudinal direction; s_x – coefficient of longitudinal wheel sliding; f_{sl} – sliding friction coefficient (φ_x if s_x=100%); f_{st} – static friction coefficient; $f_{st\,x}$ – static friction coefficient in the longitudinal direction; L_x– motion of the center of mass of a wheel in the longitudinal direction, L_y – motion of the center of mass of a wheel in the transverse direction.

For the analysis of the inaccuracy and the instability of the numerical calculation the model of the single wheel at braking (3) is used, as it is the most unstable mode. The parameters of the motion of a wheel are defined by solving the well-known system of differential equations and the constraint equations [5, 6, 8, 11] using the models of CUSF [1] for calculating the values of the friction coefficient.

Currently the researchers, who work in this field, apply the following integration steps: 0.001 – 0.005s. These steps are selected on the basis of the manifesting of the instability of the solution. Therefore, to achieve the purpose of the research the following integration steps: 0.001s, 0.002s, 0.003s, 0.004s and 0.005s are being used, and the implicit Euler method. At the present time it is the most common as being a relatively simple and accurate method.

4. Problem solving

Baseline data and constant values accepted for the solution of system (3) using the expressions (1) and (2) are presented in the table 1.

Table 1. Baseline data and constant values accepted for the solution of system (3).

Designation	Denotation	Denomination
Coefficient of a tyre radial stiffness	C_{tz}	210 N/mm
Coefficient of a tyre lateral stiffness	C_{ty}	275 N/mm
Length of a tyre contact patch with the road	l_c	223 mm
Wheel camber angle	α	0°
Normal wheel load	P_z	4000 N
Acceptable normal wheel load	$[P_z]$	6000 N
Time of braking moment change	T	0,5 s
Mass of the tyre	m_t	9,3 kg
Mass of the wheel disc	m_{belt}	7,247 kg
Linear and angular acceleration at the initial moment of time	$\dot{\omega}_0, \dot{v}_0$	0
Translational velocity of the wheel axle at the initial moment of time	v_0	10 m/s
Angular velocity of the wheel at the initial moment of time	ω_0	31,84 s-1

IOP Conf. Series: Materials Science and Engineering **315** (2018) 012004 doi:10.1088/1757-899X/315/1/012004

Table 1. (continued).

Designation	Denotation	Denomination
Coefficient of the longitudinal slip of a wheel at the initial moment of time	s_{x0}	0
Coefficient of friction between wheel and road in the longitudinal direction at the initial moment of time	φ_{x0}	0
Braking moment at the initial moment of time	M_0	0
Motions of the center of the wheel in the longitudinal and the transverse directions at the initial moment of time	L_{x0}, L_{y0}	0
Constant coefficients for calculation of CUSF	$a,b,c,d,e,$ f,g,h,i,j	a = 0.001814; b = 26.747630; c= -324.541748; d=2137.928850; e= -8375.670586; f =20260.588666; g=-30442.87972; h= 7611.479368; i=-13822.359721; j = 2929.705537

In the present study the calculations of parameters of braking of a vehicle wheel (3) for different integration steps, road surfaces and various values of lateral force (slip angle) using the implicit Euler method and using the method of average accelerations (system 4) were carried out. The equation that the system 4 differs from the system 3 is shown in grey below on the system 4.

$$\begin{cases} M_{bi} = \begin{cases} f_{st} \cdot P_z \cdot R \cdot (t_i / T) \text{ if } M_{bi} \le f_{st} \cdot P_z \cdot R \cdot (t_i/T); \\ f_{st} \cdot P_z \cdot R \text{ if } M_{bi} > f_{st} \cdot P_z \cdot R \cdot (t_i/T); \end{cases} \\ \dot{\omega}_i = \dfrac{\dot{\omega}_{i-1} + \dfrac{1}{J} \cdot \left(-M_{bi} + P_z \cdot R \cdot \varphi_{x\ i-1}\right)}{2}; \\ \omega_i = \omega_{i-1} + \dot{\omega}_i \cdot dt; \\ \dot{v}_i = -\varphi_{x\ i-1} \cdot g; \\ v_i = v_{i-1} + \dot{v}_i \cdot dt; \\ s_{xi} = 1 - \dfrac{\omega_i \cdot R}{v_i}; \\ CASF_i = f(s_{xi}); \\ \varphi_{xi} = f_{sl} \cdot s_{ix} + CASF_i \cdot f_{st\ x}(1 - s_{ix}); \\ L_{xi} = L_{xi-1} + v_i \cdot dt; \\ L_{yi} = L_{xi-1} \cdot tg(\delta). \end{cases} \qquad (4)$$

IOP Conf. Series: Materials Science and Engineering **315** (2018) 012004 doi:10.1088/1757-899X/315/1/012004

Calculations were done for the following conditions:

- surfaces: «dry asphalt concrete » (f_{sl} = 1 and f_{st}=1.28) and «wet asphalt concrete » (f_{sl} = 0.7 and f_{st} = 0.93);
- steps of integration: 0.001s, 0.002s, 0.003s, 0.004s and 0.005s;
- slip angles: 0°, 2°, 4°, 6°, 8° and 10°.

Figure 1 presents an example of the calculation of parameters of braking using the system 4 for the following conditions: the integration step 0.005s, the surface coefficient: f_{sl}=1, f_{st}=1.28 and the method of average accelerations.

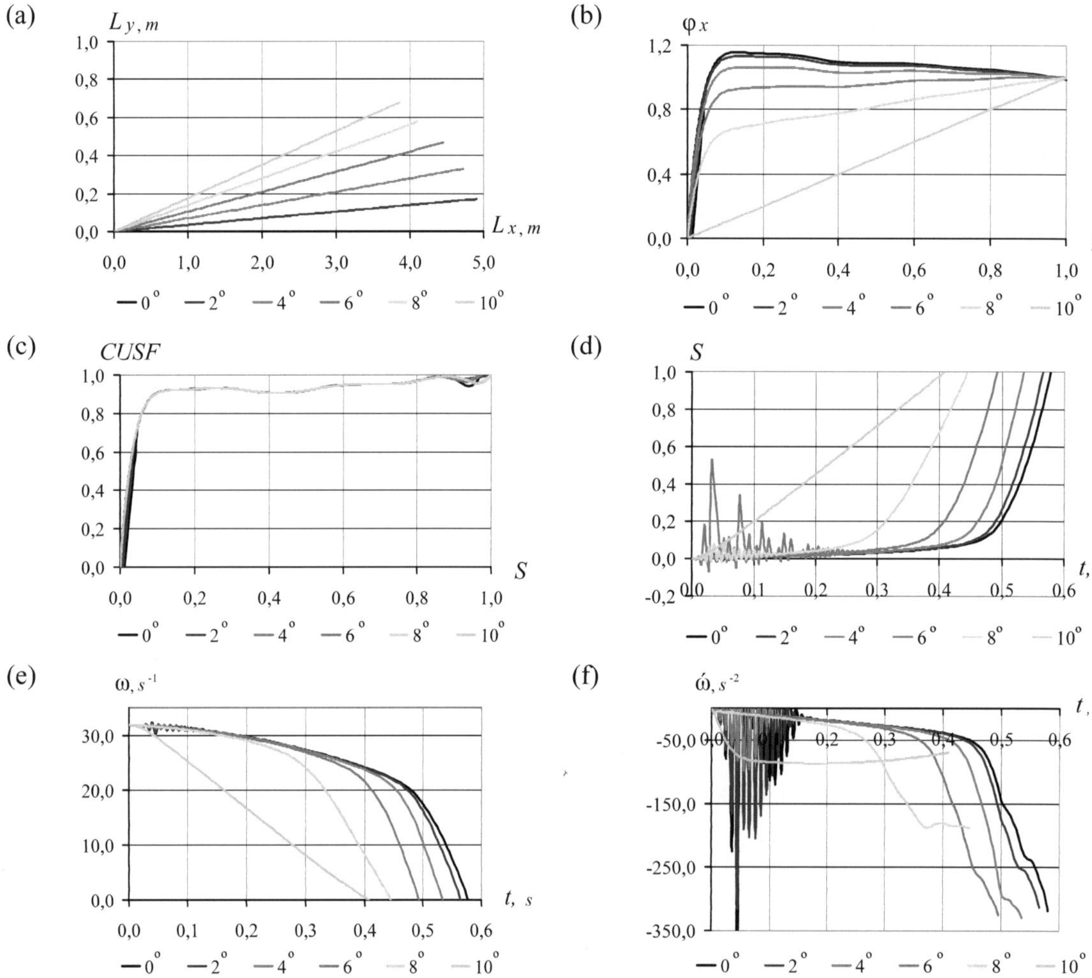

(g)

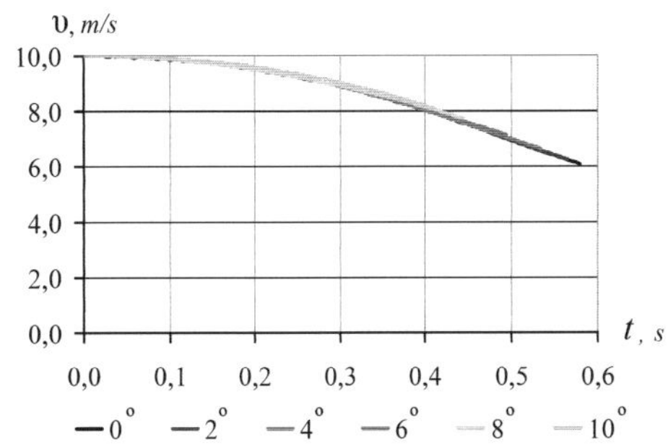

(h)

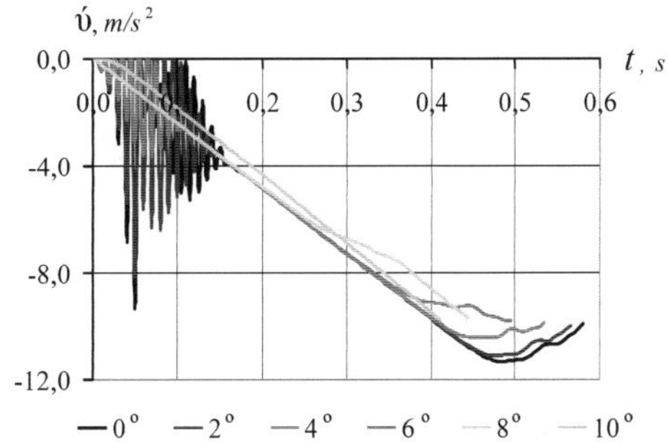

Figure 1. The motion parameters of vehicle wheel in the braking mode (integration step 0.005 s, surface coefficients: $f_{sl} = 1$, $f_{st}=1.28$, the method of average accelerations):
(a) – the dependence L_y of L_x; (b) – the dependence φ_x of s_x; (c) – the dependence *CUSF* of s_x; (d) – the dependence s_x of t; (e) – the dependence ω of t; (f) – the dependence $\dot{\omega}$ of t; (g) – the dependence v of t; (h) – the dependence $\dot{v}$ of t

On the basis of the performed analysis of the influence of calculation method on the stability of the calculation results of motion parameters of vehicle tires, overall it is found that the application of the method of average acceleration will significantly reduce the manifestation of the instability of the solution in the calculation of all motion parameters of the slowing down vehicle wheel for different conditions.

The application of this method allows to calculate the required parameters without the instability of the solution with the larger value of the integration step up to 0.004 s and more on all surfaces with different values of coefficients of static and sliding friction.

5. Practical significance

The results of research can be applied to the selection of control algorithms in vehicles motion control electronic systems and in the testing stand-simulators.

6. Conclusions

1) in the tasks, related to the real-time calculating of motion parameters of the vehicle elastic wheel during braking mode the total allowable range of integration steps is 0.001..0.005s;

2) the instability never manifests in the calculation of all motion parameters of a wheel for all slip angles when braking on all road surfaces when the integration step is 0.001s using numerical methods of first and higher orders. In other cases, the instability is manifested;

3) the strongest instability is manifested in the calculation of the angular and linear accelerations of the wheel;

4) the weakest instability is manifested in the calculation of the translational speed of the wheel and motions of the wheel center in longitudinal and transverse directions;

5) the amplitude and the duration of the interval of manifestation instability is less for large values of slip angle;

6) the amplitude and the duration of the interval of manifestation instability is less on more slippery surfaces;

7) the application of the suggested new method of average acceleration allows to significantly reduce (up to 100%) the manifestation of instability of the solution in the calculation of all parameters of motion of the elastic wheel for different braking conditions and entire range of steps of integration.

IOP Conf. Series: Materials Science and Engineering **315** (2018) 012004 doi:10.1088/1757-899X/315/1/012004

References

[1] Balakina E and Zotov N 2011 *Stability of motion of wheeled vehicles* (Volgograd: Volgograd State Technical University Press) p 464
[2] Balakina E and Zotov N 2015 Determination of the Mutual Arrangement of Forces, Reactions, and Friction Zones in the Contact Zone of an Elastic Wheel with a Solid Surface *Journal of Friction and Wear* (New York: Allerton Press) vol 36 No 1 pp 29–32
[3] Balakina E 2017 Calculation of the Geometric Position and the Sizes of the Static Friction and Sliding Friction Zones at the Point of Contact between an Elastic Wheel and a Firm Surface *Journal of Friction and Wear* (New York: Allerton Press) vol 38 No 2 pp 144–149
[4] Zarubin V and Krischenko A 2003 *Mathematical modeling in engineering* (Moscow: Bauman Moscow State Technical University Press) p 496
[5] Ivanov V 1986 Vibration of a car with anti-lock braking system during braking *Dissertation Cand. tehn. sciences* p 172
[6] Kim V and Furunzhiev R 2003 *Methodology of creating adaptive SAB ATS on the basis of force analysis* (Mogilev: Belarusian-Russian University Press) p 344
[7] Kokorev D 2004 Software simulation modeling for vibration tests *Mathematical modeling and limit problems* (Samara: Samara State Technical University Press) vol 2 No 5 pp 10-12
[8] Krantsov G 1994 Estimation of braking properties of the vehicle with automatic drive with a model method *Dissertation Cand. tehn. sciences* p 146
[9] Nikulnikov E, Kozlov Y and Balakina E 2007 Lateral force and motion stability of the vehicle at the braking mode *Automobile industry* (Moscow: Mechanical Engineering) No 12 pp 15-17
[10] Pimenov V 2001 Managed and numerical models with aftereffect *Abstract of Dissertation D-r. tehn. sciences* p 20
[11] Balakina E, Zotov N, Zotov V, Fedin A and Platonov I 2013 *Problems of modelling of dynamic processes in real time (on the example of vehicle brake dynamics)* ed S Bahmutov (Moscow: Mechanical Engineering) p 299
[12] Revin A 2002 *Theory of exploitation properties of automobiles and road-trains with ABS at braking mode* (Volgograd: Volgograd State Technical University Press) p 372
[13] Samarskiy A and Mihaylov A 1997 *Mathematical modeling: Ideas. Methods. Examples* (Moscow: Science) p 316

IOP Conf. Series: Materials Science and Engineering **315** (2018) 012005 doi:10.1088/1757-899X/315/1/012005

System analysis of vehicle active safety problem

S E Buznikov

FSUE NAMI, Moscow, Russian Federation

E-mail: sergey.buznikov@nami.ru

Abstract. The problem of the road transport safety affects the vital interests of the most of the population and is characterized by a global level of significance. The system analysis of problem of creation of competitive active vehicle safety systems is presented as an interrelated complex of tasks of multi-criterion optimization and dynamic stabilization of the state variables of a controlled object. Solving them requires generation of all possible variants of technical solutions within the software and hardware domains and synthesis of the control, which is close to optimum. For implementing the task of the system analysis the Zwicky "morphological box" method is used. Creation of comprehensive active safety systems involves solution of the problem of preventing typical collisions. For solving it, a structured set of collisions is introduced with its elements being generated also using the Zwicky "morphological box" method. The obstacle speed, the longitudinal acceleration of the controlled object and the unpredictable changes in its movement direction due to certain faults, the road surface condition and the control errors are taken as structure variables that characterize the conditions of collisions. The conditions for preventing typical collisions are presented as inequalities for physical variables that define the state vector of the object and its dynamic limits.

1. Introduction stating the purpose of the scientific work

The problem of the road transport safety affects the vital interests of the most of the population and is characterized by a global level of significance [1].

The search of ways to solve this problem is continuously conducted by the state legislative and executive bodies, public and international organizations.

However, expecting a quick solution without a clear understanding of the nature of the problem is unrealistic.

One of the conditions for improvement in the difficult economic situation is to focus on resolving the central problem of the market economy – creation of goods and services of the highest level of competitiveness.

The purpose of this scientific work is the system analysis of the issue of creation of competitive active safety systems.

One of the promising directions of resolving the problem of the vehicle traffic safety successfully developed by the leading international companies is the creation of a variety of active safety systems [2]. Such systems, including hardware and software, are designed to prevent collisions of particular types [3].

However, copying foreign technical solutions, information technologies, standards and regulations without consideration of the technological and manufacturing capabilities and without analyzing the conditions of vehicle operation in Russia will not produce a proper result.

While creation of new hardware, such as information sensors, actuators, interfaces etc. requires new materials and technologies, in the software field the range of creative activities is practically unlimited.

2. Definition of the multi-criterion optimization task in the area of technical solutions

Ensuring competitiveness of technical products is in fact directly related to solution of the problem of the multi-criterion (vector) optimization.

Competing objects are characterized by the performance vector:

$$Q = (q_1, q_2 \dots q_n)^T,$$

where: $q_i(R), 0 \leq i \leq n$ – individual performance indicators whose values depend on the technical solutions vector $R = (R_h, R_s)^T$, where R_h is the hardware vector and R_s is the software vector.

The main customer satisfaction indicators of the systems [4] in question include:

$q_0(R)$ – level of the external styling;
$q_1(R)$ – completeness of implemented functions;
$q_2(R)$ – level of monitoring and control errors;
$q_3(R)$ – level of power consumption from external sources;
$q_4(R)$ – degree of environmental impacts;
$q_5(R)$ – degree of influence on related systems;
$q_6(R)$ – level of failure resistance;
$q_7(R)$ – limitations of installation on objects;
$q_8(R)$ – operation costs;
$q_9(R)$ – system kit price.

The indicators $q_i(R)$ will be rendered to a minimized form and are dimensionless.

The special feature of the solution of the multi-criterion optimization problem under the market economy conditions is that the choice of the best option in some sense to a large extent is done not by the developer but by the end user. This means that the demand for products that do not belong to a set of compromises naturally ends up being zero. Also the fact that a product belongs to an optimum set does guarantee its commercial success with the customers as their preferences may change in unpredictable ways.

Objective properties will be acquired by technical solutions belonging to the record systems domain (R - systems) that are subject to the system of inequalities:

$$q_i(R_i) \leq q_{i\,min}, 0 \leq i \leq 9.$$

It is obvious that it is impossible to create systems with record performance using conventional development technologies and standard hardware and software, which inevitably leads to duplication of common solutions from the multitude of solutions that are not competitive.

The concept for creating record systems will be defined in the upstream research phase in terms of the consumer property indicator values, which specifies the coordinates of the development end target.

However, the process of moving to it takes certain time even when the developer knows how to implement the plans. This circumstance creates risks from the competitors who have the ability to stay ahead of developer and to enter the market with products of similar purpose soon.

As the direction of the search for new technical solutions that provide record performance one should consider the hardware domain R_h and the software domain R_s.

The analysis of the multi-criterion optimization problem for the record systems shows that its solution lies in the field of intelligent systems that are implemented in a minimal hardware configuration.

IOP Conf. Series: Materials Science and Engineering **315** (2018) 012005 doi:10.1088/1757-899X/315/1/012005

3. The results of building a structured set of typical collisions

For assessment of the set of potential collisions and ways to prevent them, a structured set of collisions is built with the aim of decomposing the main task into groups of individual tasks.

Let us consider the conditions of the collisions [5] of the controlled object and the obstacle on the road surface:

$$\begin{cases} |L_m(t) - L_{obs}(t)| \leq E_1; \\ |H_m(t) - H_{obs}(t)| \leq E_2, \end{cases} \tag{1}$$

where $L_m(t)$ and $H_m(t)$ are the coordinates of the center of mass of the controlled object in curvilinear coordinate system related to the topology of the route; $L_{obs}(t)$ and $H_{obs}(t)$ are accordingly the coordinates of the center of mass of the obstacle in the same coordinate system as the distance from the starting point of the route and the distance from the edge of the road surface; $E_1 > 0$ and $E_2 > 0$ are the constants commensurable with the dimensions of the object and the obstacle.

The collision condition (1) is the necessary and sufficient condition for the two objects to be at the moment of time t in the neighborhood of the same point on the road surface.

For the collision not to occur it is sufficient for at least one of the inequalities (1) not to hold within the travel time interval.

For solving this problem let us build a structured set of collisions. Let us assume that at moment of time $t_0 < t$, a collision does not occur and the inequalities of system (1) are not fulfilled.

Let us define what leads to a collision of the object and the obstacle at the moment of time t. For this let us put down the following system of inequalities:

$$\begin{cases} |L_m(t_0) + \Delta L_m(t) - L_{obs}(t_0) - \Delta L_{obs}(t)| \leq E_1; \\ |H_m(t_0) + \Delta H_m(t) - H_{obs}(t_0) - \Delta H_{obs}(t)| \leq E_2, where \end{cases} \tag{2}$$

$$\Delta L_m(t) = \int_{t_0}^{t} V_m(\tau)d\tau = \int_{t_0}^{t} [V_m(t_0) + a_m(\tau)\tau]d\tau;$$

$$\Delta L_{obs}(t) = \int_{t_0}^{t} V_{obs}(\tau)d\tau;$$

$$\Delta H_m(t) = \Delta\Psi_m(t)\Delta L_m(t);$$

$$\Delta H_{obs}(t) = \Delta\Psi_{obs}(t)\Delta L_m(t);$$

$$\Delta\Psi_m(t) = b_m^{-1}\int_{t_0}^{t} V_m(\tau)\psi_{cm}(\tau)d\tau + \int_{t_0}^{t} \Delta\omega_m(\tau)d\tau;$$

$$\Delta\Psi_{obs}(t) = b_{obs}^{-1}\int_{t_0}^{t} V_{obs}(\tau)\psi_{cobs}(\tau)d\tau + \int_{t_0}^{t} \Delta\omega_{obs}(\tau)d\tau;$$

$\Delta\Psi_m(t)$ and $\Delta\Psi_{obs}(t)$ are the unpredictable changes of the course angles of the controlled object and the obstacle;

$\Delta\omega_m$ and $\Delta\omega_{obs}$ are the additional components of the angular velocities of the controlled object and the obstacle occurring during wheel slippage and drift;

b_m and b_{obs} are the wheelbases of the controlled object and the obstacle;

ψ_{cm} and ψ_{cobs} are the steering angles of the object and the obstacle;

V_m and V_{obs} are the longitudinal velocities of the centers of mass of the controlled object and the obstacle;

a_m is the longitudinal acceleration of center of mass of the controlled object.

It should be noted that difference of the system of inequalities (2) for moment of time t from the inequalities at the moment of time t_0, is availability of non-zero summands $[\Delta L_m(t) - \Delta L_{obs}(t)]$ and $[\Delta H_m(t) - \Delta H_{obs}(t)]$, which, in fact, lead to collisions.

To build a structured set of collisions we use the Zwicky "morphological box" method [6]. To simplify the problem, let us assume that $V_m(t_0) \geq 0$, while unpredictable changes of the obstacle's heading do not exist $\Delta H_{obs}(t) = 0$.

IOP Conf. Series: Materials Science and Engineering **315** (2018) 012005 doi:10.1088/1757-899X/315/1/012005

We introduce three structural independent variables corresponding to driving conditions of the controlled object and the obstacle included in the equations of the additional summands $\Delta L_m(t)$, $\Delta L_{obs}(t)$ and $\Delta H_m(t)$:

$$\bar{a}_m = \begin{cases} 0, if\ a_m = 0 - driving\ at\ constant\ speed; \\ 1, if\ a_m > 0 - acceleration; \\ 2, if\ a_m < 0 - decelaration, \end{cases}$$

$$\bar{V}_{obs} = \begin{cases} 0, if\ V_{obs} = 0 - stationary\ obstacle; \\ 1, if\ V_{obs} > 0 - followed\ obstacle; \\ 2, if\ V_{obs} < 0 - oncoming\ obstacle, \end{cases}$$

$$\Delta\bar{\Psi}_m = \begin{cases} 0, if\ \Delta\widetilde{\Psi}_m(t) = 0 - \ with\ no\ destabilizing\ facts; \\ 1, if\ \Delta\widetilde{\Psi}_m(t) \neq 0 - \ due\ of\ assymetry\ of\ axle\ pairs\ of\ wheels; \\ 2, if\ \Delta\widetilde{\Psi}_m(t) \neq 0 - due\ to\ tire\ failure; \\ 3, if\ \Delta\widetilde{\Psi}_m(t) \neq 0 - due\ to\ separation\ of\ wheels\ from\ hubs; \\ 4, if\ \Delta\widetilde{\Psi}_m(t) \neq 0 - due\ of\ suspension\ or\ steering\ failure; \\ 5, if\ \Delta\widetilde{\Psi}_m(t) \neq 0 - due\ to\ longitudinal\ wheel\ slip; \\ 6, if\ \Delta\widetilde{\Psi}_m(t) \neq 0 - due\ to\ lateral\ wheel\ slip; \\ 7, if\ \Delta\widetilde{\Psi}_m(t) \neq 0 - due\ to\ asymmetry\ of\ brakes; \\ 8, if\ \Delta\widetilde{\Psi}_m(t) \neq 0 - due\ to\ road\ surface\ condition; \\ 9, if\ \Delta\widetilde{\Psi}_m(t) \neq 0 - due\ to\ driving\ errors. \end{cases}$$

The introduction of these independent structural variables implies consideration of collisions with stationary, oncoming and followed obstacles. Collisions with obstacles at the crossroads are equivalent to collisions with stationary obstacles.

Unpredictable changes of the heading in collisions may occur or not to occur as a result of unpredictable changes of the steering angle $\psi_c \neq 0$ or an additional angular frequency of revolution of the center of mass $\Delta\omega_m \neq 0$ resulting from wheel slippage or drift.

The set of states of the variable $\Delta\bar{\Psi}_m$ corresponds to the effect of certain factors that characterize the state of the object, possible faults, the road surface condition and the driving errors.

The set of typical collisions M_1 built from the above variables forms a group of 90 events in total.

The set of rollovers M_2 includes the events that may happen both on the road surface as well as due to departure from the road.

The set of rollovers $M_2(\Delta\bar{\Psi}_m)$ contains 10 events and also forms a complete group of events for rollovers.

The total capacity of sets M_1 and M_2 is 100 events and forms a model of problem suitable for its analysis by parts.

4. Analysis of a structured set of typical collisions

Introduction of a structured set of typical collisions by the decomposition of the main problem into individual problems that are more convenient for analysis and synthesis of the control actions, allows assessing the engineering level of various technical and organizational solutions aimed at alleviating the problem.

Figure 1 shows a structured set of typical collisions M_1 and rollovers M_2.

One of directions of solving the problem of collision prevention that evolved internationally in the recent decades is the construction of highways.

A highway is a multi-lane one-way road without intersections with other roads on the same level.

The highway topology enables prevention of collisions with oncoming obstacles, stationary obstacles and obstacles assumed as stationary from intersecting directions.

In this case from the structured set of collisions M_1 30 collisions with oncoming obstacles and 30 collisions with stationary obstacles are excluded. The remaining 30 collisions with followed obstacles and 10 rollovers due to driving errors, vehicle running condition and road environment stay as a potentially possible.

Implementation of this direction involves substantial costs and with current condition of the road network in Russia cannot be considered as the only possible solution.

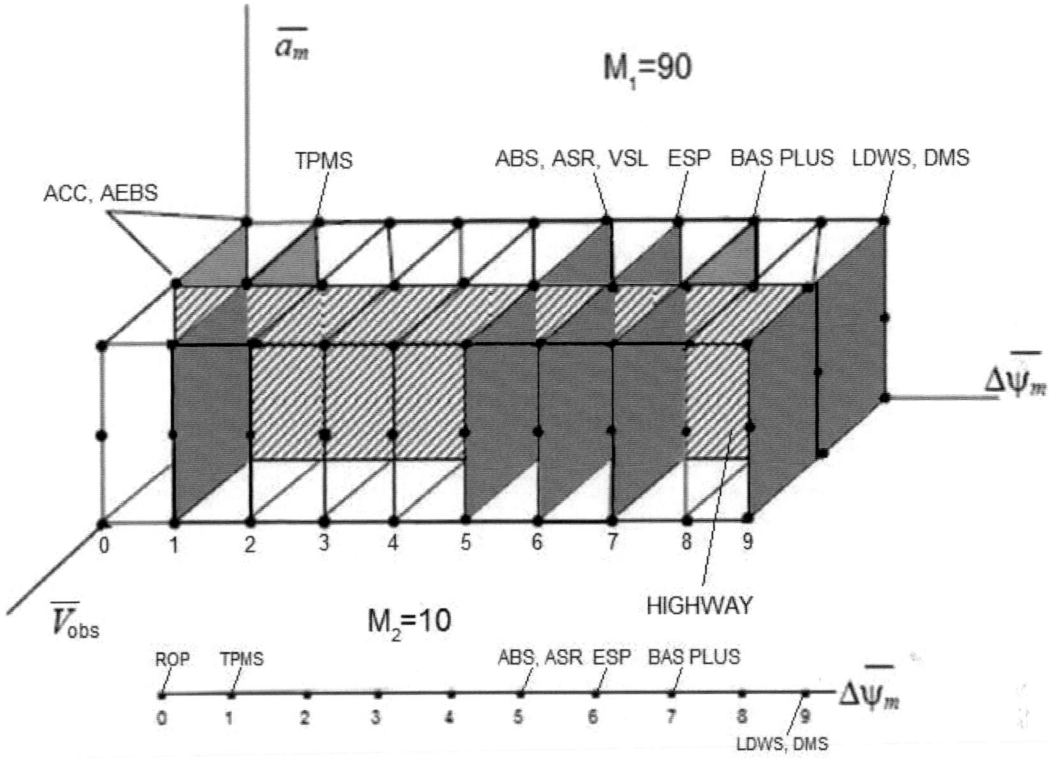

Figure 1. Structured set of typical collisions and rollovers.

Table 1. Characteristics of foreign active safety systems.

No.	Foreign active safety systems	Limitations of functioning in full configuration
1	AEBS	Braking with activation of ABS w/o consideration for distance to vehicle that follows
2	ACC	Maintaining speed and distance that are not safe
3	TPMS	Pressure in one of four wheels at $V_m > 40$ km/h; $\Delta P_i > 0.6$ bar
4	ABS, ASR	Significant reduction of braking efficiency on rough surfaces, on ice and snow
5	ESP	Inefficient for vehicles with a high center of mass and on dirt roads
6	LDWS	Needs recognizable road marking and forces driving in wheel track
7	DMS	Requires individual driver settings

Modern foreign systems (ABS, ASR, ESP, AEBS, TPMS, ACC, LDWS, DMS, etc.) designed to prevent certain groups of typical collisions of a structured set (see Figure 1) do not provide a complete solution of the general problem. At the same time, the existing active safety systems have a number of drawbacks that cannot be eliminated and significantly impair their effectiveness.

Table 1 shows the characteristics of foreign active safety systems that limit their application under Russian road and climatic conditions.

5. Task of dynamic stabilization of vehicles

The analysis of conditions of collisions of the controlled object with obstacles allows building a set of typical collisions, which allows decomposition of the general problem of collision prevention into subsets of individual problems. For defining typical collision groups, sufficient conditions for their prevention may be recorded in the form of inequalities for specific physical variables. Combining the groups of inequalities M_i for each of the variables x_i into one with upper and lower limits allows reducing the task of collision prevention to the problem of the dynamic stability controls stated most generally:

$$x_{i\,lim}^{L}(U,X,t) \le x_i(t) \le x_{i\,lim}^{U}(U,X,t), 1 \le i \le n \text{ during } U \in U_{adm}, \text{ where} \quad (3)$$

$x_i(t)$ is the i-th component of the state vector X;
$U = (U_1, U_2, \dots, U_m)^T$ is the vector of control actions with components U_j, $1 \le j \le m$;
U_{adm} is the admissible set of control actions;
$x_{i\,lim}^{U}(U,X,t) = \min[x_{i\,lim1}^{U}, x_{i\,lim2}^{U}, \dots, x_{i\,\lim r_i}^{U}]$ is the upper limit of the admissible values of x_i;
$x_{i\,lim}^{L}(U,X,t) = \min[x_{i\,lim1}^{L}, x_{i\,lim2}^{L}, \dots, x_{i\,\lim r_i}^{L}]$ is the lower limit of the admissible values of x_i.

The special feature of the task of the dynamic stability control for the class of objects in question is the fact that the overwhelming number of components of the state vector X are not measured and their dynamic limits $x_{i\,lim}^{U}$ and $x_{i\,lim}^{L}$ are not defined, which makes the control object not fully observable. Moreover, the dimension of the control vector $m \ll r$ makes the object not fully controllable. These properties are crucial in construction of modern automated control systems with operator participation. The task involves augmentation of the state coordinates that are not measured and their dynamic limits based on experience, intuition, prediction and other individual characteristics of the operator.

For automatic control of the object of the given class the solution of the most comprehensively stated problem of the dynamic stability control is necessary.

From standpoint of modern control theory, the solution of the task of dynamic stability control with the finite time interval $(t_1 \div t_2)$ will be assessed by value of the quadratic functional of the control quality with the functional limitations for the technical and economic indicators of the system implementing the control taken into account:

$$Q(t_2) = \int_{t_1}^{t_2} \sum_{i=1}^{n} C_{1i}\left[x_i(\tau) - x_{i\,lim}^{U}(U,X,t)\right]^2 d\tau + \int_{t_1}^{t_2} \sum_{i=1}^{n} C_{2i}[x_i(\tau) - x_{i\,lim}^{L}(U,X,t)]^2 d\tau \to min \quad (4)$$

with $U \in U_{adm}$, $R = (R_H, R_S)^T \in R_{adm}$; $q_i(R) \le q_{i\,adm}$, $1 \le i \le l$,

where $C_{1i} = \begin{cases} 0, if\ x_i(t) \le x_{i\,lim}^{U}; \\ C_{1i}^{*} \gg 0, if\ x_i(t) > x_{i\,lim}^{U}; \end{cases}$ $\quad C_{2i} = \begin{cases} 0, if\ x_i(t) \ge x_{i\,lim}^{L}; \\ C_{2i}^{*} \gg 0, if\ x_i(t) < x_{i\,lim}^{L}; \end{cases}$

R is the vector of technical solutions; R_H и R_S are the vectors of hardware and software solutions;
R_{adm} is the admissible set of technical solutions;
$q_i(R)$ is the i-th component of the vector of technical and economic indicators;
$q_{i\,adm}$ is the admissible value of i-th component of the vector of technical and economic indicators.

The set of the main technical and economic indicators includes the levels of power consumption, the environmental impacts, the versatility, the fault resistance, the operating costs and the cost of hardware and software.

The best solution of the problem stated as proposed is and algorithm for triggering control actions that provide dynamic stability control of the object's state vector implemented within software and hardware environment that satisfies the system of constraints for the technical and economic indicators.

IOP Conf. Series: Materials Science and Engineering **315** (2018) 012005 doi:10.1088/1757-899X/315/1/012005

6. Conclusions

Based on the analysis performed the following conclusions can be drawn:

- Creation of active safety systems represents the most promising direction for resolving the problem of the motor vehicle traffic safety.
- Modern foreign active safety systems potentially can prevent up to 57 typical collisions out of possible 100, but this is not enough to consider the problem solved.
- The augmentation of the functions of the future systems by solutions for prevention failures of tires, steering and suspension, separation of wheels from hubs along with improvement of other solutions should allow creation of a full-featured active safety system for the road and climatic conditions of the Russian Federation.
- For achievement of a significant reduction of the accident rate all vehicle in operation, including the vehicles of lowest price segments, shout be fitted with such systems.
- Attaining a highest competiveness status by such systems is possible with a minimal configuration of technical means through the use of more advanced mathematical models and information processing algorithms.

Acknowledgments

This paper was prepared based on the results of the application research performed with the financial support of the Ministry of Education and Science of Russia. Agreement No. 14.624.21.0048, unique project identifier RFMEFI62417X0048.

References

[1] Ilarionov V A 1989 The system analysis of road traffic and road accidents *Collection of Science Proc.* (Moscow:MADI) p 107

[2] Davis LC Stability of adaptive cruise control systems taking account of vehicle response time and delay *Physics Letters A*. 2012 **376** p 2658–2662

[3] Eunbi Jeong Evaluating the effectiveness of active vehicle safety systems *Accident Analysis & Prevention* 2017 **100** p 85–96

[4] Buznikov S. E The principles of construction record information control systems for commercial purposes *Materials of XIII scientific-technical Conf. "Sensor 2001"* p 171-173

[5] Buznikov S E Current state and prospects of automobile active safety systems development *Proc. of XV Int. Conf. "Control problems of sophisticated systems safety"* (Moscow:RGGU) **2** p 207-11

[6] Zwicky F The morphological approach to discovery, invention, research and construction New methods of thought and procedure (Passadena) 1967 p 273–297

IOP Conf. Series: Materials Science and Engineering **315** (2018) 012006 doi:10.1088/1757-899X/315/1/012006

The problem of the driverless vehicle specified path stability control

S E Buznikov, D V Endachev, D S Elkin and V O Strukov

FSUE NAMI, Moscow, Russian Federation

E-mail: sergey.buznikov@nami.ru

Abstract. Currently the effort of many leading foreign companies is focused on creation of driverless transport for transportation of cargo and passengers. Among many practical problems arising while creating driverless vehicles, the problem of the specified path stability control occupies a central place. The purpose of this paper is formalization of the problem in question in terms of the quadratic functional of the control quality, the comparative analysis of the possible solutions and justification of the choice of the optimum technical solution. As square value of the integral of the deviation from the specified path is proposed as the quadratic functional of the control quality. For generation of the set of software and hardware solution variants the Zwicky "morphological box" method is used within the hardware and software environments. The heading control algorithms use the wheel steering angle data and the deviation from the lane centerline (specified path) calculated based on the navigation data and the data from the video system. Where the video system does not detect the road marking, the control is carried out based on the wheel navigation system data and where recognizable road marking exits – based on to the video system data. The analysis of the test results allows making the conclusion that the application of the combined navigation system algorithms that provide quasi-optimum solution of the problem while meeting the strict functional limits for the technical and economic indicators of the driverless vehicle control system under development is effective.

1. Introduction

Currently, the effort of many leading international automotive companies is focused on creation of driverless transport for transportation of cargo and passengers. Among many practical problems arising while creating driverless vehicles, the problems of the specified path stability control and collision prevention are in the center.

It is obvious that without their successful solution optimistic forecasts of practical application of driverless vehicles are meaningless.

The purpose of this research is formalization of the problem of the specified path stability control in terms of a quadratic functional of the control quality, the comparative analysis of the possible solutions and justification of the choice of the optimum technical solution.

2. Definition of the control task

As a quadratic functional of the control quality within a finite time interval $(t_1 \div t_2)$ is proposed the use of the following expression of $Q(t_2)$ is proposed :

IOP Conf. Series: Materials Science and Engineering **315** (2018) 012006 doi:10.1088/1757-899X/315/1/012006

$$Q(t_2) = \int_{t_1}^{t_2} C_0[\Delta H(\tau)]^2 d\tau, \qquad (1)$$

where $\Delta H(\tau)$ is the deviation from the specified path; $C_0 > 0$ is the scaling factor.

Given the fact that control action $U(t)$ is restricted in absolute value and its implementation is in the software and hardware environment$R = (R_H, R_S)^T$, which should satisfy a set of limitation for the technical and economic indicators $q_i(R)$, the problem of the optimum control is defined as:

$$Q(t_2) = \int_{t_1}^{t_2} C_0[\Delta H(\tau)]^2 d\tau \to \min \qquad (2)$$

with $U \in U_{adm}$; $R \in R_{adm}, q_i(R) \leq q_{iadm}, i \leq i \leq r$, where U_{adm} and R_{adm} are the admissible sets of control actions and technical solutions.

The main technical and economic indicators $q_i(R)$ include the level of power consumption, the real environmental impacts, the failure resistance, the operating costs and the cost of the hardware and software kit.

For generating a set of variants of the hardware and software solutions the Zwicky "morphological box" method [1] in the hardware R_H and software R_S environment is used.

The minimization of the quadratic functional of the control quality when meeting the functional limitations (2) for the technical and economic indicators of the control system should be accomplished through the limited control action on the steered wheels which changes the vehicle's heading.

The best solution of the problem as defined will be a control algorithm of the heading regulator providing deviation from the specified path which is close to zero implemented in the software and hardware environment that meets the system of constraints for the technical and economic indicators.

3. The results of solving the problem of the specified path stability control

The control task is subdivided into two independent tasks: construction of the specified path and the driving path stability control.

To solving the first problem one can use: path planning on an electronic terrain map [2] and path recording based on the data of the satellite, inertial and odometric (wheel) navigation systems.

For solving the second task once can also the above presented navigation systems to determine the positioning and the orientation of driverless vehicle [3], as well as video systems allowing determination of the vehicle's positioning on the lane in relation to the visible marking.

The output data of the navigation systems based on different principles, both for specified path planning and the path stability control are the evaluations of the heading $\widehat{\Psi}_m(t)$ and the object's position $\hat{L}_y(t)$ and $\hat{L}_x(t)$ in the Cartesian coordinate system.

The evaluations $\widehat{\Psi}_m(t)$, $\hat{L}_y(t)$ and $\hat{L}_x(t)$ together with the true values contain the measurement noise of certain physical nature in the form of influence of real external factors.

4. Comparative analysis of possible technical solutions

For solving the problem of the driverless vehicle path stability control the control system can include various navigation systems. The positioning determination accuracy is influenced by various factors.

For instance satellite navigation systems (SNS) do not cope with the navigation task under the conditions of limited radio visibility (in tunnels, densely built-up areas, in areas with tall trees, etc.), as well as under the conditions of the electromagnetic noise.

The inertial navigation systems (INS) with their operating principle based on integration of accelerations and angular velocities along the orthogonal axes are characterized by error accumulation as the route length increases and the influence of road surface irregularities and off-road conditions.

The wheel navigation system (WNS) with the operating principle based on integration of the center of mass velocity V_m and the wheel steering angle FiC is also characterized by the integration error accumulation and influence of the road surface irregularities and off-road conditions on the navigation performance.

The road marking detection video system (LDW) is designed to detect deviations of the vehicle's centerline from the middle of the lane. The application of this system is limited by the lane lighting conditions, precipitation and recognizable marking.

Table 1 summarizes the results of the qualitative comparative analysis of the impact of the real environmental factors on the performance of the above navigation systems. The "-" sign marks the negative impact on the navigation system performance, and the "+» sign marks no negative influence.

Table 1. Qualitative assessment of impact of real factors on performance of navigation systems.

Exposures	Navigation system			
	SNS	INS	WNS	LDW
Limited radio visibility	—	+	+	+
Electromagnetic noises	—	+	+	+
Rough surface, off-road	+	—	—	+
Lighting level, precipitation	+	+	+	—
Long travel distance	+	—	—	+
Road marking availability	+	+	+	—
Turns with radii <20 m	+	+	+	—
Reversing	+	+	—	—

The analysis of the environmental impacts on the performance of the navigation systems shows that none of the systems in question copes with the task under the urban traffic conditions.

It appears that for obtaining an acceptable result a combination of two or more navigation systems with different operation principle is necessary. From the point of view of the additional equipment cost, the combined WNS + LDW system has significant advantage compared to other options. For driving on local paths of limited length with smooth surfaces without reversing to use of the WNS only is enough. Under more difficult driving conditions a combined SNS + WNS + VMD navigation system with higher cost could be used.

5. Differential equation system for the center of mass motion

For solution of the navigation task for the driverless vehicle in the Cartesian coordinate system the following system of equations is used:

$$\begin{cases} \dot{L}_y = V_m cos\Psi_m; \\ \dot{L}_x = V_m sin\Psi_m; \\ \dot{\Psi}_m = \omega_m + \Delta\omega_m, \end{cases} \tag{3}$$

where V_m is the longitudinal velocity of the center of mass (m·s^{-1}), Ψ_m is the heading angle (rad), ω_m is the angular speed of the center of mass while turning (rad·s^{-1}), $\Delta\omega_m$ is additional component of the angular speed of the center of mass during wheel slippage and drift (rad·s^{-1}).

The solution of the system of equations (3) for the center of mass motion using the Euler formula for $\omega_m = V_m R_m^{-1}$ and $R_m = bFiC^{-1}$, where R_m is the turn radius, b is the vehicle's wheelbase, will have the following form:

$$\begin{cases} L_y(t) = \int_{t_0}^{t} V_m(\tau) cos\Psi_m(\tau) d\tau + L_y(t_0); \\ L_x(t) = \int_{t_0}^{t} V_m(\tau) sin\Psi_m(\tau) d\tau + L_x(t_0); \\ \Psi_m(t) = b^{-1} \int_{t_0}^{t} V_m(\tau) FiC(\tau) d\tau + \int_{t_0}^{t} \Delta\omega_m(\tau) d\tau + \Psi_m(t_0), \end{cases} \quad (4)$$

where $L_y(t_0)$, $L_x(t_0)$, $\Psi_m(t_0)$ are the initial values of the coordinates and heading at the moment of time t_0.

For solving the system (4) high precision measurements of V_m, FiC and $\Delta\omega_m$ performed either by dedicated or virtual information sensors [4] would be necessary. The last technical solution option is the preferred one, as it improves almost all technical and economic indicators, including the cost and the operating expenses.

6. The path stability control system

For solving the problem of the driverless vehicle specified path stability control a double-loop computer control system was developed that uses the data of the wheel navigation system and the road marking detection video system. The first control loop must ensure maintaining of the specified heading, while the second – the reduction of the deviation from the specified path when driving on by parallel courses.

The computational portion of the system includes three computers integrated into a hybrid telemetry information and control commands exchange network. The first computer provides initial processing of the information from the wheel speed sensors and its conversion into driving parameters of movement and electronic switches for the switching of electric power steering motor. The second computer performs the navigation task, generates the commands for the electric power steering and serves as the system's control panel. The network interface between the first and second computer implemented using the Bluetooth protocol and the Bluetooth RFCOMM (Radio Frequency Communication) stack. The third computer performs the task of processing of images from the TV camera for detecting the road marking, determining the lane center and the deviation of vehicle's centerline from the lane centre. The telemetry data between the third and first computer is transmitted over the serial interface.

The equivalent schematic diagram of a double-loop path stability control system is shown in figure 1.

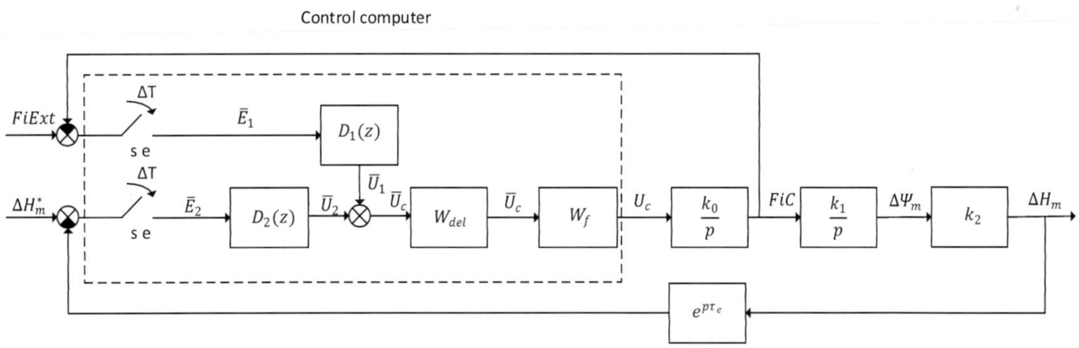

Figure 1. The equivalent schematic diagram of a double-loop computer path stability control system.

7. The single-step wheel steering angle regulator

The first control loop includes the electric motor of the power steering connected to computer-controlled reversible electronic switches. The regulator program is implemented on the navigation computer that receives information on the driving parameters, speed and wheel steering angle from the first computer, generates and sends the control commands to the first computer connected to the electronic switches with the PWM signal.

The computer portion of the system includes a pulse element that performs the quantization of the signals in time with the period of ΔT, the transfer function of the regulator $D_1(z) = k_P$, the delay element

$W_{del}(p) = e^{-p\Delta T}$ for the time of data exchange with the navigation computer and a clamper with the transfer function $W_f(p)$. The continuous part of the system in the form of the reversible power steering motor is represented by the integrating element $W_m = k_0 p^{-1}$ with a gain k_0 that relates the output signal of the steering angle FiC to the input in the form of the control voltage U_c.

The finite-difference equation of the motor in discrete time can be presented as follows:

$$FiC(k) = FiC(k-1) + \Delta T k_0 U_c(k-1); \quad (5)$$

In case a proportional regulator is used with the gain of k_p with the delay step of ΔT the following equation results:

$$U_1(k-1) = k_p[FiExt(k-2) - FiC(k-2)]; \quad (6)$$

The equation of the dynamics of $FiC(k)$ given (5) and (6) reduces to the form:

$$FiC(k) = FiC(k-1) - \Delta T k_p k_0 FiC(k-2) + \Delta T k_p k_0 FiExt(k-2); \quad (7)$$

Considering that during the pause time interval of pause the switches transfer the motor to the dynamic braking mode and $FiC(k-1) \approx FiC(k-2)$ and if $k_p = (k_0 \Delta T)^{-1}$, then $FiC(k) = FiExt(k-2)$.

To compensate the delay the setting value $FiExt$ recorded for the specified path shifts two steps forward with a linear approximation of the intermediate values and in that case $FiC(k) \approx FiExt(k)$ and the effect of the delay is completely compensated.

8. Path deviation control

The specified path stability control is performed on sections that are nearly straight $|FiExt| \leq \Psi_{lim}$ also by controlling the reversible electric power steering motor over to the second loop (Figure 1).

The equivalent schematic diagram of the computer path stability control system includes three additional elements: the heading change integrator $\Delta\Psi_m$, the proportional deviation change ΔH_m element and the element for extrapolation of its values with the coefficients $k_1 = b^{-1}V_m$; $k_2 = V_m \Delta T$ and τ_e, respectively. The continuous portion of the initial system contains two integrators connected in series and potentially may be prone to persistent oscillations. For giving to the system exponential stability a differential regulator with $D_2(z) = k_c k_{main}^{-1}(1 - z^{-1})\Delta T^{-1}$ is used:

$$U_2(k) = [U(k) - U(k-1)]\Delta T^{-1}, \quad (8)$$

where $U(k) = k_c k_{main}^{-1}(k) E(k)$; $k_{main} = b^{-1} V_m(k) \Delta T k_0$; $E(k) = \Delta H_m^*(k) - \Delta \widehat{H}_m(k)$; k_c is the factor to be set.

The assessment of the current deviation from the specified path ΔH_m is made based on the data from the WNS and LDW from condition:

$$\Delta \widehat{H}_m(k) = \begin{cases} \Delta H_{TV}, \text{if } |\Delta H_{TV}| \leq E_1 \text{ и } |\Delta H_N| \leq E_2 \text{ and } |FiExt| \leq E_3; \\ \Delta H_N \text{ otherwise,} \end{cases} \quad (9)$$

where ΔH_{TV} is deviation from specified path according to LDW (m); ΔH_N is the deviation from specified path according to the WNS (m); E_1, E_2 and E_3 are the parameters subject to setting.

The compensation of the delay introduced by the data exchange programs is achieved by extrapolating of the deviations from the lane center by defining distance at which the middle of the lane is evaluated:

$$\Delta L_d = \begin{cases} \Delta L(V_m), \text{если } \Delta L(V_m) > \Delta L_{min}; \\ \Delta L_{min}, \text{если } \Delta L(V_m) \leq \Delta L_{min}, \end{cases} \quad (10)$$

where $\Delta L(V_m) = V_m \tau_e - \Delta L_n$ (m); ΔL_n is the distance from front bumper to the center of mass (m); ΔL_{min} is the minimal range of the marking detection (m).

IOP Conf. Series: Materials Science and Engineering **315** (2018) 012006 doi:10.1088/1757-899X/315/1/012006

The deviation ΔH_N is defined as the height of the navigation triangle constructed on two nearest points of the specified path and the current position point determined from the WNS.

9. Experimental results

The research tests of the specified path stability control system were carried out at the sites of FSUE NAMI and Dmitrov proving ground.

The tests were carried out on the electric Lada Kalina vehicle. The control system set-up included the following:

- standard wheel speed sensors of the ABS system;
- standard electric power steering;
- oCam-5CR0-U video camera installed inside the car on the windscreen;
- ODROID-XU4 microcomputer ;
- primary information processing and control unit;
- a tablet computer performing the functions of the WNS and the control panel;
- autonomous control system power supply.

Figure 2 shows the frame of the video recording of the test drive on the circuit that includes a straight section with recognizable marking, a left 180 degree turn without marking and a straight section with marking that returns to the starting point. The deviation from the center of the lane ΔH_{TV} is defined as the variation of coordinates of the vehicle's longitudinal axis and the center of the lane.

Figure 2. Frame of video recording of the straight section with recognizable marking.

In figure 3 shows time diagrams of the parameters of the driving on the straight section with the recognizable marking.

The analysis of experimental results confirms the exponential convergence of the deviation from centre of lane when using the differential regulator algorithm with the deviation extrapolation.

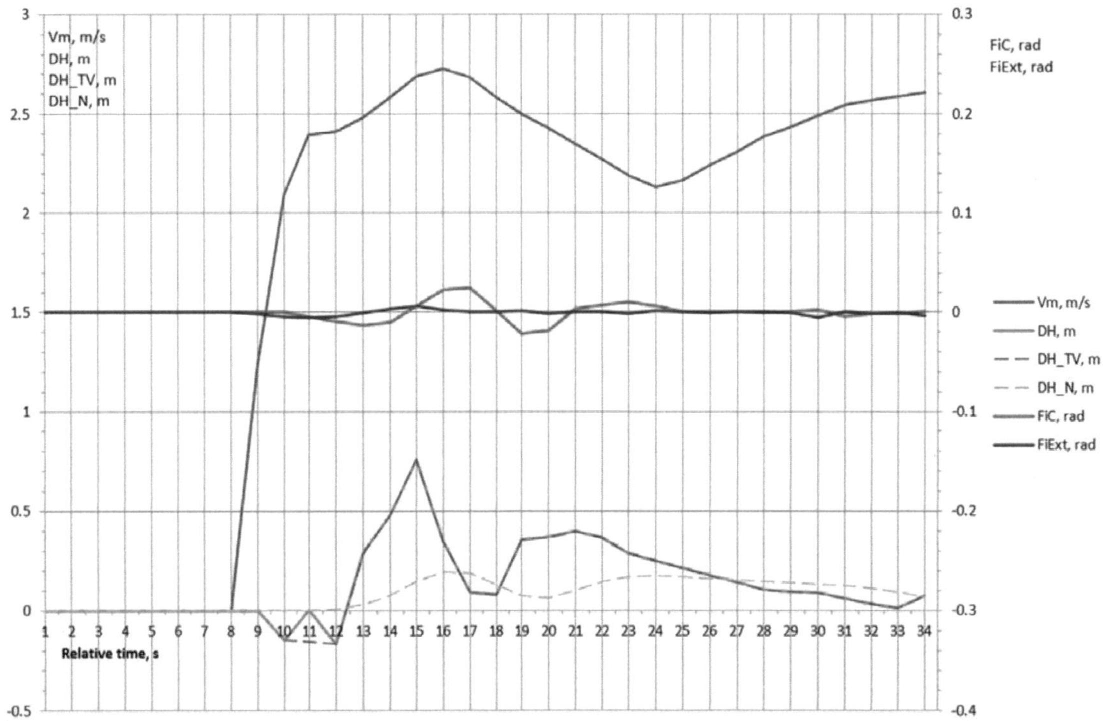

Figure 3. Time diagrams of the motion parameters on the straight section with recognizable marking.

Figure 4 shows the time diagrams of motion parameters at the turn with no visible marking.

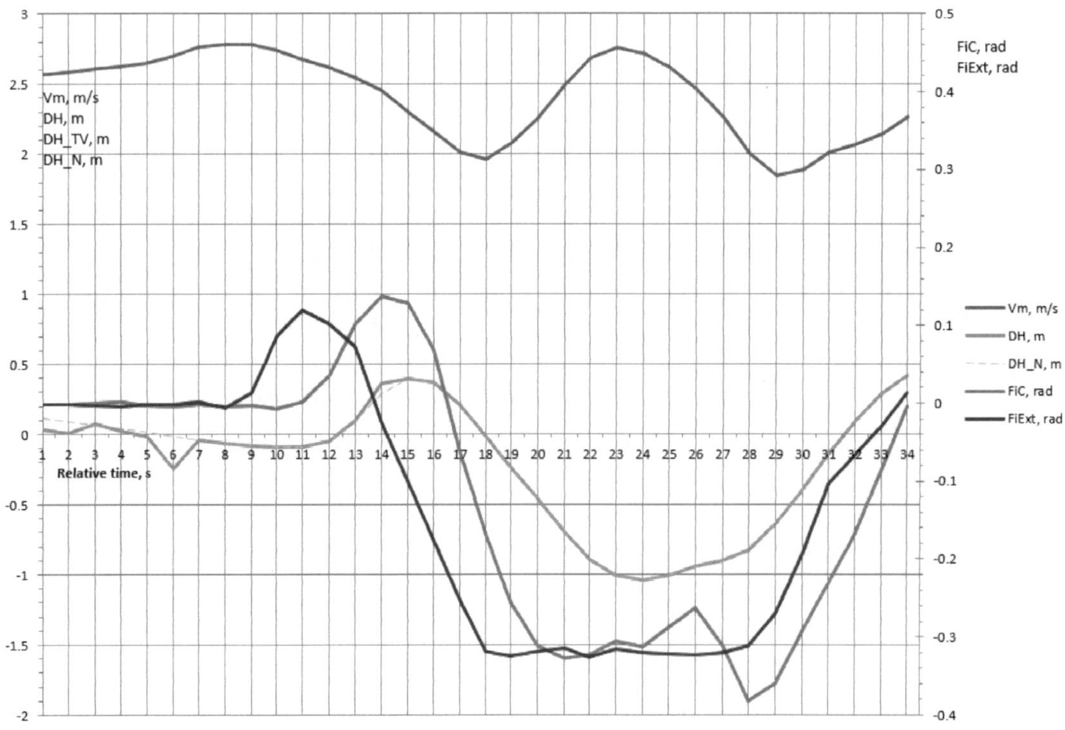

Figure 4. Time diagrams of the motion parameters at the turn with no visible marking.

The analysis of the results confirms the feasibility of the algorithm of the single-step regulator of the wheel steering angle with compensation of delay introduced by the software.

Conclusions

The analysis of the results of undertaken research of the problem of the driverless vehicle specified path stability control allows drawing the following conclusions:

- The formalized definition of the stability control problem in terms of minimization of a quadratic functional of the quality of control with consideration for the functional limitations for technical and economic indicators is adequate for the purposes of creating competitive systems;
- A comparative qualitative analysis of the navigation systems for actual operating conditions leads to conclusion that combining them is necessary;
- For circuits of limited length with sections with of recognizable the combination of the WNS and LDW gives the technical solution with the minimal configuration of additional technical means and with the optimum level of technical and economic indicators;
- The double-loop computer control system with a single step heading regulator and the virtual sensors of the center of mass velocity, the wheel steering and wheels and the additional angular speed during wheel slippage and drift and a differential deviation stabilizer that has been developed ensures compensation of the delays introduced by the software and the exponential stability with their finite step ΔT;
- The results of test driving do not contradict to the theoretical research and allow considering the achieved solution of the problem as effective.

Acknowledgments

This paper was prepared based on the results of the application research performed with the financial support of the Ministry of Education and Science of Russia. . Agreement No. 14.625.21.0043. Unique project identifier RFMEFI62517X0043.

References

[1] Zwicky F 1967 The morphological approach to discovery, invention, research and construction *New methods of thought and procedure* (Passadena) p 273-97

[2] Noreen I, Khan A and Habib Z 2016 Optimal Path Planning using RRT* based Approaches: A Survey and Future Directions *Int. Journal of Advanced Computer Science and Applications* **7** **(11)** p 97-107

[3] Kiss D and Papp D 2017 Effective Navigation in Narrow Areas: A Planning Method for Autonomous Cars *IEEE 15th Int. Symp. on Applied Machine Intelligence and Informatics (SAMI 2017)*

[4] Buznikov S E 2009 The method of constructing information virtual sensors for car's active safety systems *Proc. of XVII Int. Conf. "The management problems of safety in complex systems"* (Moscow: Russian State University for the Humanities Press) pp 420-4

IOP Conf. Series: Materials Science and Engineering **315** (2018) 012007 doi:10.1088/1757-899X/315/1/012007

Application of virtual physical modeling technology for the development of elements of the autonomous (unmanned) vehicles' systems

V Dygalo, M Lyaschenko, G Boyko and L Dygalo

Volgograd State Technical University, Volgograd, Russia

E-mail: dygalo@vstu.ru

Abstract. The article presents examples of the technical implementation devices' for developing algorithms for controlling brake system modules of autonomous vehicles.

1. Problem

Autonomous vehicle developers' has the question considering developing algorithms for controlling of wheel brake mechanism [1, 9, 14]. At the same time it's necessary to solve the problem of ensuring correct operation of the brake system units in accordance within its signals. The ways of realizing these tasks can be presented in the form of figure 1 [3].

In this process technical progress is possible only extensive use of modeling methods. There are large numbers of software tools for modeling technical systems [8]. However, along this path, there are significant difficulties that can reduce the effectiveness of modeling. In accordance to the class of problems it is an adequate description of the flow of the working fluid along the brake lines and the channels of the antilock system modulator [2, 13]. Often, the elements of the latter are products of pilot production with an unworked technology and therefore with unknown characteristics with the required accuracy. From the theory of the motion of liquids and gases along the main lines, it follows that a change in the angle of the flow axis, the shape of the channel, or the type of clearance in the control valve has a significant effect on local resistance. Aim of improving the design of the multiphase brake system modulator, this creates significant difficulties in finding its characteristics and causes a significant error in determining them for solving the modeling problem. To this, we need to add the need to take into account the transient and resistance forces in solenoids of the control valves. In the case of modeling the elements of the braking system, most often, important elements are recreated for the research, and the others are neglected, this simplifies the model. Both the reconstruction and accurate description is cumbersome or impossible at this stage and level of research. An example is the consideration of technological features of the manufacture of system elements - the shape and arrangement of the channels of the hydraulic drive system. To create a model - absolutely accurately describing the phenomena in the brake drive - is quite a difficult task at this stage.

IOP Conf. Series: Materials Science and Engineering **315** (2018) 012007 doi:10.1088/1757-899X/315/1/012007

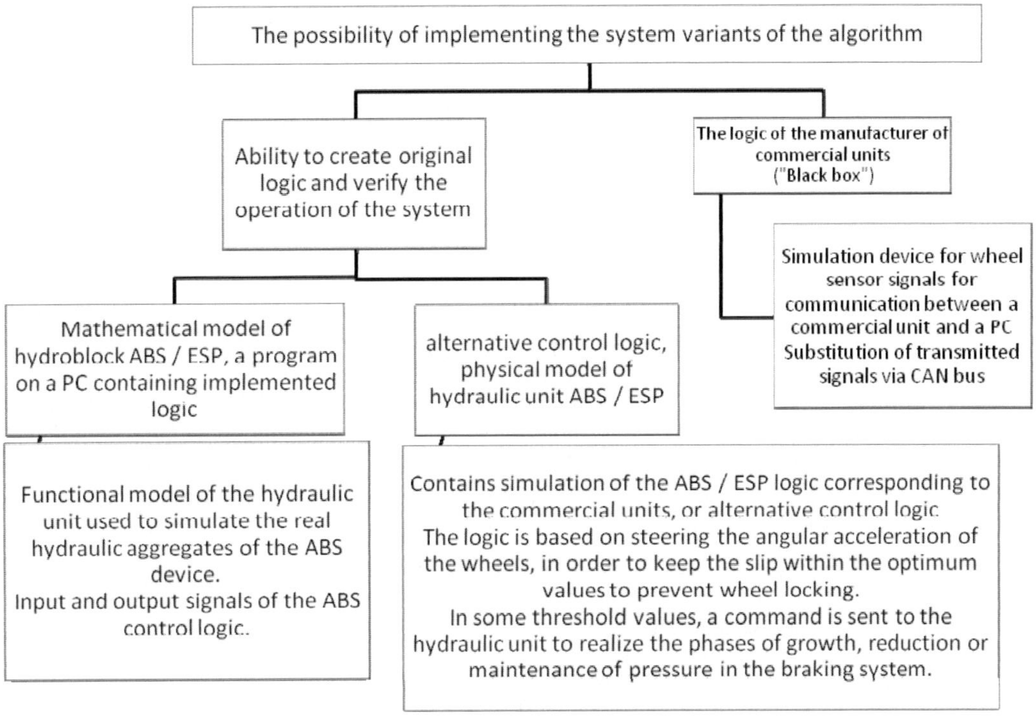

Figure 1. Block diagram of the ways of implementing variants of the system operation algorithm.

2. Solutions

One of the possible solutions of this situation may be using of virtual physical modeling technology [3, 7, 10. 11, 12], i.e. applied to this task, simulating the movement and braking of the vehicle with a virtual mathematical model, and taking into account the technological features of manufacturing the elements of the system can be taken into account by including real physical aggregates and actuating mechanisms in the system.

When applying this method, you can trigger the system and the brake mechanism in two ways:

First. Using a special mode dial that simulates the signals of the wheel speed sensor or software loaded into the control unit, which imitates signals at the program level in a special mode.

Second. Direct control of actuators. This method is more productive if, when designing an unmanned vehicle, researchers need a direct link to the system's executive mechanisms to develop alternative vehicle control logic.

3. Changing the input control signals of ECM (signal simulator)

It is necessary to know the signal parameters for device design process. But often such data are not found anywhere in the literature. This leads to the fact that researchers have to determine the input criteria for the operation of the system.

In this case several well-known vehicles of the same brand, model and model year are taken from the main characteristics of the wheel speed sensor of each wheel, as well as from the vehicle speed sensor to determine the average values. The signals with the received characteristics are supplied to the input of the system control unit with the aid of a master. In this case, a pulse generator capable of providing emulation of the signals from the wheel rotation sensors and the vehicle speed sensor can be used as the setpoint generator. The generator is controlled dynamically as from a personal computer, so it can work according to a predetermined law.

Let us consider an example of the development of a simulator of wheel sensor signals through an inexpensive National Instruments Ni 6008 USB device [4].

All elements of the electronic sensor system receive the supply voltage from a separate integrated circuit, while the "mass" of the sensor is connected by a wire, through which the signal is transmitted, to the "mass" of the computer system ABS or ESR. For Peugeot 307, the minimum sensitivity threshold of the sensor is 0.3 km / h. If the vehicle is stationary, the sensor is in standby mode. To activate the sensor, a new signal is required: a six-fold rise and fall in the amplitude. The sensor supply voltage is +12 V.

At a speed of 150 km / h, with car wheels with a radius of 0.316 meters and with 50 teeth on the rotor, the frequency of the pulses induced in the sensor is:

$$50\left[\left(\frac{150}{3,6}\right)\left(\frac{1}{0,316}\right)\left(\frac{1}{2\pi}\right)\right] = 1050 Hz$$

Sampling rate of the used ADC National Instruments Ni 6008 USB is 1000 Hz. This frequency is too low to avoid aliasing, since the consequence of the Kotel'nikov theorem (or the Whittaker-Nyquist-Shannon theorem) states that any analog signal can be reconstructed with any accuracy from its discrete readings taken at a frequency of at least in two greater maximum frequency, which is limited by the spectrum of the real signal.

Therefore, it was decided to make a circuit that converted the voltage in proportion to the gain factor G. This voltage comes to the voltage-frequency converter (hereinafter LPC), where it is converted into waves of rectangular shape with a frequency proportional to the voltage (figure 2).

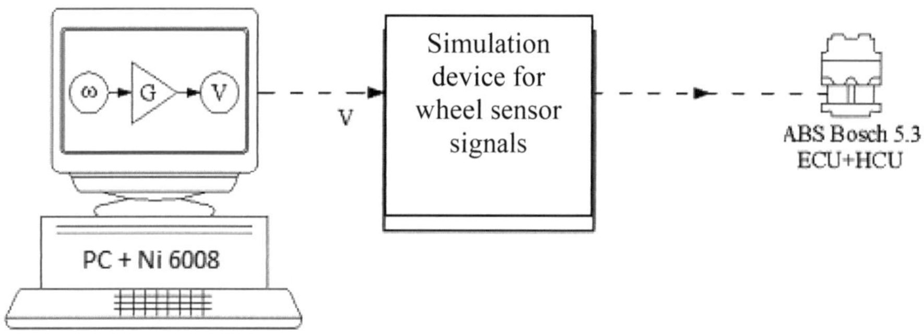

Figure 2. Connection of the device with a PC with connected to it USD and ABS Bosch 5.3.

The proportionality factor G is a voltage function that depends on the maximum angular velocity ω of the vehicle model wheel, and the maximum voltage V that supports the channels of the Ni 6008 - 5 V digital-to-analog converter (DAC). If you consider that the model must include the braking simulation at a speed of 150 km / h, it follows that the upper limit of the frequency will be 1050 Hz.

Voltage-frequency converter (VFC)

VFC is a circuit that converts the voltage into square waves of a modulated signal with a frequency proportional to the input voltage. The general principle underlying the operation of the VFC in the basic law of charging a capacitor.

Early research stages of the VSTU this task was solved on the domestic elemental basis, so in figure 3 the scheme used at the end of the last century of the VFC based on the KR1108PP1 chip is presented, which allows to reliably simulate the operation of the induction sensor, which is used as a sensor for the state of rotation of the wheels. The basis of the device created is the conversion of the voltage U_{BX}, which is an electronic analogue of the angular velocity of the wheel, into a frequency signal whose value of the period T is determined by the value of the input voltage. At the same time, the voltage source U_{BX} is a digital-to-analog converter (L-154 card manufactured by the company "L-card") installed in a personal computer, which calculates the angular velocity of the wheel in real time.

IOP Conf. Series: Materials Science and Engineering **315** (2018) 012007 doi:10.1088/1757-899X/315/1/012007

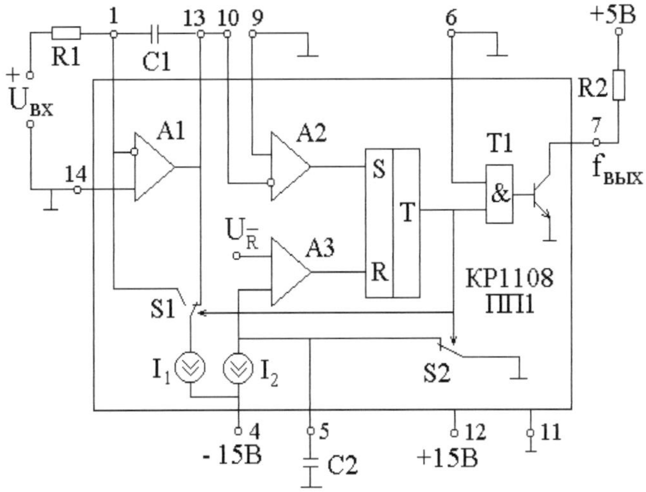

Figure 3. Functional diagram of the integrating VFC based on the KR1108PP1 chip.

At an even earlier stage in the 70s of the last century this problem was solved using electrohydraulic voltage converters [2, 3, 13] (figures 4-5).

Figure 4. External view of the hydroturbine.

Figure 5. Appearance of the block of electrohydraulic voltage converters.

At the moment, most manufacturers of converter electronic converters guarantee 0.01% linearity and thermal stability of 0.005% / °C.

The developed scheme uses the National Semiconductor VFC model LM331. The output of the LM331 VFC is an open collector transistor, which allows the output of the f_{OUT} signal with the required voltage by supplying a V_{in} voltage through a pull-up resistor. At this low voltage level will be 0 V, and high - V_{in} (figure 6).

Modulation of the output frequency is obtained by changing the high-level signal transit time. The time interval T in which the signal at a low level depends only on the nominal values of the parts in the RC circuit of the circuit represented by R2 and C2. Thus, after the choice of nominal values, this interval remains fixed. For example, the Bosch ABS 5.3 technical documentation indicates that the system operates with a signal with a duty cycle of 30% to 70%.

To solve this problem, the output of the VFC is connected to the D-trigger. Trigger is a digital circuit that preserves one of its two stable equilibrium states for an arbitrary period of time and switches in an abrupt manner from a signal from outside from one state to another.

IOP Conf. Series: Materials Science and Engineering **315** (2018) 012007 doi:10.1088/1757-899X/315/1/012007

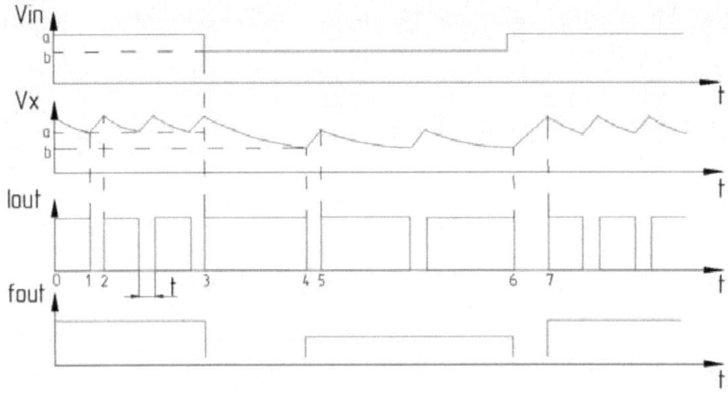

Figure 6. Dependence of the input and output characteristics of the VFC on time.

Linking the output to the input and feeding the signal from the LPCH to the input, the D-flip-flop functions as a frequency divider (figure 7), that is, it divides into two frequencies of the voltage applied to the input. The output will be a duty cycle equal to 50%. Thus, it is enough to feed the signal of the double frequency, which should be obtained at the output, and the duty cycle at the output will be equal to 50%.

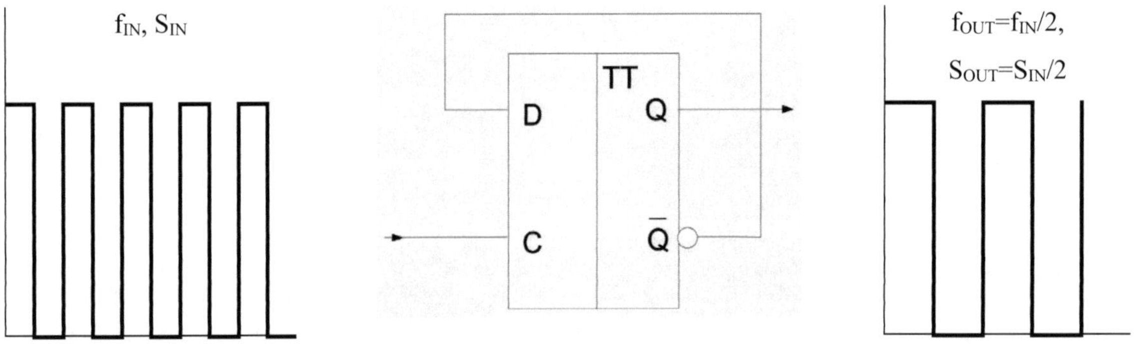

Figure 7. D-trigger as a frequency divider.

Current or voltage

To interact with ABS Bosch 5.3, the signal, in addition to the voltage level and duty cycle, must comply with the manufacturer's tolerances for the low and high current level of the signal from the wheel sensors. These tolerances are shown in figure 8.

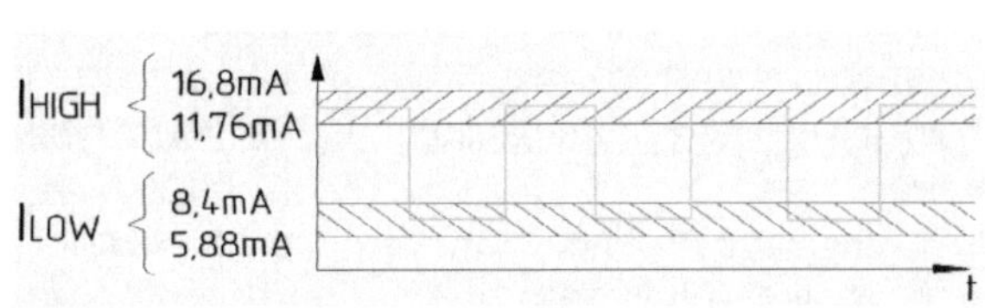

Figure 8. Technical requirements for current levels on wheel sensors.

The signal voltage of the wheel sensors, after exiting the trigger, has a high level of 12 V and a low of 0 V.

The signal comes to the base of the npn-transistor (figure 9). When the transistor is in a saturated state, the input signal rises, the current is reset directly to the "mass" via R_1. Through R_2, the current to the transistor does not go; This is prevented by the diode VD, but goes to the ABS computer. This mode is the lower threshold value I_{LOW}. When the transistor is locked, the input signal drops, the current that goes to the ABS computer must match the resistance values R_1 and R_2 in the parallel. This is the high-current mode I_{HIGH}.

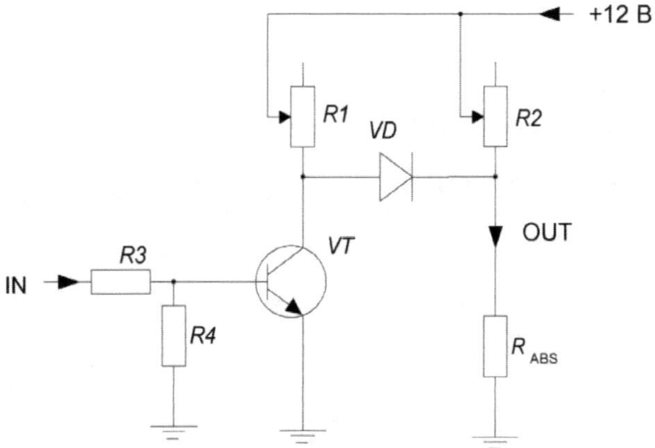

Figure 9. Schematic diagram of the current voltage.

Unlike previously used, in this improved scheme, instead of two resistances R_1 and R_2, two variable resistors were installed, in order to more accurately adjust the current level (figure 9).

This solution, if necessary, is easily scaled by the number of wheels used to control and control the movement of the vehicle (figures 10-11).

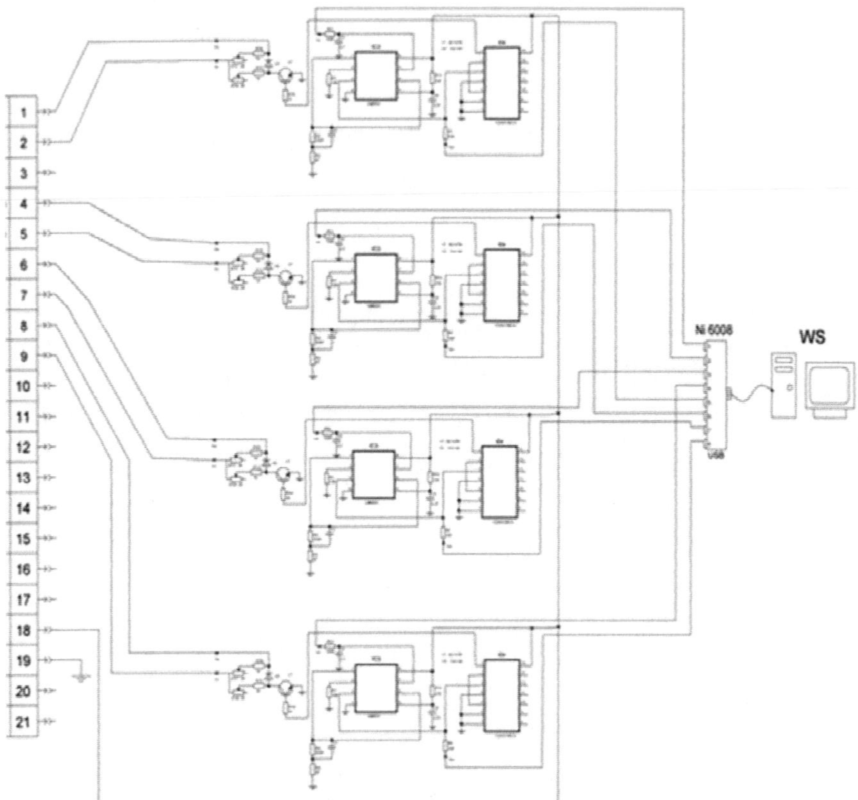

Figure 10. Diagram of connecting the device to the ABS pad LADA Kalina.

Figure 11. Appearance of the collected board.

Verification of the braking system of the car is carried out on the basis of the analysis of the oscillogram by the method of "comparing the states" of the tested vehicle and taken for the standard of a known serviceable car (figure 12).

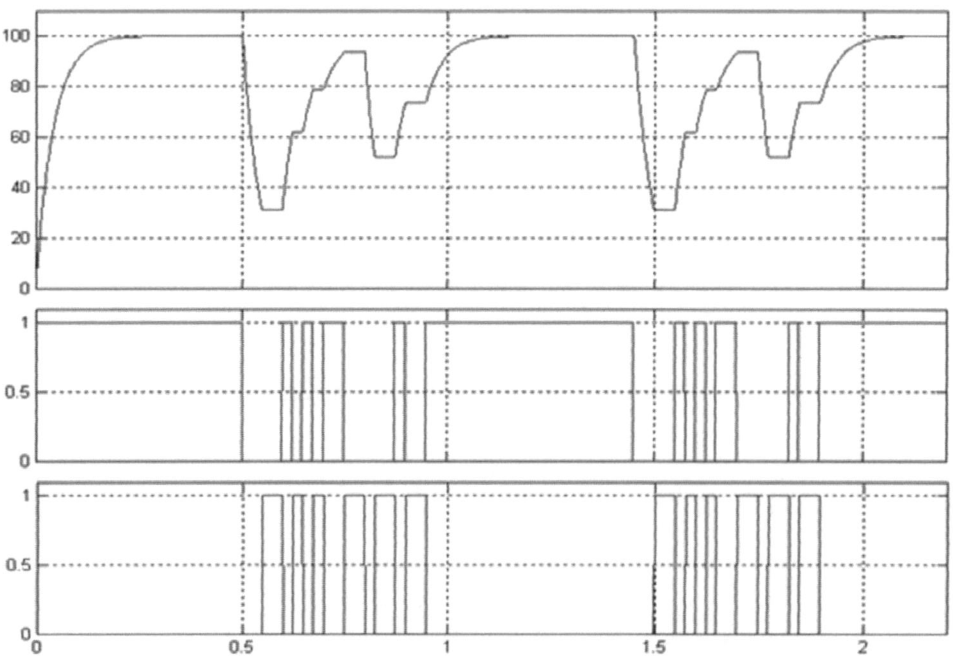

Figure 12. Generated signals.

At the present time, the researchers, in addition to the method described above, apply the method of signal substitution transmitted via the CAN. However, this method has its drawbacks, for example, the bus activity shown under certain circumstances leads to a distortion of the measurement results or a "floating break".

4. Direct control of the solenoid valves of the unit

The basic idea is to directly connect the unit to the solenoids with the appropriate communication interconnection, since almost all modern units consist of a hydraulic and electric module combined into a single unit (figure 13). The solenoids are used as a control element of the actuating mechanisms.

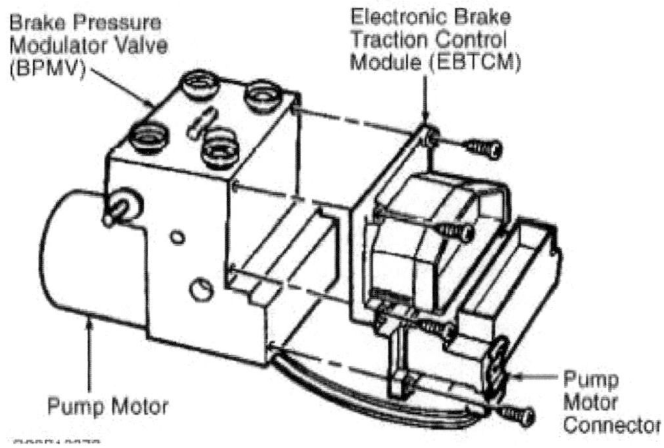

Figure 13. Hydraulic and electrical modules combined into a single unit.

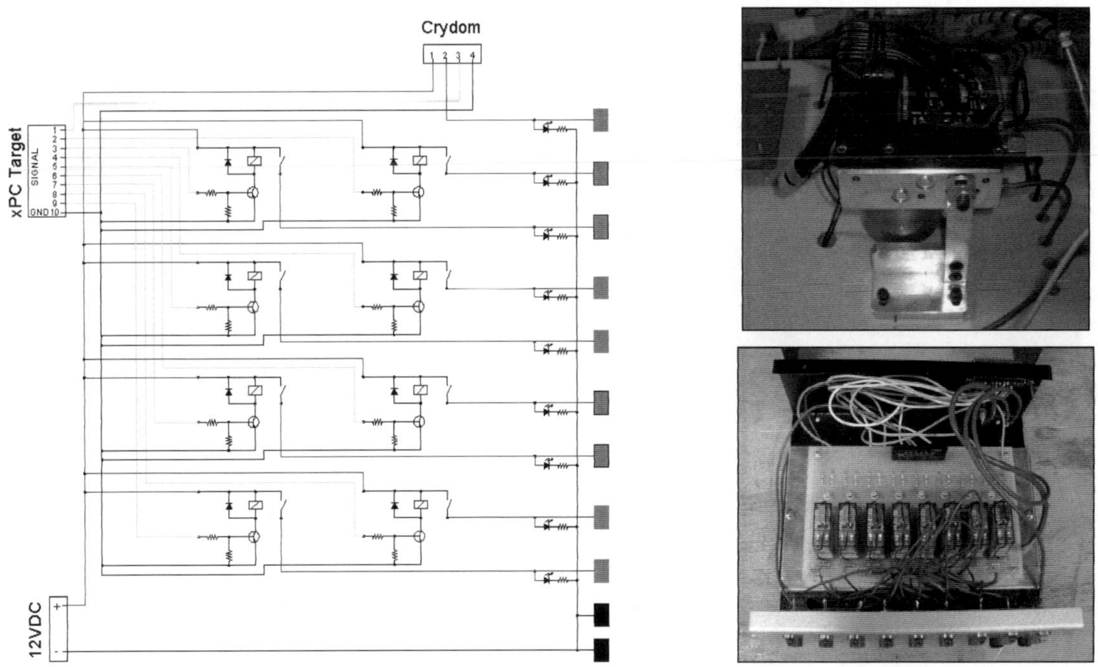

(a) (b)

Figure 14. Direct connection to solenoids (examples of technical implementation).

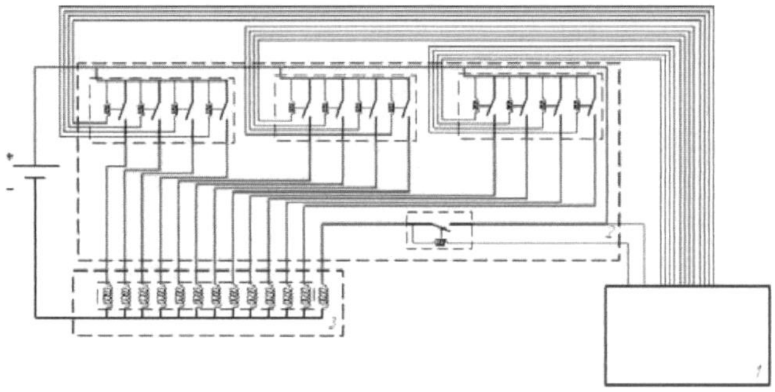

(c)

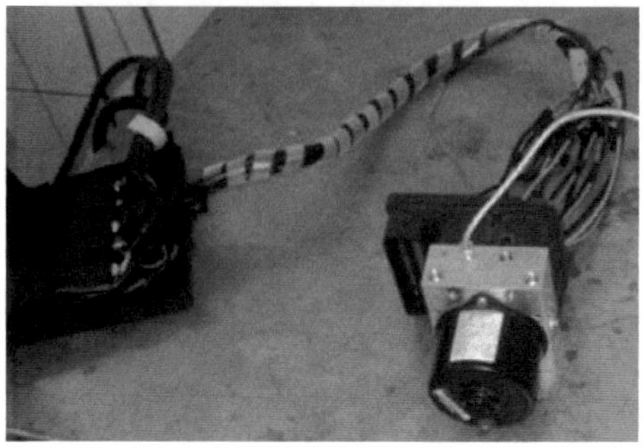

(d)

Figure 14. (Continued).

This technical solution was applied at VSTU and in the 2000s (figure 14) and was widely used in the course of the work carried out jointly with the research group of Prof. Mauro Velardochia at the Turin Polytechnic Institute [5, 6]. The technical implementation of this method has confirmed its efficiency and effectiveness. In the course of the experiments, the simulation of the ABS and ESP logic was performed, corresponding to the commercial units (figure 15), and also the algorithms for the operation of the electro hydraulic brake system have been developed, which makes it possible to apply this technique in the development of unmanned vehicle modules.

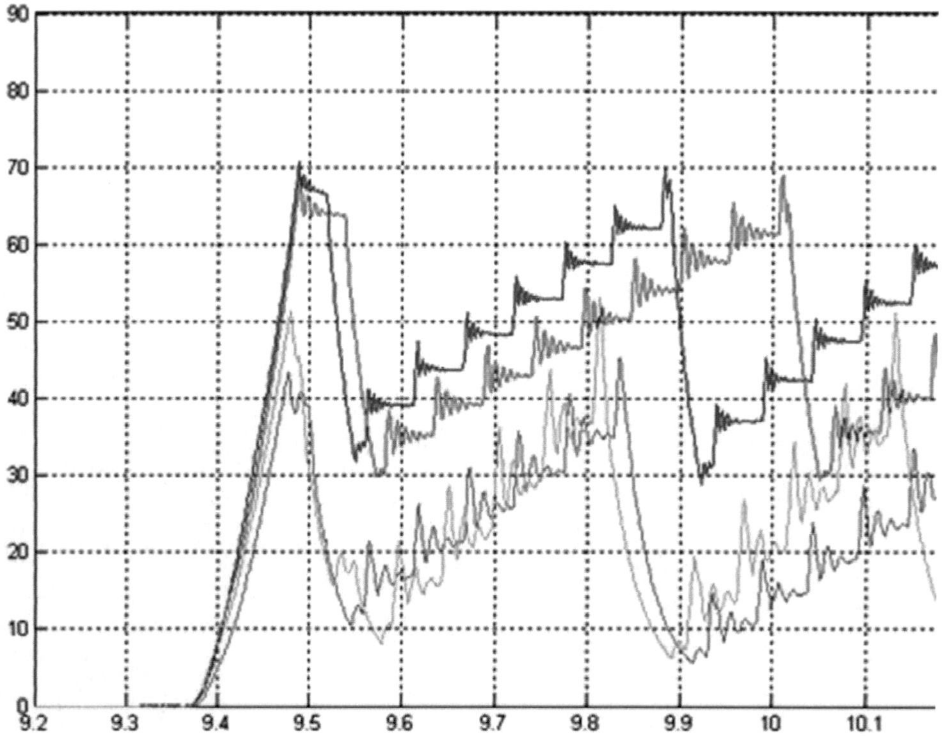

Figure 15. Oscillogram of pressure changes in wheel brake cylinders in the implementation of the algorithm of ABS and direct control of solenoids.

5. Conclusions

The application of virtual physical modeling technology that combines physical and mathematical modeling of the complex object under study, which assumes its conditional division into the full-scale and model parts connected by feedback sensors, allows developers to implement and execute algorithms of the control system of the executive mechanism of the brake system of an unmanned vehicle.

The effectiveness of Virtual-physical simulation technology is largely determined by the correctness of the accepted "division" (virtual-physical) of the research object and by the choice of the type of control computing complex.

These two approaches to the implementation of the method allow the developers to solve the tasks at different levels of integration of the unmanned vehicle control system.

References

[1] Revin A A 2014 Formation of the main operational properties of vehicles in the braking mode *Automobile industry* vol 11 pp 3–5

[2] Dygalo V G and Revin A A 2012 *Technology testing of active safety systems for motor vehicles: monograph* (Moscow: Mechanical Engineering) p 387

[3] Dygalo V G 2015 *Methodology of virtual physical tests of automated brake systems of wheeled vehicles: dis. ... Dr.Sci.Tech* (Volgograd: Volgograd State Technical University) p 348

[4] Dygalo V G 2012 Development of a simulation device for the signals of wheel sensors of the active safety system *Izv. VSTU. Series "Land transport systems"* vol 5 pp 32–5

[5] Dygalo V G, Revin A A, Sorniotti A and Velardocchia M 2005 Hardware in the loop with an electro-hydraulic brake system *Zbior referatow IX Miedzynar. Sympozjum IPMiT, (Rynia)* vol 1 (Wojskowa Akad. Techn.) pp 217–26

[6] Dygalo V G, Revin A A, Sorniotti A and Velardocchia M 2006 Stand for complex laboratory tests EGTS *Automobile industry* vol 3 pp 34–6

[7] Dygalo V G and Revin A A 2015 Alternative (virtual-physical) tests of automated brake systems of wheeled vehicles *Technology of wheel and caterpillar vehicle* vol 1 pp 37–43

[8] Smorchkov A F, Kurbanov I Sh, Dygalo V G, Dygalo L V and Boyko G V 2016 Analysis of means and test methods for active safety systems of cars *Science of the XXI century: the experience of the past - a glimpse into the future: the mother. II Intern. scientific-practical. Conf. (Omsk, Russia)* (FGBOU VO "The Siberian state. Automobile and Road Academy (SibADI)") pp 238–46

[9] Dygalo V G 2011 Virtual physical modeling technology in the design cycle of automated brake systems for multi-purpose wheeled vehicles *Bulletin of the Academy of Military Sciences* vol 2 pp 122–5

[10] Dygalo V G and Revin A A 2006 Virtual physical technologies for testing automated brake systems in the laboratory *Designing Wheeled Vehicles: Mater. Intern. n.t. Conf., cons. 70th anniversary of the cafe. "Wheeled vehicles" MSTU. N.E. Bauman (Moscow)* pp 461–3

[11] Dygalo V G and Revin A A 2005 Virtual physical testing technology for the design of automated brake systems *Mashinostroene & Electrical Engineering (Bulgaria)* vol 5 pp 26–9

[12] Tumasov A V, Groshev A M, Kostin S Yu, Saunin M I, Trusov Yu P and Gypsy V G 2011 Investigation of the properties of active safety of vehicles by the simulation simulation method *Journal of Automotive Engineers* vol 2 pp 34–7

[13] Revin A A and Dygalo V G 2001 *Research of brake dynamics of the car by methods of complex modeling technology* (Volgograd: Volgograd State Technical University) p 122

[14] Dygalo V G and Revin A A 2014 Principles of synthesis of virtual-physical models intended for the development of the brake system of a car with ABS *Automobile industry* vol 8 pp 17–9

IOP Conf. Series: Materials Science and Engineering **315** (2018) 012008 doi:10.1088/1757-899X/315/1/012008

Application of machine learning methods for traffic signs recognition

D V Filatov, K V Ignatev, A V Deviatkin and E V Serykh

V. Ulyanov (Lenin) St Petersburg State Electrotechnical University 'LETI' 5 Professora Popova Ul. St Petersburg 197376, Russian Federation

E-mail: evserykh@etu.ru

Abstract. This paper focuses on solving a relevant and pressing safety issue on intercity roads. Two approaches were considered for solving the problem of traffic signs recognition; the approaches involved neural networks to analyze images obtained from a camera in the real-time mode. The first approach is based on a sequential image processing. At the initial stage, with the help of color filters and morphological operations (dilatation and erosion), the area containing the traffic sign is located on the image, then the selected and scaled fragment of the image is analyzed using a feedforward neural network to determine the meaning of the found traffic sign. Learning of the neural network in this approach is carried out using a backpropagation method. The second approach involves convolution neural networks at both stages, i.e. when searching and selecting the area of the image containing the traffic sign, and when determining its meaning. Learning of the neural network in the second approach is carried out using the intersection over union function and a loss function. For neural networks to learn and the proposed algorithms to be tested, a series of videos from a dash cam were used that were shot under various weather and illumination conditions. As a result, the proposed approaches for traffic signs recognition were analyzed and compared by key indicators such as recognition rate percentage and the complexity of neural networks' learning process.

1. Introduction

Ensuring road safety is one of the most pressing issues in modern society. Safety issue is directly related to the 'human factor'. Driver and pedestrian inattention is one of the main causes of road traffic accidents. Among the key tasks for both active driver assistance systems (ADAS) and control systems for autonomous vehicles, the analysis of marking and traffic signs occupies an important place. Numerous computer vision systems have been developed recently for traffic signs analysis. But the characteristics of existing algorithms [1 – 5] (recognition accuracy, quantity of errors, robustness against atmospheric changes) are still not good enough to rule out a human operator.

Main tasks for computer vision systems used for the traffic signs analysis are detecting the area on a given frame that contains a road sign (localization task) and determining the meaning of the sign(s) on the selected area of the frame (classification task). The localization task can be solved in the following ways: using a deep neural network for predicting the location of the object's describing rectangle [6]; using methods for determining the parts of objects and relations between them [7 - 9].

Classification can be performed using the following methods: sliding window method [10 – 16]; gradient boosting over decision trees [17].

In the field of the development of intelligent systems for signs recognition, there are works focusing on both feedforward propagation neural networks with image preprocessing [18] and scale invariant feature transform [19] and convolution deep neural network [20].

The purpose of the work is to carry out a comparative analysis of the two methods of traffic signs recognition and identify their application features, as well as to formulate possible modifications of the given methods to achieve the best quality of recognition.

2. Materials and methods

This study utilizes a modular approach to creating a computer vision system with an option of separate optimization and adjustment. The system considered is a collection of two modules, the tasks of which are distributed as follows: module I – selecting areas that may contain a sign; module II – detecting and classifying signs in the selected area.

Schematic representation of the sequence of actions (interaction of system modules) is shown in figure 1.

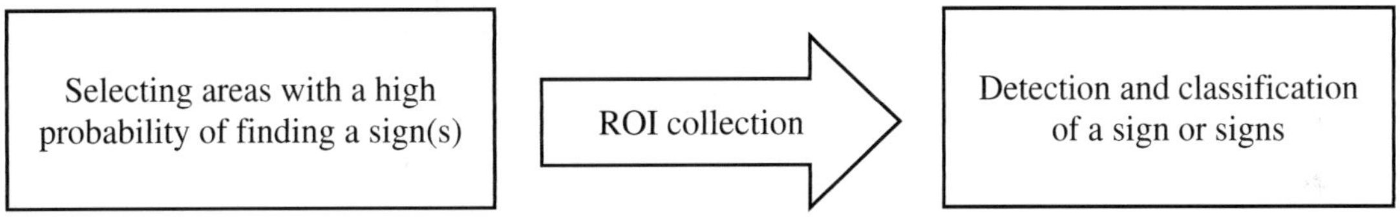

Figure 1. Schematic representation of system modules.

To use the first approach, a three-layer perceptron with a system selecting ROI on the basis of color filtering. The system of selecting the ROI consists of the following stages:

1) filtering noise using a Gaussian filter;
2) converting the image from the RGB color space to the HSV space to facilitate searching under different illumination conditions;
3) applying color filters to select areas containing basic colors of the signs in question;
4) applying morphological operations for noise filtering of the color filter (erosion) and restoring the shape of the sign with filling areas of non-basic colors (dilatation);
5) removing areas not matching in size and aspect ratio.

Areas found through this procedure are scaled to 30x30 pixels in size. For each such area, a descriptive vector consisting of 63 values is calculated. For its calculation, each of the found areas is converted into a grayscale image, then the average brightness value of all its pixels is calculated. First 30 components of the vector consist of the number of pixels in each image row, brightness of which is above the average value. The following 30 components consist of the number of pixels in each image column, brightness of which is above the average value. The remaining 3 components of the vector represent average values of color components in the HSV color space in the found area.

Classification of the detected areas is carried out with the help of a three-layer perceptron having 63 neurons in the first layer, 80 neurons in the hidden layer and the number of neurons in the output layer being equal to the number of signs considered. A sigmoid is used as an activation function in the network. A normalized describing vector of each region is sent to the input of the network.

In the second approach, in order to determine the ROI a convolution neural network with a small number of filters is used to increase performance. Neural network structure is shown in figure 2.

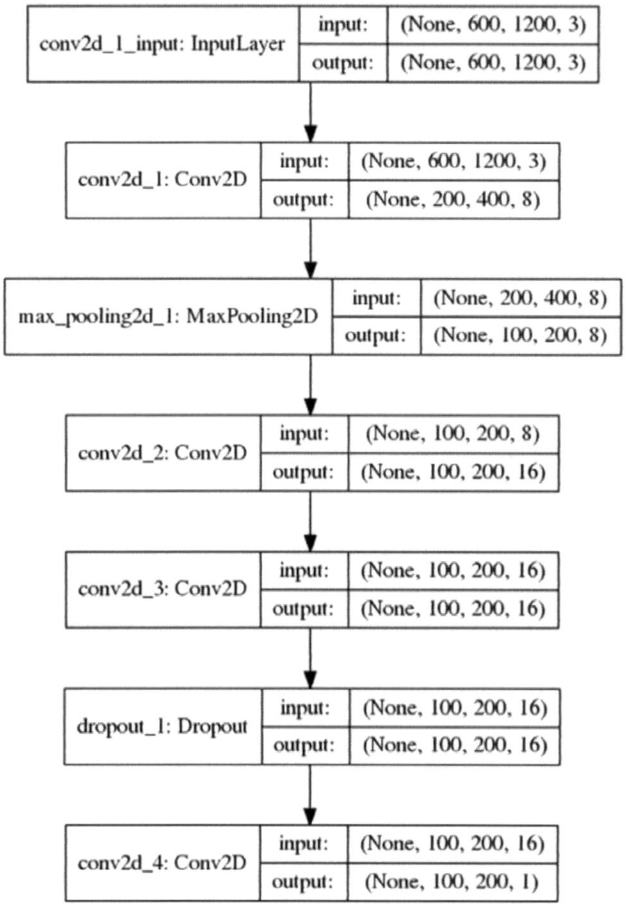

Figure 2. Convolution neural network structure for module I.

As the network input, the image from the DVR with high resolution, converted to a size of 1200x600, is used. The output of the network is an array of data of size 200x100 with a range of possible values from 0 to 1, what represents a probability scale of finding a sign in the region corresponding to the area in the original image.

As a loss log, the intersection over union function was used (1). The main optimization method was the Adam stochastic optimization approach [21].

$$IoU(A,B) = \frac{|A \cap B|}{|A \cup B|} \tag{1}$$

After receiving the probability array, the ROI's are selected from the original image (to preserve the high resolution) and transferred to the input of the module II.

To implement module II in this approach, a fully connected convolutional neural network with indirect communication between convolution layers was developed. The network structure is shown in figure 3.

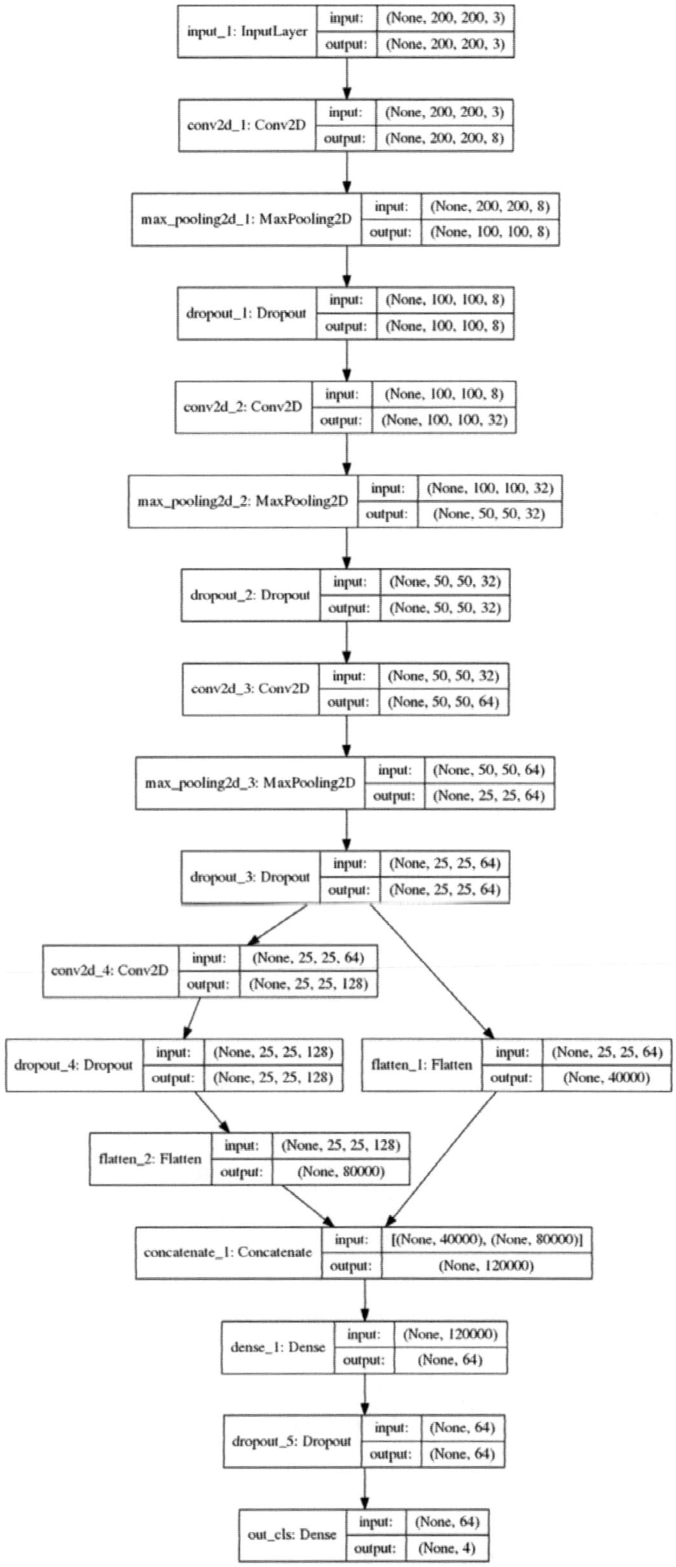

Figure 3. Convolution neural network structure for module II.

IOP Conf. Series: Materials Science and Engineering **315** (2018) 012008 doi:10.1088/1757-899X/315/1/012008

Due to the need to perform the tasks of detecting and classifying an image, the learning selection and network structure were presented in the form of multiple naming. Such a representation allows to annotate an image using an output vector elements of which are disconnected and can take a high and a low values independently of other output elements. To implement this representation, a list of signs being classified was drawn up: a stop sign, a bus stop sign, a pedestrian crossing sign and a main road sign.

Thus, the network output vector consists of four neurons. The output activation function was chosen as sigmoidal, which is a smooth nonlinear function with limiting values from 0 to 1, what corresponds to the probability of classifying the inclusion of the sign with the image.

As an estimation function, a binary crossentropy function (2) was chosen.

$$F(t,p) = -\sum_i (t_i \log(p_i) + (1-t_i)\log(1-p_i)) \quad (2)$$

Such a function is often used in logistic regression problems. Adagrad was chosen as an optimization approach [22].

3. Results

In order to compare the considered approaches, we use the basic metrics applied for the evaluation of classification problems: recall (3), precision (4), F1 (5).

$$R = \frac{TP}{TP + FN}, \quad (3)$$

$$P = \frac{TP}{TP + FP}, \quad (4)$$

$$F1 = \frac{2 * R * P}{R + P}, \quad (5)$$

TP being the number of correct classifications of the presence of a sign on the frame, FN being erroneous classification of sign's absence on the frame, FP being erroneous classification of sign's presence on a frame.

When implementing the first approach for learning a neural network, a training sample was created from 255 domains containing the whole sign or part of it, as well as areas that do not contain signs, but have a similar color. Neural network learning was carried out using the backpropagation algorithm. The results of network's test using a test selection are presented in Table 1.

Table 1. First approach test results.

Metrics	Sign			
	Stop	Bus stop	Pedestrian crossing	Main road
Recall	0.8	0.333	0.625	1.0
Precision	1.0	0.833	0.882	0.846
F1-score	0.888	0.476	0.732	0.917

In the second approach, a test set of 500 annotated frames from a dash cam was created for a convolution neural network of module I to learning. 20% of the images were randomly picked from the selection for network validation. Also, to reduce the probability of network's overtraining, data augmentation of training set was used by applying basic image processing operations: move, rotate, expand, compress.

As a result of learning, the value of estimation function reached 0.63, while the value of network recall was 0.86, what actually was the main purpose of module I learning, i.e. predicting which frame areas contain a sign. However, precision is not evaluated.

An example of image annotation using a convolution neural network is shown in figure 4.

Figure 4. Example of frame annotation using a neural network.

For the convolution neural network to learn and get configured in module II, a selection of 400 image sections with signs names depicted on frames was selected and annotated. Also, a selection of 40 images was created for testing.

Neural network learning for module II was also carried out using artificial data extension method. The selection for validation was 20% of the training set.

According to the results of learning, error function reached 0.001 in the learning selection and 0.01 in the validation selection. The results of network's test using a test selection are presented in table 2. Metric values were chosen on the basis of the largest F1 value of the metric when the threshold for sign labeling changed. The threshold of sign labeling means the smallest value of activation of the output neuron, at which the labeling of the sign value of this neuron is considered active. The value of 0.6 was chosen as the threshold for sign labeling.

Table 2. Test results for the second approach.

Metrics	Sign			
	Stop	Bus stop	Pedestrian crossing	Main road
Recall	0.8	0.866	1.0	0.8
Precision	0.88888	1.0	0.909	0.8888
F1-score	0.84	0.92	0.95	0.84

4. Conclusion

As a result of comparing two approaches for implementing the module of signs classification, the following features were revealed:

- the approach using image preprocessing is sensitive to changes in environmental conditions;
- both approaches have a high robustness in when classifying signs with noises;
- the approach using a convolution neural network requires a wider selection to enhance the ability to correctly annotate values without overtraining, what in its turn in the approach with color filtering falls on adding and modifying algorithms;
- the approach utilizing color filtering has a higher potential for scaling and implementation compared to convolution networks due to a lower requirements for computational resources;
- extension in convolution networks becomes available through increasing the number of filter layers.
- the following possible modifications are planned in future:
- module I to utilize properties of recurrent networks in view of the work with a continuous video stream;

- using a sliding frame inside the section containing a sign for localizing the sign;
- prioritizing signs for the driver (depending on the position on the image).

References

[1] Shneier M 2005 *Proc. IEEE Computer Society Int. Conf. on Computer Vision and Pattern Recognition* pp 215 – 222
[2] Ruta A, Porikli F, Li Y, Watanabe S, Kage H and Sumi K 2009 *IAPR Conference on Machine Vision Applications (MVA)* Session 15: Machine Vision for Transportation
[3] Belaroussi R, Foucher P, Tarel J P, Soheilian B, Charbonnier P and Paparoditis N 2010 *20th International Conference on Pattern Recognition (ICPR)* pp 484 – 488
[4] Houben S, Stallkamp J, Salmen J, Schlipsing M and Igel C 2013 *International Joint Conference on Neural Networks*
[5] Stallkamp J, Schlipsing M, Salmen J and Igel C 2012 *Neural Networks* **32** pp 323 – 332
[6] Erhan D, Szegedy C, Toshev A and Anguelov D 2014 *CVPR*
[7] Fergus R and Zisserman P P A 2007 *IJCV* **71** pp 273 -303
[8] Bouchard G and Triggs B 2005 *CVPR, IEEE Computer Society* pp 710 -715
[9] Felzenszwalb P, McAllester D and Ramanan D 2008 *CVPR*
[10] Bosch A, Zisserman A and Munoz X 2007 *CIVR*
[11] Chum O and Zisserman A 2007 *CVPR*
[12] Dalal N and Triggs B 2005 *CVPR* pp 886-893
[13] Ferrari V, Fevrier L, Jurie F and Schmid C 2008 *PAMI* **30** pp 36-51
[14] Rowley H A, Baluja S and Kanade T 1996 *NIPS* **8** pp 875-881
[15] Viola P and Jones M 2001 *CVPR* **1** p 511
[16] Lampert C H, Blaschko M B and Hofmann T 2008 *CVPR*
[17] Li L, Wu Y and Ye M 2014 *Informatica (2013)* **38** pp 145–153
[18] Auranuch L and Jackrit S 2017 *International Journal of Innovative Research in Computer and Comminication Engineering* **5**
[19] Hu X, Zhu X, Lia D and Li H 2010 *A special joint symposium of ISPRS Technical Commission IV & AutoCarto in conjunction with ASPRS/CaGIS 2010 Fall Specialty ConferenceNovember 15-19*
[20] Ciresan D, Meier U, Masci J and Schmidhuber J 2012 *Multi-Column Deep Neural Network for Traffic SignClassification.*
[21] Kingma D and Ba J 2015 *3rd International Conference for Learning Representations*
[22] Duchi J, Hazan E and Singer Y 2011 *Journal of Machine Learning Research* **12** pp 2121-2159

IOP Conf. Series: Materials Science and Engineering **315** (2018) 012009 doi:10.1088/1757-899X/315/1/012009

Carriageway actual parameters determination system development to form virtual road scene in the authentically recognizable road marking absence and negative obstacle presence

A M Groshev[1], S M Groshev[2], A V Tumasov[1] and A D Romanov[1]

[1] Nizhny Novgorod State Technical University named after R. E. Alekseev, Nizhny Novgorod, Russia
[2] TESA Ltd. Company, Nizhny Novgorod, Russia

E-mail: nil_st@ nntu.ru

Abstract. This paper presents the development of the carriageway actual parameters determination system to form a virtual road scene in the authentically recognizable road marking absence and negative obstacles presence. The sensors used in defining the vehicle position and the obstacles are given. The developed system will be able to function independently and also work collaboratively with the other ADAS functions increasing the functional quality of the system.

1. Introduction

The different driver aid systems such as intellectual cruise control systems, automatic parking systems and driver monitoring systems are currently implemented into the vehicle that undoubtedly exert a positive impact on road accident statistics. The video cameras combined with the other sensor types (laser range finders, radars, inertial sensors, the systems of navigation, and etc.) dominate among different touch technologies to provide high reliability (table 1). The development of a new generation of high-speed computer facilities and modern high resolution digital video sensor system employment has allowed making artificial visual devices data closer to human eye characteristics. Table 2 gives the examples of devices to collaborate incoming information during Defense Advanced Research Project Agency (DARPA) Urban Challenge competition. Since hardware opportunities are better, long-term onboard computing capacities now come closer to the human brain computing capacities for processing images [1, 2].

Table 1. Sensors.

Type	Data format	Features
Television sensor	2D - video image	Object shape definition, object environment, high resolution capacity, low cost equipment
IK sensor	2D - video image	Object shape definition, object environment, a twilight work possibility

Type	Data format	Features
IK sensor with the illumination system	2D - video image	Object shape definition, object environment, a night work possibility, area 3D model creation possibility with the use of the structured illumination
Thermovision sensor	2D-thermal video image	Object shape definition, hot spots quantity and arrangement, object environment, night work possibility
Radar	The reflected profile, polarizing images, speed field	Obstacle definition possibility in difficult meteoconditions, including in full zero view in optical and the Infrared range
Lidar	3D – video image 2D – speed field	3D - form, object arrangement, stage moving sites space distribution, the greatest object detection distance in simple meteoconditions, high resolution capacity

It should be noted that concrete is used for pavement lying during highway road construction, providing light covering color with poorly allocated white and yellow road marking lines. The lack of difference in gray color shades between pavement and road markings, especially at adverse lighting conditions, isn't enough for unambiguous marking point arrangement in the optical range. The traditional systems are able to recognize a "fresh" road markings at good visibility conditions; nevertheless the damaged marking or its absence (typical for regions) limits the system functionality.

Table 2. Road line detection (examples).

	Make of a car							
	Stanly	GM	KAT-5	Sandst orm	Google	Nvidia Car	Tesla	VW
Information obtaining	Lidar	Lidar and radar	Lidar	Lidar	Lidar, radar and video camera	Video camera	Video camera and radar	Video camera and radar
Road marking detection	–	+	–	–	+	+	+	+
Partially erased marking recognition	–	+	–	–	n/a	n/a	+	+
Carriageway recognition w/o marking	+	+	+	+	+	+	–	–
Night movement	+	+	+	+	+	+	+	+

Table 2 contains examples when component part prices exceed car prices. The use of cheap and available devices (video cameras, range-finders and thermovision cameras) becomes nowadays a key factor for civil vehicles when there are assembly cost restrictions. Object recognition issues are connected to auxiliary but extremely important algorithm training technological issues using real and model data. High quality target detection and target recognition rates can't be achieved without the training.

It should be noted that such phenomenon as road holes (negative obstacles) are well known to all car owners; the road holes have no constant sizes and specific locations on the road. Hitting a hole is always unpleasant and it can become the reason of a breakdown or an accident. To recognize negative obstacles

Ford Company developed the road holes availability / other road roughness driver warning system. The system represents a digital map on those holes and other roughnesses will be marked at real time. The satellite navigation system with double-sided communication is in its core. Passing a certain road area having pavement defects, onboard electronics and sensor read out of roughness characteristics by means of measuring wheel forces and transfer the processed information to a cloud service. Other cars download the information from the cloud server and display it on the on-board computer screen. GOOGLE Company's system is arranged similarly. It is based on the vertical accelerometer. The main advantage of the system is driver notification about invisible holes during bad weather conditions (under snow or water layer). Undoubted advantage of the system is availability of the information to cars not equipped with the obstacle detection system. The main disadvantage of the system is that the information on obstacles is transferred certain time after the first car has already collided with an obstacle and is useful only to other cars. Obstacle definition accuracy is limited with the vehicle coordinates determination accuracy. The system information isn't of any value for the first driver hitting the open hatch.

Besides global positioning satellite signals aren't available everywhere. Even if system signals are available, the user can face another problem. Coordinate determination accuracy for civilian receivers is 15 meters, but navigation system carrier geometrical sizes are much smaller than the positioning error.

The purpose of the work is the carriageway actual parameters determination system development to form a virtual road scene in the condition of authentically recognizable road marking absence. In particular it determines carriageway limits when road markings are being absent or damaged; the system draws a virtual road marking using road signs requirements, navigation data, roadside width, and road conditions. The system also detects and emphasizes driver attention on small-sized obstacles in a way of the movement, especially on negative obstacles (an open hatch, a hole, etc.).

An integrated approach to interconnected technical sight system component interaction allows implementing new decisions in roadside detecting using additional structured illumination.

The system includes several sensors and is able to adapt under specific operation conditions, including virtual route laying possibility under road network absence (permafrost, steppes, etc.).

Information shown in the table 3 is displayed onto the dashboard or windshield after being collected, processed and analyzed.

Table 3. The output information.

Type	Features
Driving lane	Along a driving lane two green strips are highlighted. In case of forbidding marking detection the border is highlighted in red color
Negative or small-sized obstacle	The obstacle is allocated in a red rectangle and the obstacle symbol is removed
The distance to an obstacle in a driving lane	The obstacle is allocated in a red rectangle and the distance is specified below
Emphasis attention on the remote obstacle found by the radar	The obstacle found at a distant area and posing potential hazard is brought onto the separate screen in "zoom"

2. Principle of operation

Road Vehicle Position Assessment algorithm by means of optical sensors is based on image processing algorithms. These algorithms allow allocating carriageway borders on the image accepted by the observation system using this information to solve an observation system external orientation problem. The road marking lines allocation algorithm is developed within the event image analysis offered earlier [4]. The technical sight system consists of minimum two optical-electronic sensors providing binocular technical sight; is established on the vehicle; and is oriented toward driving direction. At different time points working scene image stereo pairs are formed, identical objects are allocated on images and their three-dimensional coordinates are measured. Object movement trajectories in a shot are determined by each object coordinates. If any object movement trajectory crosses a vehicle movement trajectory, the

IOP Conf. Series: Materials Science and Engineering **315** (2018) 012009 doi:10.1088/1757-899X/315/1/012009

warning signal forms. Sensor optical axes are oriented under equal angles ($\alpha/2$) normally to the axis of the vehicle movement. The sensor arrangement provides the enhanced object space coordinates measurement accuracy at the increased base expense and smaller minimum working distance because of the main optical axes orientation at an angle to each other.

Obstacle Detection on the Plane algorithm described in [5] (a way of three-dimensional object detection on rather smooth surface) bases on the "differential orthophoto" method. Stereo pair left and right images are orthogonally projected onto the auxiliary (horizontal) plane taking into account spreading surface analytical model. Object availability leads to deviation appearance from the surface model. Therefore an object is "projected" on the area blocked by it. There are specific angular brightness and geometrical structures due to the difference between stereosystem camera positions on the left and right orthophoto difference (a differential orthophoto) in the object border field. Thus, the object detection problem can be reduced to simpler to the search of angular two-dimensional structure on a differential orthophoto.

Infrared Range Obstacles and Road Marking detection algorithm is completely developed by the authors. In modern computer and in particular technical sight systems the main role is played by lighting. Creation of a uniform luminous flux on all video camera vision fields in the absence of light patches on controlled objects is a separate difficult scientific and technical task.

Currently rather realistic images of any specific area, including all necessary infrastructure facilities in the visible range waves can be made by means of computer simulating. Unfortunately, the software allowing carrying out high-quality and naturalistic modeling of images in the other ranges is practically absent or is extremely inaccessible in the market.

Obstacle detection using infrared range sensors, both with illumination and without it generally is similar to optical range sensor detection. The essential difference of the system is ability to work with the small-sized, negative obstacles and also with roadsides detected with the use of the structured illumination [6 - 8]. There is an analysis of an illuminated image part and the hypothesis of its geometrical sizes is checked. The use of the specialized structured illumination allows significantly simplifying and reducing the cost of all system generally.

The thermal image helps to find road limits to detect coming turns earlier, etc. Thermal Range Obstacles detection algorithm based on road infrastructure specific elements identification to have different temperatures and on optical range sensors analysis simultaneously. The performance check is necessary as in some cases where there is a shadow from a foreign object on a vehicle driving lane that actually isn't on a driving lane (figure 1) leading to a collision warning system false drop.

Figure 1. The road and a tree shadow (thermovision and optical ranges).

Figure 2. The crossroad (thermovision and optical ranges).

Figure 3. The straight road part (thermovision and optical ranges).

Figures 1 – 3 show the photographs of roads sections in the thermovision and optical ranges. Measurements were made in non-optimal days for thermal imager (summer, afternoon, + 30 C air temperature, + 42 C road surface temperature) [7 - 8]. The specific road elements were found. At night pavement detection is possible due to different heat capacity of an asphalt/concrete surfacing and the surrounding areas. Optimum conditions are provided right after the sunset (the most dangerous period) when the driver hasn't yet adapted to new night traffic conditions.

Detection of large obstacles such as vehicles is based on detecting specific car elements (figure 1), especially a radiator or the exhaust system (at a foreshortening 0/4) and wheel naves (at a foreshortening 4/4). If there is any uncharacteristic thermal anomaly detecting (a hot spot) in a driving lane or moving in the direction of a driving lane the system warns the driver and helps him to avoid collision with people and animals coming out onto the unlighted carriageways at night.

Detecting negative obstacles (figure 4), thermal road surfaces fields, with abnormal temperatures are evaluated.

Figure 4. Manhole and barriers.

The algorithms mentioned in the paper supplement such distance measurement subsystems as "radar" and "lidar" and are used as auxiliary elements in normal weather conditions to measure the distance to the objects already found.

The radar becomes the main obstacle detection system during poor conditions (rain, snowfall, fog, etc.). Optical, IK and thermal range systems help to define and recognize targets at the shortest distance. The main principle is "we don't see an obstacle, but we know it is and are ready to classify it".

Acknowledgements
The research has been done with the financial support of small form scientific and technical sphere enterprise development assistance fund (Innovation Assistance Fund).

References

[1] Zheltov S Yu and Wisilter Yu V 2009 Perspectives of intellectualization of control systems of aircraft due to application of computer vision technology *MIPT* **1** (4) pp 164 - 181

[2] Visilter Yu V, Vishnyakov B V, Vygolov O V, Gorbatsevich V S and Knyaz V A 2016 Intelligent Information Processing Technologies for Navigation and Control Problems of Unmanned Aerial Vehicles *SPIIRAS Proc.* **2** (45) pp 26 - 44

[3] Visilter Yu, Zheltov S and Stepanov A 1996 Events-based Image Analysis for Machine Vision and Digital Photogrammetry *SPIE Proc. International Archives of Photogrammetry and Remote Sensing* **31** p 5

[4] Zheltov S Yu, Sybiryakov A V and Vygolov O V 2002 Car collision avoidance system based on orthophoto transformation *Intern. Archives of Photogrammetry, Remote Sensing and Spatial Information Sciences* (Greece) **34** pp 125-130

[5] Xin L, Wenguang H and Haoliang S 2008 The Theory and Application of Structured Light Photogrammetry with Known Angle *XXIst ISPRS Congress Technical Commission* pp 101-106

[6] Lanman D, Crispell D and Taubin G 2009 Surround structured lighting: 3-D scanning with orthographic illumination *Computer Vision and Image Understanding* **113** (11) pp 1107-17

[7] Tarasov V V and Yakushenko V G 2004 *Infrared systems of the looking type* (Moscow: Logos) p 443

[8] Gosorg G 1988 *Infrared thermography. Fundamentals, technique, application* (Moscow: Mir) p 399

IOP Conf. Series: Materials Science and Engineering **315** (2018) 012010 doi:10.1088/1757-899X/315/1/012010

Estimation of light commercial vehicles dynamics by means of HIL-testbench simulation

A Groshev[1], A Tumasov[1], E Toropov[1] and P Sereda[2]

[1] Nizhny Novgorod State Technical University n.a. R.E. Alekseev, Minin st., 24, 603950, Nizhny Novgorod, Russia
[2] GAZ Group, Light Commercial Vehicles Division, Nizhny Novgorod, Russia

E-mail: anton.tumasov@nntu.ru

Abstract. The high level of active safety of vehicles is impossible without driver assistance electronic systems. Electronic stability control (ESC) system is one of them. Nowadays such systems are obligatory for installation on vehicles of different categories. The approval of active safety level of vehicles with ESC is possible by means of high speed road tests. The most frequently implemented tests are "fish hook" and "sine with dwell" tests. Such kind of tests provided by The Global technical regulation No. 8 are published by the United Nations Economic Commission for Europe as well as by ECE 13-11. At the same time, not only road tests could be used for estimation of vehicles dynamics. Modern software and hardware technologies allow imitating real tests with acceptable reliability and good convergence between real test data and simulation results. ECE 13-11 Annex 21 – Appendix 1 "Use Of The Dynamic Stability Simulation" regulates demands for special Simulation Test bench that could be used not only for preliminary estimation of vehicles dynamics, but also for official vehicles homologation. This paper describes the approach, proposed by the researchers from Nizhny Novgorod State Technical University n.a. R.E. Alekseev (NNSTU, Russia) with support of engineers of United Engineering Center GAZ Group, as well as specialists of Gorky Automobile Plant. The idea of approach is to use the special HIL (hardware in the loop) -test bench, that consists of Real Time PC with Real Time Software and braking system components including electronic control unit (ECU) of ESC system. The HIL-test bench allows imitating vehicle dynamics in condition of "fish hook" and "sine with dwell" tests. The paper describes the scheme and structure of HIL-test bench and some peculiarities that should be taken into account during HIL-simulation.

1. Introduction

There are a lot of papers describing the positive usage experience of simulation results for prediction of active safety of vehicles equipped by ECS [1…4]. At the same time it is easy to see that different vehicles (passenger cars, vans, trucks and heavy trucks) are simulated with using of different methodologies.

This paper describes the approach, proposed by researches from Nizhny Novgorod State Technical University n.a. R.E. Alekseev (NNSTU, Russia) with support of engineers of United Engineering Center GAZ Group as well as by specialists of Gorky Automobile Plant (Russian light commercial vehicles (LCV) producer).

The main purpose of current research is to develop methods for justification and selection of the simulation model parameters that are necessary for providing acceptable modeling results that would properly match the road tests results.

2. HIL-testbench description

The general scheme of HIL-test bench developed in NNSTU shown on figure 1. The proposed HIL-test bench consists of a computer (3) with installed software that allows creating dynamic models of simulated vehicle, a real-time computer (1) that interacts between the hardware (2) and virtual parts of the test bench. The hardware part consists of the electro-hydraulic control unit (ECU) of the ESC system with the possibility of switching it off, the brake pedal unit including the brake master cylinder, brake fluid nozzle and vacuum brake booster assembly. For creating a vacuum, a vacuum pump is connected to the vacuum brake booster by a pneumatic conduit. Hydraulic pipelines connect the pedal assembly to the ECU and equipped with pressure sensors. It also connects the actuators of the front and rear axles, respectively, to the ECU and equipped with pressure sensors.

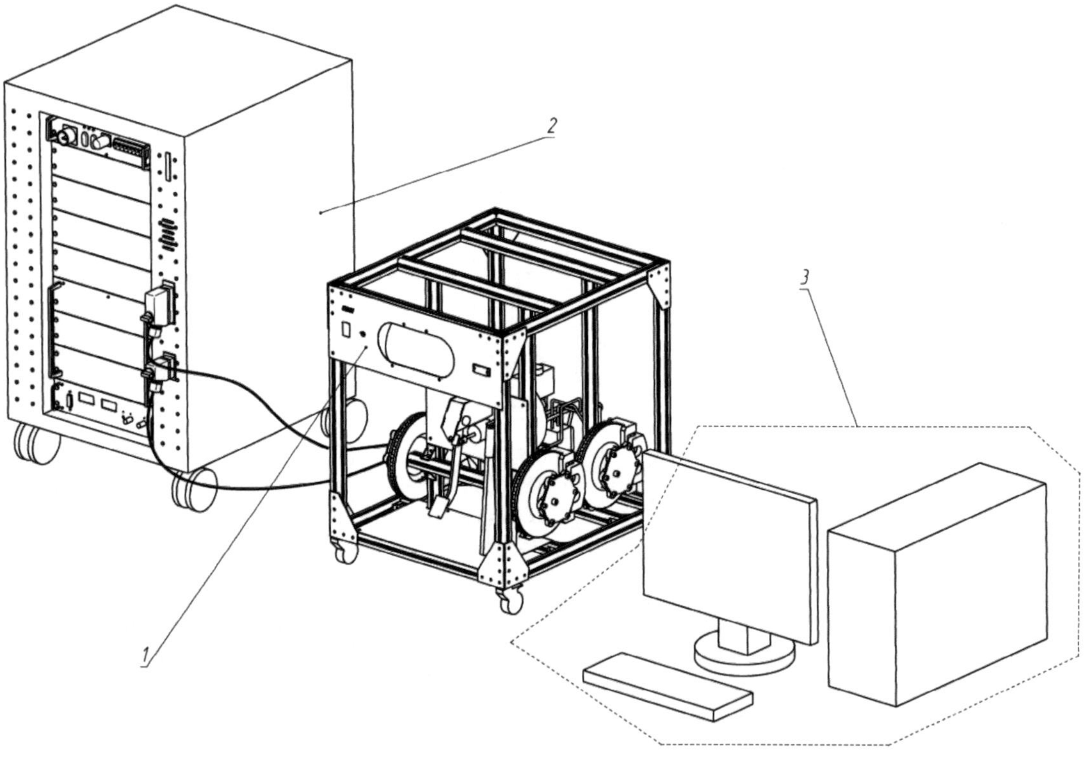

Figure 1. The general scheme of HIL-testbench.

The principle of HIL-test bench operation is based on real time communication between virtual and physical parts. In special real-time software the virtual vehicle model and conditions of all necessary dynamic tests are implemented. Such virtual parameters appearing during high speed maneuver as the speed of wheels rotation, the angle and angular speed of rotation of the steering wheel, the rotational speed of the internal combustion engine, longitudinal and transverse accelerations, and the yaw rate are transmitted to a real-time computer that realizes the interaction of the virtual and physical parts of HIL-test bench, sending signals on the physical part of the stand and vice versa at each time step. Depending on the conditions of the virtual tests, the braking system operates in a way as it would do if it were installed on a real vehicle: either by the brake pedal robot or directly by the electro-hydraulic control unit of the ESC ECU actuating the brake mechanisms. The signals generated from the brake force sensor on the brake pedal, the pressure sensors of the hydraulic pipelines emulate signals that are necessary for the operation of the ESC ECU. Real time PC converts pressure values to the brake moments that applied to a virtual vehicle model, thereby adjusting the speed and trajectory of the movement. The results of the simulation performed on a real time PC are visually reproduced at each time step of the simulated process and demonstrates vehicle dynamics under the given conditions. Thus, the result of simulation of each time step is available online in a real time mode, as if the process was started in a natural environment. Switcher allows disabling ESC ECU to simulate vehicles dynamics without ESC support.

Figure 2. Photo of HIL-testbench assembled in NNSTU.

Figure 3. Visualisation of HIL-Simulation of LCV behaviour in a real-time mode.

It is possible to point out several important problems that should be solved for precise simulation:

- determination of vehicles axial moments of inertia;
- determination of tires characteristics;
- an adequate simulation of engine processes.

3. Estimation of vehicles axial moments of inertia

Axial moments of inertia can be determined analytically by presenting of vehicles design as a set of individual elements: wheels, axles, cabin, frame, body and etc. It is possible to determine the moment of inertia of the entire vehicle by using its detailed CAD model, specifying the density of the material and determining the mass and moments of inertia of individual vehicles parts. Nevertheless, an experimental verification of this approach is required, since it is difficult to take into consideration liquids (fuel, coolant and others); interior materials of the cabin and others "fine points", which can affect the value of the moment of inertia.

The method of calculation where the vehicle is approximated by a set of simplified geometric shapes with the use of empirical coefficients increasing the accuracy of calculations are described in [5]. The main advantage of calculation methods is the physical absence of the researching object. However, the actual vehicle differs from the ideal calculation model. Therefore, experimental methods have higher measurement accuracy.

The analysis of the previous scientific works shows that a new trend of the vehicle moments of inertia calculation is being observed. In the paper [6], the authors do not try to measure the mass and yaw moment of inertia of the vehicle using any kind of test benches they present the possibility of the yaw moment of inertia calculation, during the vehicle motion. The disadvantage of this approach is using a bicycle model of the vehicle in the calculation algorithm. This model has a number of assumptions. There are many effective developments among the existing measuring methods as well as test bench structures. The most universal is the original method and the experimental facility, presented in [7]. There are test benches of the moments of inertia measuring, with the use of three string [8] and bifilar [9] pendulums. In these studies, the measurement objects are the scale models, which are much smaller than their originals in dimensions and mass.

The authors of current paper propose the method of oscillating platform on the basis of the review and analysis of existing experimental methods of calculation of the vehicle moment of inertia. The essence of this method lies in the implementation of free angular oscillations of the platform with a vehicle located with respect to a certain axis of rotation (movement force caused by a spring). In this case the moment of inertia of the vehicle with the respect to the Y axis passing through its center of mass parallel to the axis of rotation can be determined as follows:

$$J_{YY} = \frac{(T_s^2 - T_{pl}^2)\cdot c \cdot l^2}{4\cdot\pi^2} - m_a \cdot d^2$$

where: m_a - the mass of the vehicle; c - spring rate; d - the distance between the mass center of the vehicle and the axis of rotation; T_{pl} - oscillation period of the platform without vehicle; T_s - oscillation period of the platform with vehicle; l - the distance between the attachment point of the spring and the axis of rotation.

4. Determination of tires characteristics

There are different methods of tire characteristics determination. The most reliable is the determination of tires characteristics by results of special skid-trailer tests on real roads [10].

According to the test results, it is possible to have a set of curves (dependencies) for different loads appeared on the wheel as well as:

- dependence of the longitudinal slip ratio from longitudinal force taking into account different values of vertical force and wheel camber angle;
- dependence of the lateral force from the wheel slip angle taking into account different values of vertical force and wheel camber angle;

- dependence of the lateral force, longitudinal force and stabilizing moment from the longitudinal force taking into account different values of vertical force and wheel camber angle.
- The received data with the help of special software allows to create a separate file with the characteristics of the tire which is subsequently used in real-time simulation.

5. Operating modes of the engine

Reproduction of correct engine operation at the HIL-Test bench (an adequate response of the engine to requests from the ESC ECU) is one of the most difficult tasks.

Vehicles power unit is used to achieve the target lateral acceleration. Thus, the engine model should provide the torque for transmission and wheels, taking into consideration appropriate transmission configuration (front / rear wheel drive, four-wheel drive, electric motor wheels, differential configurations, etc.).

Such parameters of engine as power, speed of response to the rotation of the throttle, clutch parameters, etc., may not be taken into account in the HIL-Test bench. However, when the chassis control system is activated, any parameters of the power unit affecting the controller's performance of the system must be included in the model.

As an example, figure 4 presents a diagram of engine operation during the maneuver "fish hook" (real road test data). This maneuver begins at a speed of 80 ± 2 km/h, with the steering robot acting on the vehicle's steering wheel (the gas pedal always remains pressed). Because the speed of a vehicle is high enough, the successful maneuvering (without rollover) is possible only for a condition when the ESC system "strangle" the engine to reduce (regardless of the desire of the driver) the torque. The ESC ECU sends the corresponding signal to the engine control unit that "executes the appropriate instructions".

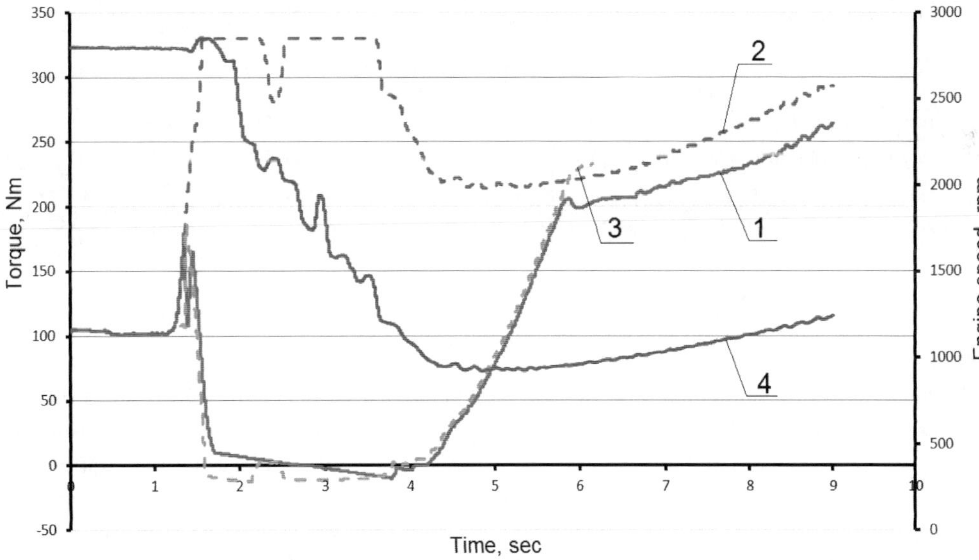

Figure 4. Engine operation during the maneuver "fish hook": 1 – torque, generated by engine ECU; 2 – driver demand; 3 – torque, required by ESC system; 4 – engine speed.

It is obvious that such "effect" should be fully reproduced on the HIL-Test bench during the simulation. The engine model must "obey" the ESC ECU commands. For this purpose, information about the CAN-matrix of the engine ECU is put into the HIL-Test bench software, which allows reproducing the correct operation of the engine.

6. Validation of HIL-Testbench

The requirements of ECE 13-11 Annex 21 – Appendix 1 "Use Of The Dynamic Stability Simulation" has some recommendations regarding to validation of HIL-Test bench which can be considered as suitable for virtual testing. At the same time it is reasonable to use others methodologies that are aimed at validation of vehicles dynamic simulation:

- ISO/DIS 19364: Passenger cars. Vehicle dynamic simulation and validation. Steady-state circular driving behavior.
- ISO/DIS 19365: Passenger cars. Vehicle dynamic simulation and validation. Sine with dwell stability control testing.

For example, one method is proposed to set values of lateral accelerations in accordance with the results of road tests that registered at a given speed at a certain angle of rotation of the steering wheel. From each point, the region of permissible deviations is plotted (a "corridor" of values). The results of simulation must be within this area. If the simulation data of the HIL-Test bench is inside the corridor, then the simulation results should be considered as adequate.

With the support of Russian State Scientific Research Center FSUE "NAMI", NNSTU engineers performed a range of road tests including "fish hook" and "sine with dwell" tests (figure 5) – all tests were conducted with front and rear outriggers, designed and produced by NNSTU. The experimental data will be used for test-bench validation that will be the subject of further papers.

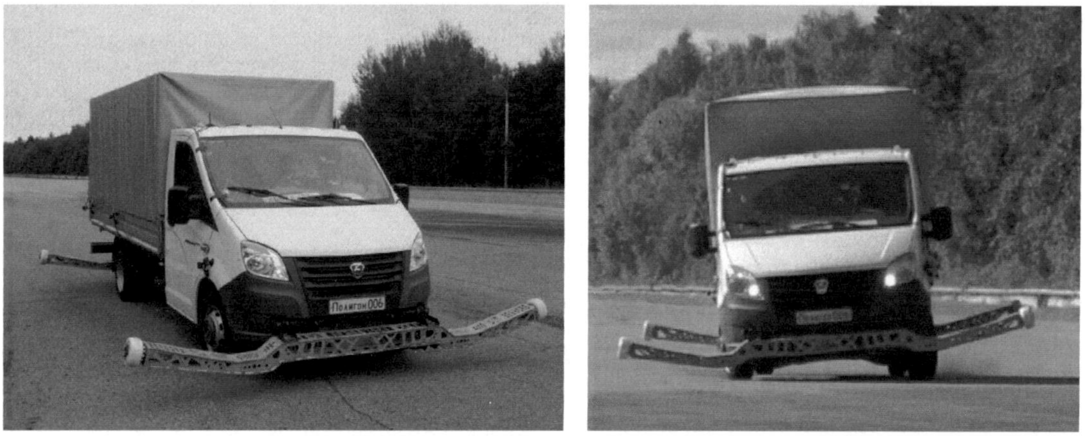

Figure 5. Road tests of LCV equipped by ECS.

7. Conclusions

- The design of HIL-Test bench for simulation of vehicle dynamics equipped by ECS proposed.
- The detailed attention must be paid to estimation of vehicles axial moments of inertia that are necessary for adequate simulation. The estimation of the impact of the platform options on the error in the calculation of the vehicle moment of inertias by oscillation method was presented. It is discovered that the error caused by the distance between the platform axle and springs can be compensated by increasing the rigidity of the springs. There was a significant influence of the platform rigidity on the accuracy of results.
- Some proposals for determination of tires characteristics and adequate simulation of engine processes as well as methods of simulation validation are described.

Acknowledgements

This research was made with the financial support of Ministry of Education and Science of the Russian Federation in the frame of the complex project "The establishment of the high-tech manufacturing of GAZ commercial vehicles family, equipped with intelligent driver assistance systems" under the contract №02.G25.31.0193 from 27.04.2016 (Governmental Regulation №218 from 09.04.2010).

References

[1] Wüst K and Lutz A 2012 Simulation-aided Homologation of ESP Systems for NAFTA Vans by Software-in-the-Loop Simulation *Daimler AG, Robert Bosch GmbH. IPG Technology Conference*

[2] Hahn K, Holzmann H, Weyer F, Roemer M et al. *2012* Simulation-based Certification of ESC Systems for Passenger Vehicles in Europe *SAE Int. J. Passeng. Cars - Electron. Electr. Syst.* **5** (1) pp 89-99, *doi:10.4271/2012-01-0235*

[3] Horak J, Pfitzer S, Keckeisen M, Neumann C and Wüst K 2012 Simulation-Based Homologation of Truck ESC Systems *TWT GmbH Science & Innovation, Daimler AG. 21st Aachen Colloquium Automobile and Engine Technology* pp 1621-33

[4] Rao S 2013 Development of a hardware in the loop simulation system for heavy truck ESC evaluation and trailer parameter and state estimation *PhD Dissertation* Ohio State University p 228

[5] Maclnnis D 1997 A comparison of moment of inertia estimation techniques for vehicle dynamics simulation *SAE Technical Paper 970951 doi:10.4271/970951*

[6] Yu Z 2015 A least-squares regression based method for vehicle yaw moment of inertia estimation *American Control Conference* (Chicago, USA) pp 5432–37

[7] Okuzumi H Identification of the rigid body characteristics of a power plant by using experimentally obtained transfer functions *Int. J. of Vehicle Design* vol 15 Nos 3/4/5 pp 425-34

[8] Witaya W 2009 Scaled vehicle for interactive dynamic simulation *Proceedings of the 2008 IEEE International Conference on Robotics and Biomimetics* (Bangkok, Thailand) pp 554-9.

[9] Jardin M 2007 Optimized measurements of UAV mass moment of inertia with a bifilar pendulum *AIAA Guidance, Navigation and Control Conference and Exhibit* (South Carolina, USA) pp 1-23

[10] Skid Trailer For Tire Characterisation & Labelling *DUFOURNIER Technologies presentation*

IOP Conf. Series: Materials Science and Engineering **315** (2018) 012011 doi:10.1088/1757-899X/315/1/012011

Development of autonomous vehicles' testing system

A M Ivanov and S S Shadrin

MADI, Russia

E-mail: ivanov-am@madi.ru

Abstract. This article describes overview of automated and, in perspective, autonomous vehicles' (AV) implementation risks. Set of activities, actual before the use of AVs on public roads, minimizing negative technical and social problems of AVs' implementation is presented. Classification of vehicle's automated control systems operating conditions is formulated. Groups of tests for AVs are developed and justified, sequence of AVs' testing system formation is proposed.

1. Introduction

In recent years, improvement of car design is concentrated in the area of control automation. There are cars on the market with electronic driver assistance systems (DAS/ADAS), such as dynamic stability control systems, lane-keeping systems, advanced emergency braking systems (AEBS), adaptive cruise control, automatic parking and others. Some manufactures have developed cars with automatically commanded steering function technology (ASCF), available at high speed driving. The issues of complete automation of vehicle control, that is, the production of fully autonomous cars (5th SAE level) and their operation on public roads are actively discussed all over the world [1]. Thus, the following advantages of cars transition to AVs [2] can be mentioned:

- road accidents reduction;
- increase of road capacity;
- increase of average driving speed;
- traffic jams reduction;
- reduction of exhaust emissions;
- reduction of driver's workload;
- useful free time usage by driver;
- transportation cost reduction.

Automatic control systems progress and preparation of fully AVs for use on public roads have caused the change of Vienna convention on road traffic [3]. Accepted corrections allow to control and change parameters of vehicle motion not only by driver, but also by vehicle's automatic control systems.

At the same time, construction and control algorithms of vehicle's automatic control systems analysis has shown, that transition to automated and, in perspective, AVs could lead to the next negative technical and social problems [2]:

- inadequate car's behavior due to imperfection of control algorithms;
- inadequate car's behavior in the events of failures, malfunctions, connection with infrastructure lost, after road accidents, vehicle's parameters change (tires change, software modification);
- malfunction due to conflict of driver's actions and automatic systems operation;

- unemployment of professional drivers;
- operation in conditions, which are not specified by car manufacturer;
- lack of cybersecurity;
- moral and ethical problems;
- contradiction with existing national legislation.

Solving of the above problems is actual and well-timed task.

2. Autonomous Vehicles' Testing System

So that to minimize listed implementation risks, set of activities offers to be done prior getting of legal permission for automated systems and AVs to access public roads (see Table 1). The proposed actions relate to car manufactures, for example, in the case of service manual content, car's compliance verification according to national legislation aspects, as well as AVs' testing system before the use on public roads improvement, and process of periodic technical inspections enhancement. Fundamentally, the testing system should cover the whole life cycle of AV, beginning from the stage of design and ending with utilization. At the same time, methods of testing consist of simulations, proving ground tests, public road tests and further data monitoring of automated systems in use.

Table 1. Set of activities for automated (autonomous) car's systems implementation risks minimizing.

Problem	Solution for risks reduction
1. Inadequate behavior due to control algorithms.	Testing of functional safety (ISO 26262), as well as operational safety with simulation of typical Russian road conditions and situations.
2. Inadequate behavior in use:	
- due to failures, malfunctions, connection with infrastructure lost;	Testing with simulation of abnormal system operation (hazard identification and risk assessment).
- after road accident;	Virtual and full-scale crash tests conduction with activated system (elements of the system) of autonomous driving.
- due to vehicle's parameters change.	It is necessary to ensure: - legislation requirements for car manufactures; - testing of typical conditions (e.g., car on winter studded tires); - monitoring during periodic technical inspections.
3. Wrong or late driver's actions.	It is necessary to develop: - requirements for the human-machine interface; - testing methods for conflict situations; - requirements for service manual content; - change of driver training process.
4. Drivers' unemployment.	Step-by-step AVs implementation.
5. Operation in conditions, which are not specified by car manufacturer.	Requirements for service manual content.
6. Poor cybersecurity.	Parameters of cybersecurity testing.
7. Moral and ethical problems.	Development of the principles of system operation in conflict situations. Requirements for implementation of these principles in control algorithms.
8. Contradiction with existing legislation.	Access to public road after design improvements or legislation changes.

Consider a common approach to the content of automated systems, as well as AVs, testing system before their driving on public roads. One of the main questions in testing system development is the next – in what road situations (typical, conflict, emergency) should AVs or automatic control systems be tested?

Research works, that were held in USA [4, 5], have allowed to formulate those road situations to which car's automatic control systems should respond adequately, namely:

- detect and respond to speed limit changes;
- perform high speed freeway merge;
- perform low speed merge;
- park on the shoulder or transition the vehicle to a minimal risk state;
- detect and respond to dangerous oncoming traffic;
- detection of permitted and restricted zones for driving on the road, and maneuvers performing in permitted areas;
- perform car following including 'stop and go' and emergency braking;
- detect and respond to stopped vehicles;
- detect and respond to intended lane changes;
- detect and respond to static obstacles in roadway;
- detection of traffic light signals, road signs of stopping and giving the way;
- respond to traffic light signals, road signs of stopping and giving the way;
- navigate intersections and perform turns;
- perform driving at roundabouts;
- detection of free parking slots and parking;
- detect and respond to prescriptive and prohibitory road signs;
- detect and respond to traffic-controller's signals;
- follow the priority rights of driving;
- respect other requirements of national traffic rules;
- follow other road requirements (temporary road signs, marking, etc.);
- detect and drive safely in road repair areas;
- react to the situation in the area of road accident;
- detect and respond to emergency vehicles;
- detect and respond to cars with activated light alarm;
- give way to pedestrians and cyclists on pedestrian crossings;
- ensure safe distance between vehicles, pedestrians, cyclists on roadside;
- detect and respond to temporary changes in traffic patterns.

This list can be broaden out with situations, actual for Russian operating conditions:

+ perform overtaking on two-lane roads with oncoming traffic;
+ choose safe cornering speed;
+ detect and respond to dangerous defects of road pavement (roughness, ruts, etc.), as well as road areas with poor adhesion characteristics.

Actual set of road situations for automated control systems testing pattern depends on functional purpose and application conditions of tested system. The following grading of operating conditions is suggested:

- geographic location (country, region, city, etc.);
- road types (e.g., highways, one-way or dual road with oncoming traffic, parking areas, etc.);
- weather conditions (temperature range, relative humidity, possible road surface conditions, possible atmospheric phenomena, etc.);
- time of a day (day, night, time ranges);
- available speed ranges;
- other conditions (traffic type, requirements to infrastructure and so forth).

Car manufacturer in service manual handbook should provide the above listed information, and this information defines quantity and specification of conducted tests.

At present, there are some standardized methods for testing of automated control systems, namely:

- UN Regulations 131 [6]:
    - braking with stationary target;
    - braking with a moving target.

- Standard ISO 22839 [7] with requirements and test procedures for emergency braking systems:
  - braking with slow target;
  - motorcycle beyond the duplicated target begins braking;
  - duplicate target appearance in the nearest lane;
  - target drives in road lane with lateral offset;
  - duplicate target appearance in the nearest lane while cornering;
  - target above the road.

Draft of automatic steering systems' testing procedures that is intended to be included in UN Regulations 79 is of attentive interest [8]:

+ Functional tests:
- FU 1 - lane keeping;
- FU 2 - lane change process interruption;
- FU 3 - lane change.

+ Provocative tests:
- TR 1 - lateral acceleration limit exceeding;
- TR 2 - road marking disappearance;
- TR 3 - driver loss;
- TR 4 - sensor failure;
- TR 5 - driver's force in control process during automated minimal risk maneuver performing.

+ Emergency tests:
- EM 1- braking in front of moving target;
- EM 2 - braking in front of stationary target.

It should be noted that most of the tests simulate driving on a highway during the day in dry weather, which is a cause of objective criticism.

In general, combination of possible external conditions, vehicle conditions, actions of driver and other traffic participants, and variability of traffic situations form infinite number of possible scenarios for automated (autonomous) vehicle control systems' functional testing. Thus, nobody can guarantee 100% safety of autonomous vehicle operation, as well as operation of its driver assistance subsystems. In this case, we suppose the following sequence of necessary actions to be logically reasoned: 1) extended requirements development for product certification at type approval level; 2) requirements implementation to car manufacturer to apply the system of functional safety ensuring at all stages of product's life cycle; 3) independent control of autonomous vehicles' safety parameters in use.

We consider that automated (autonomous) vehicles' testing system should include four main groups of tests:

1) functional tests – tests in which system's performance is checked in accordance with intended purpose in normal conditions, special attention is paid to the interaction with other vehicle systems;

2) provocative tests – tests in which the probability of false positives is examined, as well as noise immunity, system's behavior in the event of malfunctions, and exceeding the limits of acceptable conditions of use;

3) emergency tests - test of system response to critical situations, not related to the functional purpose, including impact of random events assessment;

4) interaction with a driver – tests in which system reaction is checked when driver intervenes or does not interfere with control actions, including the presence of deliberately inadequate actions.

In view of the foregoing, the following sequence of actions is proposed for AVs' testing system formation:

- ascertaining permissible operating conditions and functionality of individual automatic control systems (performs by car manufacturer);
- choice of functional, provocative, emergency tests and interaction with a driver testing procedures for individual subsystems and/or complete autonomous vehicle with respect to permissible operating condition;
- tests conduction.

Example of proposed test procedures for advanced emergency braking system (AEBS) with operating conditions of highway driving in dry or rainy weather is shown below (see Table 2).

Table 2. Proposed AEBS test procedures.

Tests	Method sample
Functional tests	
1. Braking in front of stationary target:	
- dry road surface;	UN Reg. 131, Reg. 79 Amend. (EM2).
- wet road surface.	Test procedure has not been developed.
2. Braking in front of moving target:	
- dry road surface;	UN Reg. 131, Reg. 79 Amend. (EM1), ISO 22839.
- wet road surface.	Test procedure has not been developed.
Provocative tests	
1. Motorcycle braking beyond the target.	ISO 22839
2. Duplicate target in the nearest lane.	ISO 22839
3. Target drives with lateral offset.	ISO 22839
4. Duplicate target in the nearest lane while cornering.	ISO 22839
5. Target above the road.	ISO 22839
6. Car-leader overtakes stationary target.	ISO 22178
Interaction with a driver tests	
1. Steer attempt of collision avoidance with target after AEBS activation.	Test procedure has not been developed.

3. Conclusion

It is necessary to provide for set of activities to be done prior getting of legal permission for automated systems and AVs to access public roads, so that to reduce negative implementation risks. The main technical step is testing system development with respect to permissible operating conditions and automated control systems functionality. It is necessary to develop test procedures for specific road situations, taking into consideration typical weather conditions, proving grounds improvement is also of important tasks.

The following sequence of actions is proposed for AVs' testing system formation at type approval level:

- ascertaining permissible operating conditions and functionality of individual automatic control systems (performs by car manufacturer);
- choice of functional, provocative, emergency tests and interaction with a driver testing procedures for individual subsystems and/or complete autonomous vehicle with respect to permissible operating condition;
- tests conduction.

References

[1] Saikin A M, Bakhmutov S V, Terenchenko A S, et al Tendency of creation of "driverless" vehicles abroad *Biosciences biotechnology research Asia* **11** (Spl. Edn.) 2014 pp 241-246

[2] Shadrin S S Methodology of autonomous road vehicles, integrated in intellectual transport environment, driving control systems design: dissertation doctor of technical science (Bauman State Technical University Moscow) 2017 p 400

[3] Annex to report of the sixty-eighth session of the Working Party on Road Traffic Safety (Geneva) 2014 ECE/TRANS/WP.1/145

[4] Christopher Nowakowski, et al Development of California Regulations to Govern the Testing and Operation of Automated Driving Systems (California PATH Program University of

California, Berkeley) 2014 Available at http://docs.trb.org/prp/15-2269.pdf.
[5] Federal Automated Vehicles Policy Accelerating the Next Revolution in Roadway Safety U.S. Department of Transportation NHTSA 2016 12507-091216-v9 p 116
[6] UNITED NATIONS: Regulation 131. Uniform provisions concerning the approval of motor vehicles with regard to the Advanced Emergency Braking Systems (AEBS)
[7] ISO 22839:2013 Intelligent transport systems – Forward vehicle collision mitigation systems – Operation, performance, and verification requirements
[8] Proposal for amendments to Regulation No 79 to include ACSF > 10 km/h. (Informal Document: ACSF-05-03) The proposal is based on document ACSF 04-20 Submitted by the experts of Germany and Japan

Normative legal regulating of vehicles with a high degree of automation of control: strategy and tactics for implementation in Russia

B V Kisulenko, A V Bocharov and V V Pugachev

Federal State Unitary Enterprise «Central Automobile & Automotive Engine Research Institute «NAMI» (FSUE «NAMI»), Moscow, Russia

E-mail: a.bocharov@nami.ru

Abstract. The article discusses the risks specific to vehicles with a high level of automation of control, and conditions the limits on the operating conditions of such vehicles. The article determines existing legal barriers to the implementation of autonomous vehicles. The article contains an analysis of foreign practice of regulating in the European Union, Japan and the United States and information about the UNECE activities aimed at enabling operation of vehicles with a high degree of automation control. Basing on the results of the analysis, the authors made proposals for removal of legal barriers. The article also contains proposals for the development of specific requirements for autonomous vehicles associated with their specific features of design.

1. Introduction

Modern digital technologies are increasingly embedded in transport sector, creating high economic and social effect. Automation of transport is becoming the main trend of the global market, creating competitive advantages to the owners of advanced technologies. The most promising areas, which can be implemented while achieving the highest level of automation of vehicle control, include:

- ensuring the transport of goods in a convoy of vehicles using communication system between themselves and with the elements of road infrastructure;
- providing operating of autonomous vehicles in complex and dangerous road and weather conditions;
- providing shuttle transportation of passengers in restricted areas (airports, exhibitions, sporting events);
- providing without additional assistance transportation of people belonging to the disabled social groups (people with physical disabilities, elderly).

The level of automation in control is a combination of conditions described in SAE J3016 [1], which presently is commonly recognized in the world. A high level of automation refers to a technology level at which the driver is not required to monitor executing control tasks by an automatic system under identified conditions or permanently (that is, when the driver, in principle, is not present).

Currently, all developed countries are characterized by the situation when the development of the regulatory framework is behind the implementation of innovative ideas in automation of vehicle control. At the same time, significant efforts are made to overcome that at the national level and at the level of the UN Economic Commission for Europe (UNECE).

The authors propose an approach towards creating a regulatory framework that would ensure the implementation of highly automated vehicles in Russia.

2. Risks specific for vehicles with a high level of automation

Vehicles with automatic control functions are characterized by the same risks as conventional vehicles. The first and foremost is the risk of a road accident and the severity of its consequences. The level of this risk depends on the active and passive vehicle safety. Other important risks: risk of fire and burning intensity, the risk of harm to human health and the environment, the risk of unauthorized use, the risk of interference in the work of the electronic equipment, the risk of electric shock.

Partial (or full) transfer of the vehicle control functions to an automated system creates a risk of accidents due to incorrect operation of that system. Experts widely discussed the incident with a deadly outcome with the Tesla vehicle, which, being in autopilot mode, was not recognized on a light background ambient light train moving along crossing trajectory.

It is possible to reduce the risk of accidents with automated vehicles the likelihood of their intersection with other objects could be reduced. For that the preferred approach would be organization of operations of such vehicles on lanes, dedicated or fenced from other vehicles or pedestrians.

While for automated vehicles the risk of accidents can be reduced through the use of automated control systems, there are other risks specific to this group of vehicles associated with the use of computer technologies: the error of geolocation and, as a result, the risk of accidents with elements of road infrastructure, loss of data, including data about the parameters of a vehicle motion, as well as personal data of a user; improper interaction with a driver (remote operator); loss of control of a vehicle due to technical failure or in connection with an unauthorized intervention in the control system. These risks require the development of special measures to safeguard the data, including those relating to the private life of a user, cyber security interface for operator's interaction with an automatic control system.

Currently, the situation is complicated by the lack of practice of widespread use of vehicles with a high-level automation of control. At the same time:

- there is not enough experimental data on the mutual influence of the work of a large number of sensory systems in traffic conditions on public roads;
- the intelligent techniques to analyze traffic are not well developed, which is especially important in the urban transport environment;
- there are no standardized and massively proven communication systems between vehicles and between them and the infrastructure.

In addition, the existing road infrastructure including existing road traffic control systems and road environment are focused on a human driver.

3. Legal barriers to the implementation of vehicles with a high degree of automation

The 1968 Convention on Road Traffic [2] stipulates in Article 8: «Every moving vehicle or combination of vehicles shall have a driver… Every driver shall possess the necessary physical and mental ability and be in a fit physical and mental condition to drive. Every driver of a power-driven vehicle shall possess the knowledge and skill necessary for driving the vehicle. Every driver shall at all times be able to control his vehicle…». Article 13 of the said Convention stipulates: «Every driver of a vehicle shall in all circumstances have his vehicle under control so as to be able to exercise due and proper care and to be at all times in a position to perform all maneuvers required of him. He shall, when adjusting the speed of his vehicle, pay constant regard to the circumstances, in particular the lie of the land, the state of the road, the condition and load of his vehicle, the weather conditions and the density of traffic, so as to be able to stop his vehicle within his range of forward vision and short of any foreseeable obstruction. He shall slow down and if necessary stop whenever circumstances so require, and particularly when visibility is not good».

If an automated control system takes over part or all of the functions of a driver, the cited provisions of the 1968 Convention stop triggering. Without amending this Convention, the automation control systems, especially of high level, are illegitimate. The problem should be resolved at the international level in the framework of the UNECE and the certain steps, which will be discussed below, have already been made in this direction.

The actual in the Russian Federation Road Traffic Regulations based on the 1968 Convention, do not allow the lack of a driver in a vehicle A very extensive national law relating to the possession of vehicles and their operation, is not adapted to the situation when a driver, who is traditionally responsible for the behavior of his vehicle on the road and for damages caused by it, is not privy to the actions of his vehicle or may not be present. In this case, who should bear the responsibility for damage: a vehicle manufacturer, a manufacturer of an automated control system, a vehicle owner or an insurance company? The question currently remains open.

In addition, we note that in respect of vehicles with a high level of control automation at the regional (Eurasian Economic Union) or at the national level the relevant technical regulations are not developed and technical standards are absent. That is, such vehicles are currently, in fact, are in a legal vacuum. This situation, of course, has a chilling effect on the dissemination of technologies of automated control.

4. Foreign experience

In most economically developed countries, a number of automated control systems for vehicles is already offered on the market. Basically, those are systems with a low level of automation, however, the examples of vehicles with a high level of automation are already known. However, the experience of regulating in relation to such systems at present is not sufficient for the development of legal acts. Therefore, at the national level the guidance documents containing recommendations on the functioning and effectiveness of such systems are published. This is due to the fact that the rate of appearance in the market of new technologies of automated control significantly outperforms procedural terms of development, agreement and approval of regulatory acts.

The experience of the European Union (EU) is of interest. EU is developing a roadmap for the deployment of vehicles with a high degree of automation [3]. The road map GEAR 2030[4] aims at ensuring a consistent approach to the deployment of automated vehicles and the relevant legal framework of the EU. This roadmap includes:

1) general strategy development of automation technologies of vehicle control that should proceed in stages, with priority attention to traffic safety, for example, automated driving on motorways without intersections;

2) the list of activities, including: (a) analysis of the existing legal base defining conditions for the introduction of vehicles with automation control; (b) funding for coordinated research, innovative developments, large scale trials, etc.; (c) actions to ensure international cooperation and competitiveness;

A part of the general strategy is the inclusion of physical and digital traffic infrastructure in the overall technology of vehicles automated control with specific design parameters in terms of ensuring information exchange with other vehicles and traffic control. Social issues (for example, driver qualification, use of the personal data, interaction of a driver with a vehicle and other road users, ethical issues, environmental issues, social inclusion) and economic issues (impact on economic activity) should also be considered.

Document [3] states that at the current stage in EU legislation there are no significant legal obstacles for an upcoming appearance of conventionally automated vehicles (third level of automation [1]). Article 20 of the framework Directive 2007/46/EC of the European whole vehicle type approval [4] provides for the special compliance assessment procedure in respect of new technologies, which have not yet covered by developed regulatory requirements. It is noted that for higher levels of automation,

[4] The program GEAR 2030 (Competitive and sustainable growth of the automotive industry in the European Union up to 2030) focuses on three topics: adaptation of the production chain to the new global challenges; automated and connected vehicles; trade, international harmonization and global competitiveness. The program is developed by the eponymous high level working group formed on 19 October 2015.

as well as for ensuring information exchange with vehicles, changes in EU legislation may be required (in particular, traffic rules, traffic signs, driving license, road-worthiness, legal liability, insurance, protection against theft and cyber security, privacy and personal data protection, conformity assessment, etc.). The primary regulatory measures should include the introduction of systems for registration of parameters of vehicle movement (also known as «black boxes») for further investigation of road incidents. And the traffic rules should contain a provision that the driver must not misuse or confuse the automated control system.

Japan issued the guidelines for testing of automated vehicles on public roads (published in May 2016) and the criteria for permitting the use of public roads for testing of automated driving technologies with remote control (published in June 2017) [5]. According to these documents, a vehicle must meet the safety requirements, including specially approved for this vehicle. A person performing the functions of a driver must be on the driver's seat, observe the condition of the vehicle and the environment and, if necessary, in case of an emergency, take over the control. Testing of fully automated vehicles (without driver) is allowed to carry out on limited short sections of roads. Such vehicles shall remain under the supervision of a remote operator who, if necessary, must maintain control himself. The report [5] notes that by 2020 Japan should develop measures allowing for the motion of vehicles in a convoy when a driver actively controls just the first vehicle in a convoy.

The U.S. National Highway Traffic Safety Administration (NHTSA) published the «Federal policy on automated vehicles» [6]. This document sets out an ambitious approach to accelerate the uptake of vehicles with high level of automation. The speed of appearance of concerns that accompany the proliferation of such vehicles leads to the use of new approaches to ensuring the successful adoption of new technologies without creating new significant risks for safety, while providing new advantages in terms of safety with the disclosure of the entire existing capabilities in the future. For meeting this challenge, the necessary experience and knowledge have to be quickly acquired to keep pace with events, to expand the potential in the field of regulating, and increase the speed of making decisions. It is noted that this policy is a first step in regulating of automated vehicles, and is more of a guide rather than a technical rule intended for distribution to manufacturers and others the best practices of design, testing and implementation of such vehicles. In addition to the description of the mechanism of decision-making and allocation of responsibility under the laws of the United States, the policy contains a section concerning guidance on the providing for effectiveness of automated vehicles on the stages from their design and development to sales and operation on public roads. The «implementation» refers to operation of vehicles by persons who are not employees or agents of their designer, developer or manufacturer.

NHTSA expects that manufacturers and suppliers of highly automated vehicles will immediately begin to be guided by the adopted policy. Data obtained as a result of their activities, should be available for authorities, industry and the public to increase their knowledge and understanding of developing technologies. For that the policy recommends to manufacturers and other organizations on a voluntary basis provide to NHTSA reports on the observance of the provisions contained in the policy. With that, the legitimate interests of confidentiality and competitive advantages should be protected. Reports should contain safety assessment of each system of control of automated vehicles from the perspective of the measures contained in the policy, and the results of the tests performed to assess safety. The safety assessment should be conducted in the following areas:

- safety of control system;
- detection of objects and events and response to them;
- passive and post-accident safety;
- transition to the non-operational mode (minimal risk condition);
- cyber security (protection against unauthorized access and protection against failures of electronic systems);
- the human-machine interface;
- software;
- data recording and sharing;

- ensuring confidentiality measures;
- registration and certification;
- compliance with Federal, state and local laws;
- adherence of ethical measures;
- verification techniques;
- education and training of consumers.

5. Regulating at the international level

In actual conditions of globalization, the harmonization of policies is undertaken by developed countries for promotion of technologies of vehicle automated control for the purpose of minimizing non-tariff barriers. To optimize resources, a unified global approach is implemented to the formation of the regulatory framework. The leading role in that in the field of vehicles is played by the World Forum for Harmonization of Vehicle Regulations (WP.29), as well as the Global Forum for Road Traffic Safety (WP.1), within the framework of the Inland Transport Committee (ITC) of the UNECE. The Russian specialists are involved in activities of those forums.

The organizational structure of forums and working groups of the UNECE associated with intelligent transport systems (ITS) and the automation of vehicle control, interaction between them and the issues they work on is shown on figure 1.

The World Forum WP.29 and its subsidiary working groups develop new technical regulations (UN Regulations, UN Global Technical Regulations (GTR) and UN Rules), harmonize existing technical regulations and their amendments and updates. WP.29, due to the nature of its activities on the development of internationally agreed safety requirements to motor vehicles, became to the center of the discussions of issues related to the technologies of automated control. WP.29 established in its structure a special working group on intelligent transport systems and automated driving (ITS-AD), which takes care of strategic issues that support implementation of new technologies. This group has developed a number of guidance documents that form the common understanding on ITS, principles of providing assistance to a driver from ITS and it deals with issues of terminology, classification of automated vehicles and the development of general principles of cyber security. The perspective topics for consideration, which ITS-AD would begin, are shown in table 1 [7].

IOP Conf. Series: Materials Science and Engineering **315** (2018) 012012 doi:10.1088/1757-899X/315/1/012012

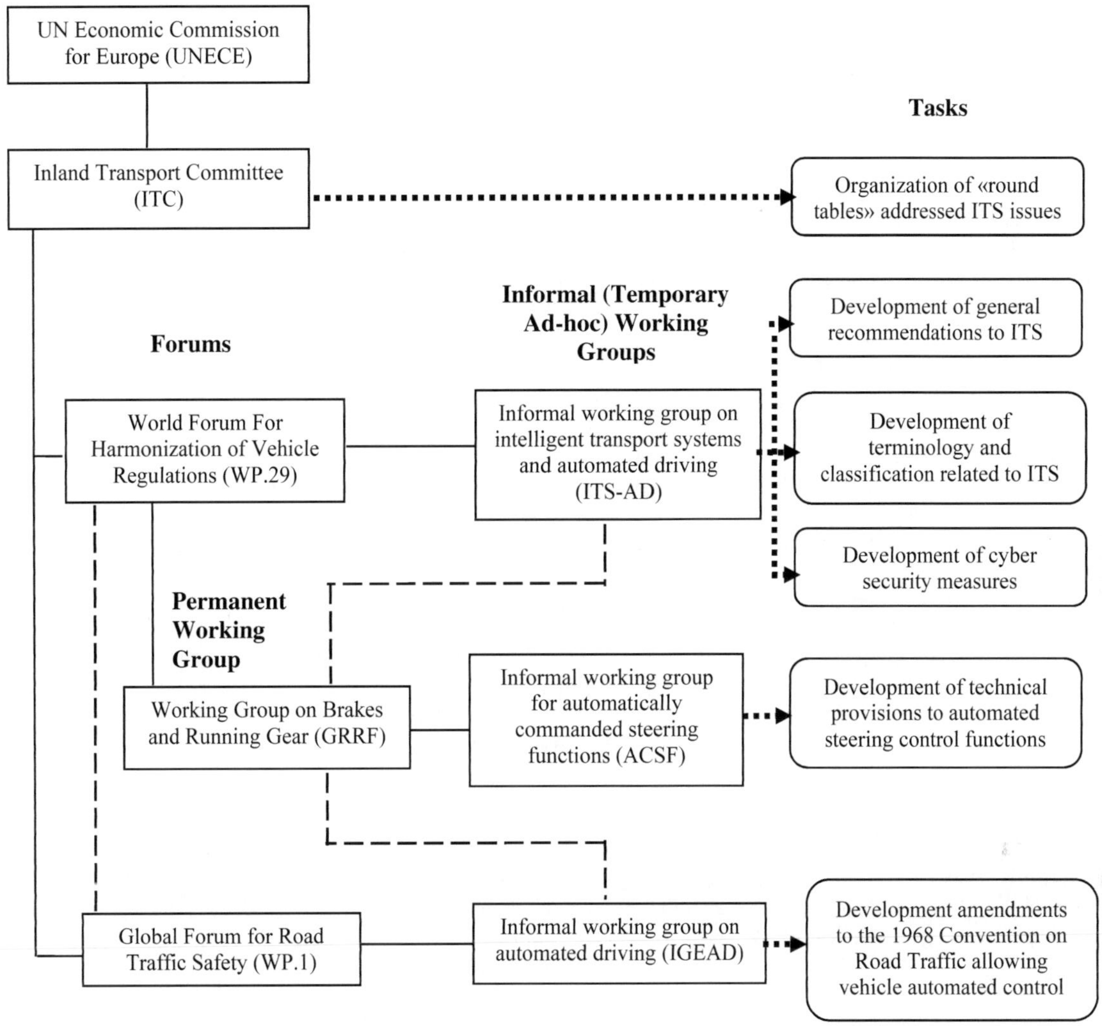

Figure 1. The organizational structure of UNECE forums, participating in discussions of ITS.

The World Forum WP.29 and its subsidiary working groups develop new technical regulations (UN Regulations, UN Global Technical Regulations (GTR) and UN Rules), harmonize existing technical regulations and their amendments and updates. WP.29, due to the nature of its activities on the development of internationally agreed safety requirements to motor vehicles, became to the center of the discussions of issues related to the technologies of automated control. WP.29 established in its structure a special working group on intelligent transport systems and automated driving (ITS-AD), which takes care of strategic issues that support implementation of new technologies. This group has developed a number of guidance documents that form the common understanding on ITS, principles of providing assistance to a driver from ITS and it deals with issues of terminology, classification of automated vehicles and the development of general principles of cyber security. The perspective topics for consideration, which ITS-AD would begin, are shown in table 1 [7].

Table 1. The perspective topics for consideration by the informal working group ITS-AD.

	Vehicle Category		
	Advanced passenger vehicle	Shuttle, Pod	Platoon for Trucks
Objective	• To provide a more comfortable driving environment • To improve safety	• Alternative of public transport	• CO2 reduction in truck transport • Improvement of working environment
Use case (Operational design domain)	Alternate driving or driving support: • for traffic congestion at low velocity • on motorway at high velocity • on urban road (driver must exist)	• Mobility services within a fixed (limited) area such as geo-fence • No Driver is necessary at the final stage	• Platoon running on motorway • Keeping the lane as much as possible after selecting a specific lane • No driver is necessary on the following vehicles at the final stage
Examples of concrete approach	To establish the regulation for level 3 technology under human driver: • 1st step: for traffic congestion at low velocity [30/60km/h or less] • 2nd step: on motorway at high velocity • 3rd step: on urban road	Considering that type of vehicles to be used and how to use them are different in each country, minimum guidelines should be issued assuming usage conditions. Considering the consistency with Vienna and Geneva Convention: • 1st step: On vehicle supervisor or remote control • 2nd step: Unmanned system	Considering the difference in traffic rules such as safe distance regulation in each country, minimum guidelines should be issued. • 1st step: With drivers on the following vehicles • 2nd step: Without drivers on the following vehicle
Possible major safety requirements	• Driver monitoring • Minimal risk maneuver • Data storage • Transition process	• (1st step: On vehicle supervisor or remote control) • Clarification of geo-fenced driving environments (as French proposal) • Requirements for emergency brake • Cyber security • Maximum velocity • (2nd step: Unmanned system) • Maximum velocity • Minimal risk maneuver • Data storage	• (1st step: With drivers on the following vehicles) • Referring the technical requirement for ACC and CACC • (2nd step: Without drivers on the following vehicles) • Minimal risk maneuver • Data storage • Safety requirements for electronic combination

A separate ad hoc working group is developing provisions for automatically commanded steering functions (ACSF).

The Global Forum WP.1 administers the 1949 Geneva and 1968 Vienna Conventions on Road Traffic, which are the basis for the national traffic regulations.

In cooperation with WP.29, WP.1 at its 68th session in March 2014, adopted the amendments to Article 8 of the 1968 Convention, according to which, it is considered that vehicle systems that affect the way of driving a vehicle, comply with the provisions of the Convention, if they meet the conditions

IOP Conf. Series: Materials Science and Engineering **315** (2018) 012012 doi:10.1088/1757-899X/315/1/012012

of manufacture, installation and operation according to the international legal instruments concerning wheeled vehicles, equipment and parts which can be fitted and/or be used on wheeled vehicles (those are the regulations developed by WP.29). It is also considered that such systems not meeting the conditions outlined above, comply with the provisions of the Convention, if a driver can at any time switch those to manual mode or disable.

A similar amendment was included in the 1949 Convention.

6. Proposals to eliminate legal uncertainties

In Russia, technologies of vehicle automated control practically are not standardized for the time being. The relevant work on standardization, based on existing and developing international standards and national standards of technologically advanced countries are needed to be carried out.

Standardization should cover:

- Terminology;
- The technical level of vehicles with automated control that ensure safe operation within the established boundary conditions;
- The order of admission to operation for vehicles with automated control as innovative ones (in accordance with the provisions of paragraph 16 of the Customs Union Technical Regulation «Concerning Safety of Wheeled Vehicles» (CU TR 018/2011);
- Requirements to road infrastructure;
- Information exchange between vehicles and road infrastructure;
- Protection of vehicles with automated control from unauthorized interference in their operation.

In respect of vehicles with a high level of automation of control the operation rules should be developed, providing:

- The restriction of territory of their operation;
- The rules of motion of vehicles with automated control;
- The transport security.

FSUE "NAMI" carried out the analysis of the regulatory legal base aimed at the establishment of legal acts subject to change in connection with the introduction of vehicles with automated control. The list of legal acts requiring analysis and subject to revision is shown in table 2.

Table 2. The list of legal acts requiring analysis and subject to revision.

No.	Legal Act	Necessity of changes
1.	Federal Law «Concerning traffic safety»	The need to establish common requirements for the operation of vehicles with a high level of automation
2.	Federal Law «Concerning transport security»	The need to establish responsibility and procedures, emerging and regulated under this law
3.	Federal Law «Concerning compulsory insurance of civil liability of vehicle owners»	The need to define the insurance event and the liability in a traffic accident involving a vehicle with highly automated control
4.	The Civil Code of the Russian Federation	The need for defining responsibilities and procedures in cases regulated by the civil legislation associated with damage from vehicles with a high level of automation
5.	The Code of the Russian Federation About Administrative Offences	The necessity of defining liability for administrative offences related to the operation of vehicles with a high level of automation
6.	The Criminal Code of the Russian Federation	The necessity of defining liability for criminal offences related to the operation of vehicles with a high level of automation
7.	Federal Law «The Charter of road transport and urban surface electric transport»	The need to establish responsibility and procedures, emerging and regulated under this law

No.	Legal Act	Necessity of changes
8.	Traffic Regulations	The need to establish conditions and limitations for operation of vehicles with a high level of automation
9.	Regulations for the carriage of goods by road transport	The need to establish responsibility and procedures, emerging and regulated under these regulations
10.	The Customs Union Technical Regulation «Concerning Safety of Wheeled Vehicles» (CU TR 018/2011)	The need to establish the safety requirements for vehicles with a high level of automation in their release for circulation and operation

7. Development of technical requirements to automated vehicles

In recent times it is said a lot at various levels about the need for quick establishment of technical requirements to vehicles with automated control.

However, the experience of other developed countries, as well as the results of international activities within the framework of the UNECE, the development of the requirements will be processed gradually, with the accumulation of sufficient experience.

When developing requirements the following principles are proposed to be used as guidance:

1. Automated systems solve individual tasks of a vehicle control for a driver. The higher the automation level, the more tasks are solved. But it is very difficult to cover a significant number of control tasks solved by a driver, therefore, at least at the present stage of development of automation technology, that number needs to be limited one way or other, for example, by means of the organization of designated lanes on roads, minimizing intersection with other road users, etc.;

2. The above-cited Article 13 of the 1968 Convention on Road Traffic on the need to maintain control over a vehicle under any circumstances is relevant in the case of automated vehicle control. Such vehicle speed, which will provide the desired system performance should be determined as well as (1) the procedure for the transfer of control from a system to a driver, if the boundary conditions for the operation of a system would be exceeded, and (2) the procedure for safe vehicle stop, if the transfer of control to a driver would not happen (so-called a minimum risk maneuver);

3. The following general requirements associated with the specificity of automated control system functioning and its interaction with the human operator (driver) should be specified:

a) To the reliability and functional safety of a system;

b) To the interface that enables interaction with a human operator (driver) and an automated control system;

c) To the record of data relating to happened events and actions of the system, so that, if necessary, the course of events could be reconstructed; the technical and legal ability to transfer these data to interested persons or authorities should be provided;

d) To the measures ensuring cyber security: protection from unauthorized access («eSecurity») and protection against failures, for example, the systems outage, the data redundancy in a system etc. («eSafety»);

e) To the information of other road users that the automated control system is in operating condition;

f) To the knowledge of an operator (driver) on the peculiarities of a system and about functions, which it performs.

Before the vehicle requirements with automated control will be developed in the form of normative documents (international and national standards, technical regulations), which CU TR 018/2011 will refer to, such vehicle in accordance with paragraph 16 of CU TR 018/2011 will be qualified as innovative. In such vehicle new design solutions, qualitatively changing its major performance indicators, which cannot be evaluated in accordance with CU TR 018/2011, are realized. In the case of innovative vehicle safety requirements shall be established by decision of the authorized body of a member state of the Customs Union for technical regulating, which carries out compliance assessment. This provision of the CU TR 018/2011 gives the opportunity to put into circulation a vehicle with automated control, without waiting for the adoption of the standards. This does not mean that the safety

IOP Conf. Series: Materials Science and Engineering **315** (2018) 012012 doi:10.1088/1757-899X/315/1/012012

of such vehicles will not be checked. The individually developed list of requirements should consider the risks associated with the operation of automated vehicles, including those described in this article.

As an example, a list of special requirements proposed by the authors is considered in relation to autonomous vehicles designed to operate on closed areas on specific routes on a dedicated lane, which does not presume the presence of other vehicles (other than vehicles of a similar type) and pedestrians (table 3).

Table 3. The list of special requirements for autonomous vehicles for operation in closed areas.

No.	Element or property of the object of technical regulating	The document, compliance with which ensures compliance with the requirement	The applicability of the requirements for autonomous vehicles for operation in closed areas
1.	Braking performance	The base document is UN Regulation No. 13-11	Partial compliance with Regulation. The testing methods need to be changed. The performance of braking system and ABS should be checked Should be designed to determine the for speed 25 km/h excluding control action from a driver on the brake pedal. 1. The function of the service, secondary and parking brake systems without control action from a driver's side; 2. The performance of the ABS; 3. Safety inspection for complex electronic systems.
2.	Vehicle protection from unauthorized use	The legal document is not available	The measures for protection against unauthorized access to the electronic control unit should be taken (cybersecurity).
3.	Speed measuring devices	The legal document is not available	The correct determination of a vehicle speed should be confirmed. The requirements and test method should be developed.
4.	Requirements to ensure safe operation of autonomous vehicles	The legal document is not available. The specific provisions are contained in UN Regulations Nos. 79-02, 107-06 and 131-01.	The numerical targets in relation to the following requirements and the relevant test methods should be developed, including the following items: 1. Providing motion within a dedicated lane; 2. Providing stops according to the route. 3. Providing opening and closing the service doors at the bus stop; 4. Providing a possibility for the emergency stop for passengers («emergency brake»); 5. Providing voice communication between passengers and the vehicle operation manager; 6. Ensuring passenger comfort with regard to accelerations; 7. Ensuring detection and detour of fixed and moving obstacles, also in the dark. 8. Providing self-diagnosis of faults.

8. Conclusions

For the time being, all developed countries can be characterized by the situation when the development of the regulatory framework is behind the implementation of innovative ideas of automation of vehicle control. At the same time, significant efforts are applied to overcome such situation at the national level and at the level of UNECE.

Vehicles with automatic control functions can be characterized by the same risks as conventional vehicles. At the same time, there are new risks associated with the use of computer technologies: the error of geo-location and, as a result, the risk of accidents with elements of road infrastructure, loss of data, including data about the parameters of vehicle motion, as well as personal data of a user; improper interaction with a driver (remote operator); loss of control over a vehicle due to technical failure, and in connection with the unauthorized intervention in the control system. These specific risks require the development of appropriate regulatory measures.

Currently, for vehicles with a high level of control automation neither at the regional (Eurasian Economic Union) nor at the national level no measures for technical regulating and technical standards are developed.

In economically developed countries, the experience of regulating in relation to automated control systems is not sufficient for the development of legal acts. This is due to the fact that the speed of the introduction on market of the new technologies of automated control significantly outperforms procedural terms of development, consideration and approval of legal acts. Therefore, at the national level guidance documents containing recommendations on functioning and effectiveness of such systems are published.

The experience of economically developed countries and the cooperative regulatory activities under the auspices of the UNECE are the basis for the development of regulatory requirements for vehicles with automated control. The participation of the Russian Federation in international agreements leads to the use of technical regulations developed and accepted in international forums under the auspices of the UNECE. In this regard, rulemaking at the national level in relation to vehicle automated control should be carried out in concert with activities at the UNECE level. This approach, in addition to effects associated with the facilitation of promotion of Russian products in foreign markets, would also help to save on scientific and technical developments, without repeating the research already been conducted in other countries.

The authors propose an approach towards building a regulatory framework that will ensure the implementation of highly automated vehicles in Russia. Proposals to eliminate legal uncertainties include the development of changes in national legislation as well as standards and regulations for operation of vehicles with automated control. The formulated principles are proposed to guide the development of technical requirements for automated vehicles. Until that those requirements would be developed in the form of regulatory documents (international and national standards, technical regulations), such vehicles in accordance with paragraph 16 of CU TR 018/2011 could be qualified as innovative. In the case of innovative vehicles safety requirements shall be established by a decision of the authorized body for technical regulating of a member state of the Customs Union, which carries out compliance assessment. An individually developed list of requirements should consider the risks associated with the operation of automated vehicles, including those described in this article. As an example, the list of special requirements proposed by the authors for the autonomous vehicle designated for traffic in closed areas.

References

[1] SAE Standard J3016: Taxonomy and Definitions for Terms Related to On-Road Motor Vehicle Automated Driving Systems (SAE International Inc.) 2014
[2] Convention on Road Traffic Done in Vienna 8 November 1968 (United Nations) 1968
[3] Roadmap on Highly Automated vehicles *GEAR 2030 Discussion Paper* (Brussels. European Commission Directorate-General for Internal Market, Industry, Entrepreneurship and SMEs) 2016
[4] Directive of the European Parliament and of the Council of 5 September 2007 establishing a framework for the approval of motor vehicles and their trailers, and of systems, components and separate technical units intended for such vehicles (Framework Directive) 2007/46/EC (Brussels. European Commission) 2007
[5] Sano Y 2017 Guidelines for Testing of Automated Vehicles *Workshop on Governance of*

Automated Vehicles 28-29 June 2017 (Geneva, UNECE)
[6] Federal Automated Vehicles Policy *DOT HS 812 329* (Washington D.C. U.S. Department of Transportation. National Highway Traffic Safety Administration) 2016
[7] Expansion of consideration target of ITS-AD IWG *Document ITS-AD-12-08* (Geneva, UNECE) 2017
[8] Report of the Working Group for Road Traffic Safety about its sixty-eighth session *Document ECE/TRANS/WP.1/145* (Geneva, UNECE) 2014

IOP Conf. Series: Materials Science and Engineering **315** (2018) 012013 doi:10.1088/1757-899X/315/1/012013

Concept of intellectual charging system for electrical and plug-in hybrid vehicles in Russian Federation

A Kolbasov[1], K Karpukhin[1], A Terenchenko[1] and I Kavalchuk[2]

[1] Federal State Unitary Enterprise Central Scientific Research Automobile and Automotive Institute "NAMI" (FSUE «NAMI»), Avtomotornaya Street, 2, Moscow, Russia

[2] School of Science and Technology, RMIT University Vietnam, 702 Nguyen Van Linh, District 7, Tân Phong, Ho Chi Minh City, Hồ Chí Minh, Vietnam

Email: aleksey.kolbasov@nami.ru

Abstract. Electric vehicles have become the most common solution to improve sustainability of the transportation systems all around the world. Despite all benefits, wide adaptation of electric vehicles requires major changes in the infrastructure, including grid adaptation to the rapidly increased power demand and development of the Connected Car concept. This paper discusses the approaches to improve usability of electric vehicles, by creating suitable web-services, with possible connections vehicle-to-vehicle, vehicle-to-infrastructure, and vehicle-to-grid. Developed concept combines information about electrical loads on the grid in specific direction, navigation information from the on-board system, existing and empty charging slots and power availability. In addition, this paper presents the universal concept of the photovoltaic integrated charging stations, which are connected to the developed information systems. It helps to achieve rapid adaptation of the overall infrastructure to the needs of the electric vehicles users with minor changes in the existing grid and loads.

1. Introduction

Ecology of the vehicles has become the main concern of the engineers and researchers in the last twenty years. The sustainable electric vehicles (EV) and plug-in hybrids have become the key solution to meet the strict ecological requirements with reasonable cost and technical development. Zero emission vehicles improve the air quality of the cities as well as reduced noise level.

However, modern EVs still face the problem of limited driving range on a single charge and availability of the charging stations within the driving distance. Long charging time has also become an issue for the users.

Developed countries have improved their infrastructure to meet the novel challenges in the power consumption and its distribution. In addition, several countries have introduced special tax excemptions and specific changes in road rules for the users of EVs. It allows them to rapidly increase the number of sustainable transport on the roads.

Nowadays, official electric vehicles market in Russia is limited to a single brand Renault. Another manufacturer of EVs, Mitsubishi, has left the market after four years of presence. The main reasons were the low development of the infrastructure and the absence of the special taxation polices for EVs and their users.

Figure 1. EV market in Russia in 2016.

As it can be seen on figure 1, Nissan Leaf and Tesla Model S are the leading EVs in terms of sale, but they have been imported unofficially or as second-hand vehicles. The only available new EVs are Renault Twizy and Renault Kangoo ZE.

According to the proposed plans, PJSC "Russian Grids" has started Russian-wide program of infrastructure development and adaptation to meet EVs requirements. In addition to the program, Minister of Energy of Russia, A. Novak, announced in 2016, that the number of EVs in Russia will reach 200 thousand units by 2020. As a part of this development, wide integration of EVs is assigned to the private and commercial transport.

Past researches have shown that the charging infrastructure in Russia is located in the large agglomerates and it is not presented in local centres. In addition, existing charging solutions are provided by various companies. That situation leads to the lack of unified information supply for the drivers and customers.

This paper presents the initial analysis of the existing infrastructure in Moscow regions, as the region with wide adaptation of Electrical transport. Next section presents the developed information system for information collection and sharing within the concept of Connected Car system. Next part of the paper discusses the design and implementation of the universal charging station for EVs based on photovoltaic (PV) system with integration of the solution in the existing electrical grid and developed information solution. It also presents the preliminary results of the system integration. Last section discusses the main results of the developed system and shows the key direction of the future work.

2. Analysis of the existing infrastructure solutions in Moscow region

Moscow region is known as a region with the most developed infrastructure for EVs due to the high number of vehicles-in-use and existing intelligent transportation solutions. Due to these factors, Moscow region was analysed to see the available charging solutions for EVs.

Currently, there are three providers of the charging stations for EVs in Moscow region: PJSC "Mosenergo", PJSC "RUSSIAN GRIDS" and LLC "Revolta". They have different charging standards, different combination of the charging cables and accessing polices. These differences create a challenging environment for the drivers of EVs. In addition, charging access card is different for all three solutions, as it can be seen in table 1.

Table 1. Comparison table of charging companies in Moscow region.

Company Name	Number of Charging Stations	Access Card Type	Cost of Charging, Rub/hour	Web-service
LLC "Revolta" emi ЭЛЕКТРО-МОБИЛЬНАЯ ИНФРАСТРУКТУРА	25	«EMI» Card 	10	Yes
PJSC "RUSSIAN GRIDS" РОССЕТИ МОЭС	11	«MOESK-EV» Card 	Free	No
PJSC "Mosenergo" МОСЭНЕРГО	Unknown	«Troika» Card 	Free	No

Current number of charging stations creates several problems. First problem is availability of the charging ports due to the high demand. Second problem is distance between charging stations, which leads to the overrun of EVs, higher power consumption and, as a conclusion, lower satisfaction for the users. Third problem is availability of the specific charging plugs on the specific stations.

Because of these issues, a typical EV driver in Moscow has to spend more than half-an -hour to find suitable charging station. It leads to the decreased sustainability of the overall transportation system and low satisfaction rate.

Based on the observation and analysis, couple more challenges have been identified:

- Hard to identify location of the charging stations;
- Impossible to identify availability of the charging ports;
- Various charging providers have different access cards;
- Pricing and payment for the specific company;
- Availability of a specific charging plug on the nearby station;
- Absence of the centralised information centres for EVs.

3. Developed centralised information platform based on connected car concept

Connected car concept includes all possible connection between a specific vehicle and its environment. The initial connection protocols combine two key channels Vehicle-to-Vehicle (V2V) and Vehicle-to-Infrastructure (V2I). Further development of the Internet of Things and smart devices, provided another communication channel Vehicle-to-Portable Device (V2D).

In addition, EVs can be used as an energy storage. Due to the rapid integration of unreliable renewable distributed generators, EVs have become elements in the power distribution network through the protocol Vehicle-to-Grid (V2G). This protocol allows EVs to create bidirectional power flow. Within this protocol, EV can get the energy from the grid, if the excess if presented in the distribution network, or operate as an external battery storage for the peak-consumption times.

Developed solution is based not only on the technical ideas and existing solutions within the concept of Connected Car, but also based on the discussions with EV-users. New system combines the demand from individual vehicles and creates a general pattern of the electrical demand within specific regions and matches it with the combined supply. Centralised nature of the solution allows to get information about charging capacity from all providers and supply it to the users.

Intellectual charging scheme is shown on figure 2.

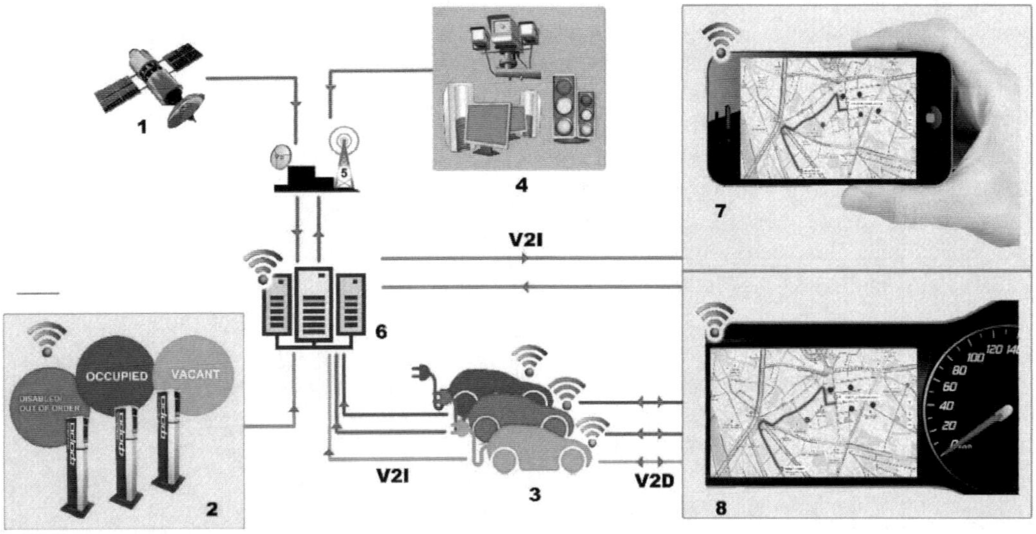

Figure 2. Connection scheme of developed solution with infrastructure Elements and EVs; where 1) Navigation satellites; 2) Charging stations with Internet connectivity (4G/5G, ZigBee); 3) EVs with V2V and V2D connectivity; 4) Traffic Management Centre; 5) Navigation Service Provider; 6) Centralised Data and Control Centre; 7) Mobile devices with installed application from centralised service provider; 8) In-car information system with GPS and Centralised system connectivity.

Developed solution is based on intellectual data analysis concept. It analyses the data about available charging stations, vacant stations and number of EVs in the designated area. At the same time, each car supplies information not only about current destination and optimal route, but also about remaining charge in the battery and charging technical information. Charging information includes model of the car. Required charging port and availability of fast charging technologies. These parameters are sent to the centralised control centre by the owner, when he registers his EV for the first time in the system. In addition, key parameters for the most common EVs can be assigned automatically, based on the manufacturer's data, supplied to the centre.

Collected data can adjust the routes, according to the availability of the charging stations on the route, traffic congestions and planned arrival time. These algorithms help to achieve higher efficiency from individual vehicle as well as decrease the load on the selected charging stations and spread the load evenly between all charging stations, as system can analyse and predict demand for the stations, based on the current shared routes.

Furthermore, developed system constantly monitors state of charge of the battery due to the V2D protocol. It helps to analyse the state of health of the battery (SoH), detect the issues with temperature or degradation and supply technical data to the manufacturers for further improvements. Optimisation algorithms in the route planning allows users to charge their EVs in scenarios, when the traffic congestions are heavy, so the battery can be drained fast and not efficiently, which mean more benefits for the users.

4. Design of universal charging station for intellectual system

The main problems with rapid integration of fast charging stations in current grid is the peak load, which is created by these stations. Outdated grid cannot handle high power charging in the peak loading times, which leads to the slower charging time, decreasing of the power quality and even power outage in the selected regions. As a result, it slows the integration of EVs in the society.

On the other hand, the outdated infrastructure with limitations in the powerflow directions also limits integration of sustainable energy sources, such as photovoltaic systems (PV). The key challenge in integration of PV systems is inoptimal generation time, when the peak of generation for the households is not matching the period with the peak demand. As a result, many researchers have proposed to use additional energy storage systems to store generated energy for later use.

Similar concept was proposed in the design of the universal charging stations. Solar energy allows to increase the capacity of the power generation with limited interventions to the grid due to the applications of the energy storage system, associated with the generation unit. Current power capacity of powerplants cannot meet increasing demand of power by EVs. In addition, long distribution and transmission lines decreases the sustainability of the integration processes of EVs.

Developed solution is shown on figure 3.

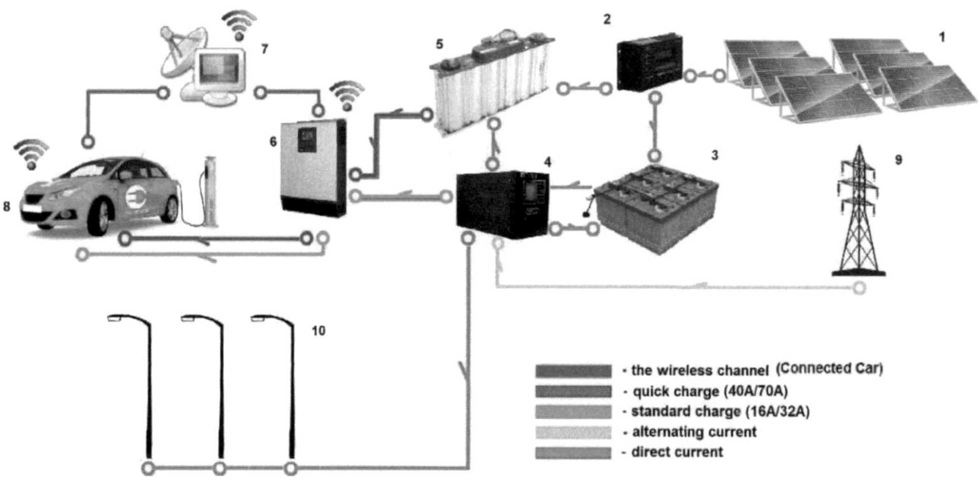

Figure 3. Developed architecture of charging stations.

Developed solution includes set of solar panels (1). Power generation capacity of the solar panels can be calculated based on the current demand in the area and number of registered EVs in the neighbour area. Inverter (2) operates as charge controller for the energy storage elements (battery (3) and ultracapacitors (4)). In addition, charge controller plays the role of maximum power point tracking (MPPT) unit for solar generation under partial shading conditions. Integrated MPPT unit operates with the PV panels based on the partial swarm optimisation concept to achieve immediate response and high possible variables.

State of charge of energy storage devices and generation capability of the system is continuously monitored by centralised control centre (7). This centre communicates with vehicle via V2D and V2I protocols to exchange information about charging demand of the vehicle and provide information about power supply in the nearby stations for further analysis by driver, if required. AC/DC inverter (4) allows charging station to take power from the main grid and, if the infrastructure allows, send extra generated power back to the grid.

In addition, developed solution powers the local light system (10) to improve safety of the areas and increase efficiency of the solution. In Russia, the requirements for the street light varies from 5 to 16 hours per day depending from the season.

Initial calculations were performed based on the assumption, that charging station will accommodate parking space for one car. As a result, the area of PV panels in the calculations was limited to 20 m^2. In the analysed solution, monocrystalline PV elements were used due to the high performance in high range of weather conditions. Analysis showed, that developed solution can generate up to 16kWh per day and 4007 kWh per year, as can be seen in figure 4. Generated energy provides the distance of 24000 km for an average EV. At the same time, it allows to decrease the amount of CO_2 emissions by 1963 kg.

IOP Conf. Series: Materials Science and Engineering **315** (2018) 012013 doi:10.1088/1757-899X/315/1/012013

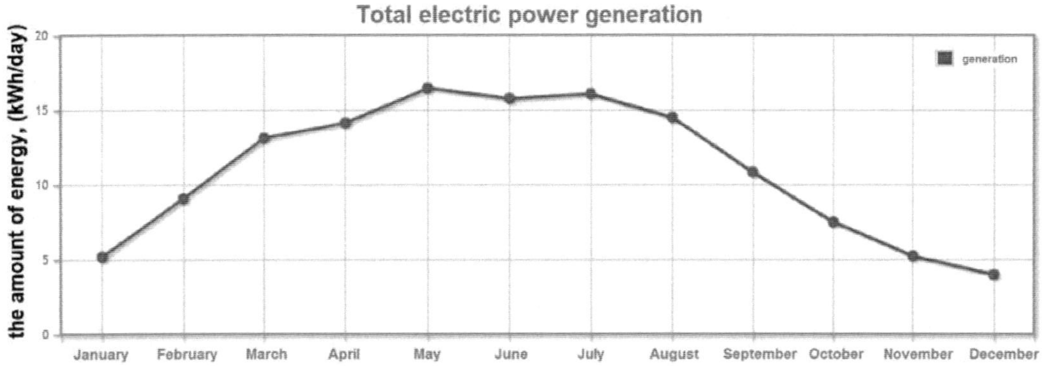

The monthly average electricity production kWh/day

January	February	March	April	May	June	July	August	September	October	November	December
5.18	9.06	13.09	14.13	16.45	15.76	16.07	14.49	10.81	7.49	5.21	3.95

Annual electricity generation: 10.97 kWh/day **Total electric power generation for the year: 4005.77 kWh/day**

Figure 4. Analysis of generated energy capacity for the developed charging system.

5. Conclusion and future work

This paper presented and discussed the concept of the intellectual information solution for the drivers of EVs. Developed solution helps to combine power requirements from the drivers and charging infrastructure elements and create sustainable environment. The foundation of the developed solution comes from the concept of Connected Car.

Second section of the developed solution allows to update the charging infrastructure using small investments. Separated self-powered solar-based charging stations were designed and analysed as a part of developed infrastructure concept. Monocrystals PV panels allows to generate enough energy throughout the year in the middle part of Russia for an average EVs. This solution helps to further improve sustainability of the electric grid and EVs integration policies.

As a future work, the research will lead to the hardware manufacturing of the system and implementation in several regions for further testing of the system.

Acknowledgement

This article was supported by the Ministry of Education and Science of the Russian Federation, agreement No. 14.624.21.0047; unique identifier of the project: RFMEFI62417X0047.

References

[1] Karpukhin K E and Terenchenko A S 2016 Features of Creation and Operation of Electric and Hybrid Vehicles in Countries with Difficult Climatic Conditions, for Example, in the Russian Federation *IOP Conference Series: Materials Science and Engineering* Vol. 157 **1** 012014. DOI: 10.1088/1757-899X/157/1/012014

[2] Ipatov A A, Kutenjov V F, Luksho V A et al. 2010, *Motor transport and ecology of metropolises* (Moscow: Mashinostroenie Publishers) 254

[3] Kolbasov A F, Karpukhin K E and Debelov V V 2017 Research infrastructure for personal electric transportation: current problems, possible solutions *Žurnal AAI* **2** 36-45

[4] Kavalchuk I, Nguyen H L, Pham T and Stojcevski A, 2016 Methodology of Intelligent Energy Management System Simulation for Electric Vehicle Applications with Asynchronous Logic Controller, *7th International Conference on Intelligent Systems, Modelling and Simulation (ISMS)*, 225-230 DOI: 10.1109/ISMS.2016.35

[5] Varlamov D O, Skvortsov A A, Hortov V P 2016 Matlab/simulink simulation of an active

balancing battery system of an electric vehicle in the power redistribution mode *Žurnal AAI* **5** 26-29

[6] Wolfram P, Lutsey N 2016 Electric vehicles: Literature review of technology costs and carbon emissions *International council on clean transportation* **14** 123

Energy efficient motion control of the electric bus on route

G O Kotiev, D O Butarovich and B B Kositsyn

Bauman Moscow State Technical University,
Ul. Baumanskaya 2-ya, 5, Moscow, 105005, Russian Federation

E-mail: kotievgo@yandex.ru

Abstract. At present, the urgent problem is the reduction of energy costs of urban motor transport. The article proposes a method of solving this problem by developing an energy-efficient law governing the movement of an electric bus along a city route. To solve this problem, an algorithm is developed based on the dynamic programming method. The proposed method allows you to take into account the constraints imposed on the phase coordinates, control action, as well as on the time of the route. In the course of solving the problem, the model of rectilinear motion of an electric bus on a horizontal reference surface is considered, taking into account the assumptions that allow it to be adapted for the implementation of the method. For the formation of a control action in the equations of motion dynamics, an algorithm for changing the traction / braking torque on the wheels of an electric bus is considered, depending on the magnitude of the control parameter and the speed of motion. An optimal phase trajectory was obtained on a selected section of the road for the prototype of an electric bus. The article presents the comparison of simulation results obtained with the optimal energy efficient control law with the results obtained by a test driver. The comparison proved feasibility of the energy efficient control law for the automobile city electric transport.

In the modern metropolis, a significant role in passenger transportation is played by land road transport. With an intensive increase in the city's population, urban boundaries are being expanded, the transport network is enlarged and the traffic schedule for road transport is consolidated. At present, the city is widely used by the city automobile electric transport - the electric bus. For this type of electric transport, the task of improving the energy efficiency of traffic on the urban route (which involves the presence of stops and a restriction on the time of travel between them) is especially important, since with a reduction in energy costs it will be possible to:

- increase the passenger capacity of the electric bus by rationally choosing the parameters of the energy storage device (the number of batteries) on board, and, consequently, reducing the curb weight;
- reducing the cost of the electric bus by reducing the required number of batteries to ensure the necessary power reserve;
- increase the power reserve of the electric bus with the same capacity of the battery.

The essence of the proposed method consists in software control of the speed of the electric bus on the way between stops for landing and disembarking passengers at a given time in order to minimize energy consumption during movement. In this case, the driver determines the beginning of the movement. The control system must provide the calculated speed of the electric bus as a function of the traveled path, by means of the traction moment of the electric machine supplied to the wheels. The braking torque of the regenerative brake and the braking torque produced by the service braking system.

Thus, the task of an energy-efficient law governing the speed of an electric bus between stops on a city route should be formulated as an optimization problem. Determine the law of change in speed $v(s)$ from the traversed path of the electric bus on the route when driving from a waypoint with phase coordinates $s_1,\ v_1 = 0$ to the waypoint $s_2,\ v_2 = 0$ during $T = t_2 - t_1$ with the minimization of the energy consumed by the motion determined by the objective function:

$$J = \int_{t_1}^{t_2} (N_a - N_{rb} + N_b)dt, \qquad or \qquad J = \int_{s_1}^{s_2} \frac{(N_a - N_{rb} + N_b)}{v} ds$$

where

N_a – the capacity of the discharge of the accumulator, developed during the acceleration of the electric bus;

N_{rb} – power charge of the drive with regenerative braking;

N_b – power dissipated when using the service braking system.

To determine the energy-efficient control law, we describe the rectilinear motion of an electric bus on a horizontal reference surface as a motion of a rigid body. To do this, let's assume that the electric bus moves along the city route without slipping of its wheels, which means that r_{w_i}(rolling radius of the i-wheel) $\approx r_{w0_i}$(rolling radius without sliding of the i-wheel) $\approx r_{d_i}$ (dynamic radius of the i-wheel). In addition, we will assume that the rolling radii of all the wheels of an electric bus without slipping are equal. Then the equation of dynamics [1] electric bus will take the form:

$$m\delta \frac{dv}{dt} = \frac{M_{el}\eta_{tr}u_{tr} - M_f - M_b}{r_{w0}} - P_w,$$

where

m – mass of the electric bus;

$\frac{dv}{dt}$, v – acceleration and speed of the electric bus;

r_{w0} – rolling radius without sliding of the electric bus wheels;

M_{el} – tractive /braking torque generated by the electric machine;

M_f – moment of rolling resistance of an electric bus [1];

M_b – reduced braking torque of the service braking system;

P_w – force of aerodynamic resistance;

η_{tr} – efficiency of the mechanical part of the transmission;

u_{tr} – transmission ratio;

δ – rotational inertia coefficient.

$$\delta = 1 + \frac{\sum_{i=1}^{4} J_{w_i}}{m{r_{w0}}^2},$$

where

J_{w_i} – the moment of inertia of the mechanical part of the transmission and the wheel reduced to the i-th wheel.

For further convenience, we write the equation of the dynamics of the electric bus in the coordinates path $s - v$:

$$m\delta \frac{dv}{ds} = \frac{1}{v}\left(\frac{M_{el}\eta_{tr}u_{tr} - M_f - M_b}{r_{w0}} - P_w\right),$$

This simplified model of the motion of the electric bus as a rigid body is used to reduce computation time and increase the speed of the algorithm.

To determine the relationship N_a and N_{rb} with M_{el} and v we apply the following sequence of power conversion in the rectilinear motion of the electric bus (figure 1):

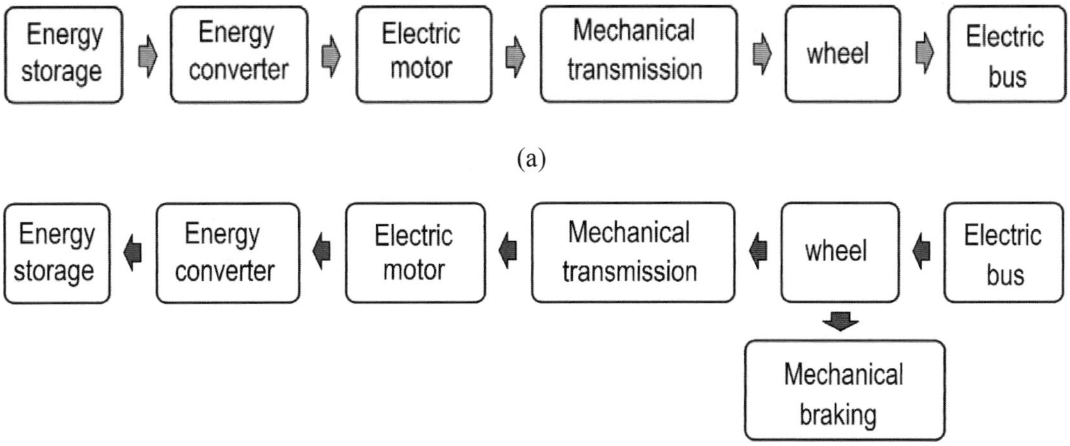

(a)

(b)

Figure 1. Power conversion in rectilinear motion during:
a) – energy storage unit discharge;
b) – energy storage unit charge.

Then we get:

$$\begin{cases} N_a\eta_{cP}\eta_{em_p} = \sum_{i=3}^{4} M_{w_i}\omega_{w_i} = \frac{v\cdot u_{tr}}{r_{w0}}M_{el}, & \text{traction mode} \\ N_{rb} = \eta_{tr}\eta_{c_T}\eta_{em_T}\sum_{i=3}^{4} M_{w_i}\omega_{w_i} = \eta_{c_T}\eta_{em_T}\frac{v\cdot u_{tr}}{r_{w0}}M_{el}, & \text{energy recuperation mode} \end{cases},$$

where

M_{w_i} – traction/braking torque on the i-wheel created by the electric machine;

$\frac{d\omega_w}{dt}$, ω_{w_i} – angular acceleration and angular velocity of the i-th wheel;

η_{cP} – efficiency of the electric power converter at dispersal and speed maintenance;

η_{cT} – efficiency of the electric power converter at regenerative braking;

η_{em_p} – efficiency of the electric machine during acceleration and holding speed;

η_{em_T} – efficiency of the electric machine with regenerative braking.

At the same time, the maximum power dissipated during braking of the electric bus by the service braking system is independent of the speed of movement and defined as:

$$N_b = \sum_{i=1}^{4} M_{b_i}\omega_{w_i} = \frac{v}{r_{w0}}M_b,$$

where

M_{b_i} – braking torque generated by the service braking system on the i-wheel.

Thus we have provided a possibility to determine the discharge / charge power of the drive and the braking power by the working braking system and, consequently, to find the value of the minimized functional at each moment of time by solving the equations of dynamics.

Let us describe the law governing the change in the driving and braking moment of the electric bus. The driver, by pressing the accelerator pedal, determines the "desire" to move according to the calculated traffic law. Proceeding from this, the control system must provide the desired speed of the electric bus, by means of the traction moment of the electric machine supplied to the wheels, the braking torque of the regenerative brake and the braking torque created by the service braking system. For convenience of calculations we combine M_{el} and M_{b_i} into one variable $M_h = M_{el}\eta_{tr}u_{tr} - M_b$ and determine its value depending on the magnitude of the control action h. In addition, it must be taken into account that during braking, the energy recovery mode is first switched on, and after reaching the control parameter of a certain value, the service braking system is turned on.

$$M_h = \begin{cases} \begin{cases} M_{em}u_{tr}\eta_{tr}(h/h_{em})\ , \text{ if } \dfrac{v\cdot u_{tr}}{r_w} < \dfrac{N_{em}}{M_{em}} \ and\ v > 0 \\ \dfrac{N_{em}(h/h_{em})}{v/r_w}\eta_{tr}\ , \text{ if } \dfrac{v\cdot u_{tr}}{r_w} > \dfrac{N_{em}}{M_{em}} \ and\ v < v_{em} \\ 0\ , \text{if } v > v_{дв} \end{cases} , \text{ if } h > 0\ and\ h < h_{em} \\ \begin{cases} \dfrac{M_{rec}u_{tr}(h/h_{rec})}{\eta_{tr}}\ , \text{ if } \dfrac{v\cdot u_{tr}}{r_w} < \dfrac{N_{rec}}{M_{rec}} \ and\ v > 0 \\ \dfrac{N_{rec}(h/h_{rec})}{\eta_{tr}\cdot v/r_w}\ , \text{ if } \dfrac{v\cdot u_{tr}}{r_w} > \dfrac{N_{rec}}{M_{rec}} \ and\ v < v_{rec} \\ 0\ , \text{if } v > v_{рек} \end{cases} , \text{ if } h > -h_{rec}\ and\ h < 0 \\ \begin{cases} -\dfrac{M_{rec}u_{tr}}{\eta_{tr}} + M_b\left(\dfrac{h+h_{rec}}{h_т - h_{rec}}\right), \text{ if } \dfrac{v\cdot u_{tr}}{r_w} < \dfrac{N_{rec}}{M_{rec}} \ and\ v > 0 \\ -\dfrac{N_{rec}}{\eta_{tr}\cdot v/r_w} + M_b\left(\dfrac{h+h_{rec}}{h_т - h_{rec}}\right), \text{ if } \dfrac{v\cdot u_{tr}}{r_w} > \dfrac{N_{rec}}{M_{rec}} \ and\ v < v_{rec} \\ M_b\left(\dfrac{h+h_{rec}}{h_т - h_{rec}}\right), \text{if } v > v_{rec} \end{cases} , \text{ if } h < -h_{rec} \end{cases} ,$$

where
M_{em} – the maximum tractive moment created by the electric machine in the traction mode;
M_{rec} – maximum braking torque generated by the electric machine in the energy recovery mode;
M_b – maximum braking torque produced by the service braking system;
N_{em} – maximum tractive power developed by the electric machine in traction mode;
N_{rec} – the maximum power developed by the electric machine in the mode of energy recovery;
v_{em} – the maximum speed of the electric bus, to which the electric machine can operate in traction mode;
v_{rec} – the maximum speed of the electric bus to which the electric machine can operate in the energy recovery mode;
h_{em} – the amount of control at which the electric machine operates in the mode of maximum traction power;
h_{rec} – the control value at which the electric machine operates in the maximum recovery power mode;
$h_т$ – the control value at which the electric machine operates in the maximum recovery power mode and the maximum braking torque of the service braking system is used.

Thus, a system of equations is obtained describing the driving traction / braking torque realized by the motion control system as a function of the speed of the electric bus v and the magnitude of the control action h. value h may vary from $-h_т$ to h_{em}, introducing into the equation of dynamics a limitation on the driving traction/braking torque.

The search for optimal control is carried out using Bellman's dynamic programming method [2]. The algorithm for implementing the Bellman method is as follows. At the first stage, it is necessary to determine the phase space, that is, the state space of the electric bus $v(s)$, in which it can be physically located under the condition that the boundary conditions are met. To do this, it is necessary to integrate the equation of motion of the electric bus while fixing the initial boundary conditions $(s_1, v_1 = 0)$ and feed the traction moment to the driving wheels according to the external characteristic of the electric motor, when moving from the initial state to the final state. The magnitude of the control will then be equal to h_{em}. After this, it is necessary to obtain solutions of the equation of motion for the fixed finite boundary conditions (s_2, $v_2 = 0$) and the braking torque the regenerative brake taken from its external characteristic, as well as the maximum torque of the service braking system M_T, moving from the final state to the initial state. The magnitude of the control will then be equal to $-h_T$. Then, if necessary, consider the restriction of the speed of the electric bus on the route in accordance with the traffic rules. Thus, curves are obtained that limit the space in which the electric bus can be located without violating the traffic rules, provided that the driving traction / braking torque when applying the permissible control action will ensure that the boundary conditions are met. A graphic illustration of the space obtained is given in figure 2.

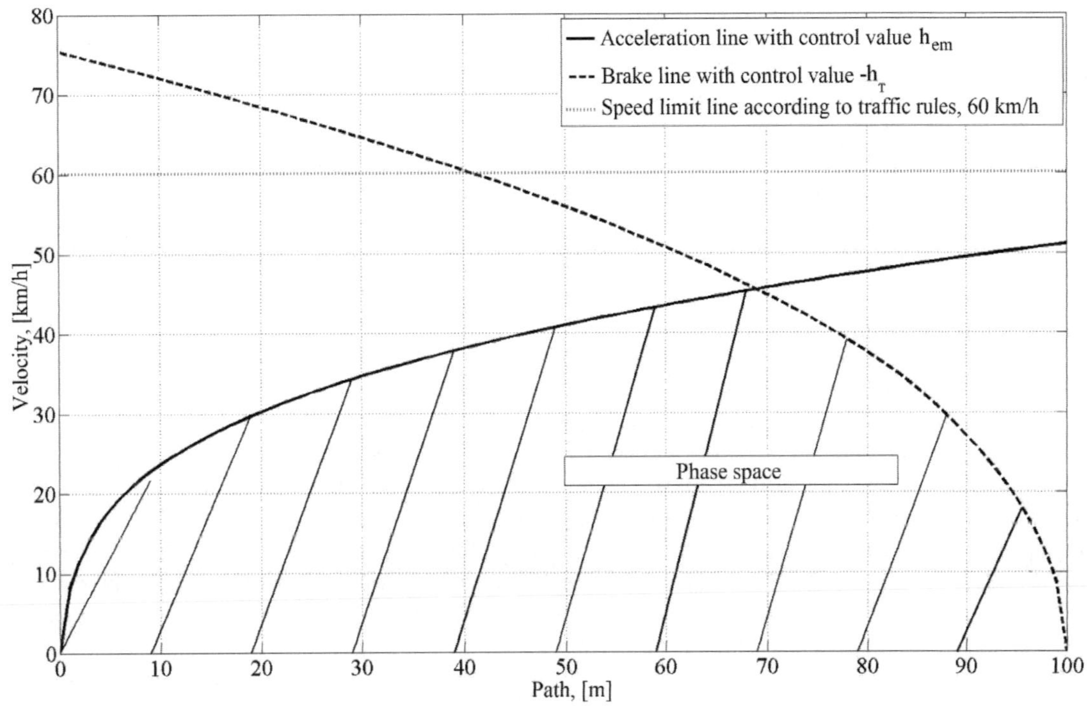

Figure 2. Phase space of electric bus states.

The next stage is the discretization of the resulting space, that is, the partition of the phase surface onto the grid of the states considered. We divide the section of the path under consideration into N points, and the allowable speed for L. Since for each point of the considered route the range of permissible speeds can be different, the number of points of the velocity partition at a given point of the path in question will be equal to L_k, where k – the number of the point under consideration in the route under study. The calculation time and the accuracy of the obtained control law directly depend on the frequency of the partition.

Then it is necessary to compile the Bellman function for the system under consideration. For convenience, we will use a functional with a path dependence:

$$J = \int_{s_1}^{s_2} \frac{(N_a - N_{rb} + N_b)}{v} ds$$

Then in our case the Bellman function will have the form:

$$Z_{N-k}(s(N-k), v(N-k,l)) = \\ = \min_{h(N-k,l)\epsilon H} \Big\{ \big(N_a(s(N-k), v(N-k,l), h(N-k,l)\big) \\ - N_{rb}\big(s(N-k), v(N-k,l), h(N-k,l)\big) + N_b\big(s(N-k), v(N-k,l), h(N-k,l)\big) \\ + \lambda) \cdot \frac{\big(s(N-k+1) - s(N-k)\big)}{0{,}5\big(v(N-k+1,p) + v(N-k,l)\big)} \\ + Z_{N-k+1}(s(N-k+1), v(N-k+1,p)) \Big\},$$

where

$Z(s,v)$ – the minimum value of the Bellman function at the point of phase space with the length of the traversed route s And speed v;

N – number of points of subdivision of the path under study;

H – area of admissible controls;

k – number of the point in question in the path under study $k \in [1 \dots N]$;

l – the number of the point in question in the array of permissible velocities for $(N-k)$ point of the path$l \in [1 \dots L_{N-k}]$;

p – the number of the point in question in the array of permissible velocities for $(N-k+1)$ point of the path$p \in [1 \dots L_{N-k+1}]$;

λ – indefinite Lagrange multiplier.

The value of the functional, at a point with phase coordinates k, l, is the minimum value of the sum of the value of the functional, at a point with phase coordinates k, p and the amount of energy expended for the transition from the state k, l to k, p, provided that the control applied to the system h be in the range of acceptable values. Thus, by calculating the value of the Bellman function at each point of the phase space, moving from the final position to the initial position, we obtain a two-dimensional array of minimum energy values that must be expended to move from the point of space to the finite point. The sequence of calculations occurs precisely from the end point to the initial one because the value of the function $Z(s,v)$ At the end point of the route does not affect the very process of going to this point.

For the comfortable transportation of passengers, it is necessary to provide a restriction on the acceleration developed by the electric bus (1 m/s^2). To do this, when calculating the values of the Bellman function, one must choose a control that is not only in the range of admissible values (from $-h_т$ to h_{em}), but also creates a rooting that does not exceed the permissible limits.

The next step is to write down the equation of state. Since the position of the electric bus on the phase surface depends on the previous state and control at this step, then:

$$Z_k\big(s(k), v(k,l')\big) = \\ = Z_k\big(s(k-1), v(k-1,p')\big) \\ + \big(N_a(s(k), v(k,l'), h(k,l')\big) - N_{rb}\big(s(k), v(k,l'), h(k,l')\big) \\ + N_b\big(s(k), v(k,l'), h(k,l')\big) + \lambda) \cdot \frac{\big(s(k) - s(k-1)\big)}{0{,}5\big(v(k,l') + v(k-1,p')\big)}$$

where

l' – the number of the point in question in the array of permissible velocities for k point of the path$l \in [1 \dots L_k]$;

p' – the number of the point in question in the array of permissible velocities for $(k-1)$ point of the path$p \in [1 \dots L_{k-1}]$.

IOP Conf. Series: Materials Science and Engineering **315** (2018) 012014 doi:10.1088/1757-899X/315/1/012014

Now moving from the initial state to the final state, solving the equation of state, we obtain the optimal phase trajectory $s'(v')$ movement of the electric bus in the area under consideration, as well as control h', Providing the movement of the electric bus along the received trajectory.

In the case when the indeterminate Lagrange multiplier λ Is equal 0 The method gives a trivial solution: to move with the minimum possible speed. In this case, the time to reach the final state tends to infinity. Thus, in order to ensure a given time of passage of the route, it is necessary to vary iteratively the value of the indeterminate Lagrange multiplier λ, Repeating the above actions.

Similar calculations can be made for various options for congestion of the electric bus, thus obtaining various traffic laws depending on the number of passengers in the cabin. In addition, if it is possible to solve the equation of state on board the electric bus, it is possible to find the optimal control law for moving from the current state of the electric bus to the end point. Thus, you can take into account, for example, the intervention of the driver to prevent a traffic accident or stop the electric bus at the traffic light [[2]].

In order to estimate the energy the developed method would save in comparison with the energy spent by the electric bus on the route now, simulation results with the use of the energy-efficient control law obtained by the method were compared with the results of the road tests.

As the test object, we used the electric bus developed by the "GAZ" group in cooperation with the Bauman Moscow State Technical University (figure 3).

Figure 3. Electric bus built on the LiAZ 5292.30 chassis.

The electric bus was bus built on the LiAZ 5292.30 chassis (figure 4). The curb weight of the electric bus is 13265 kg, the gross weight of the electric bus is 18000 kg.

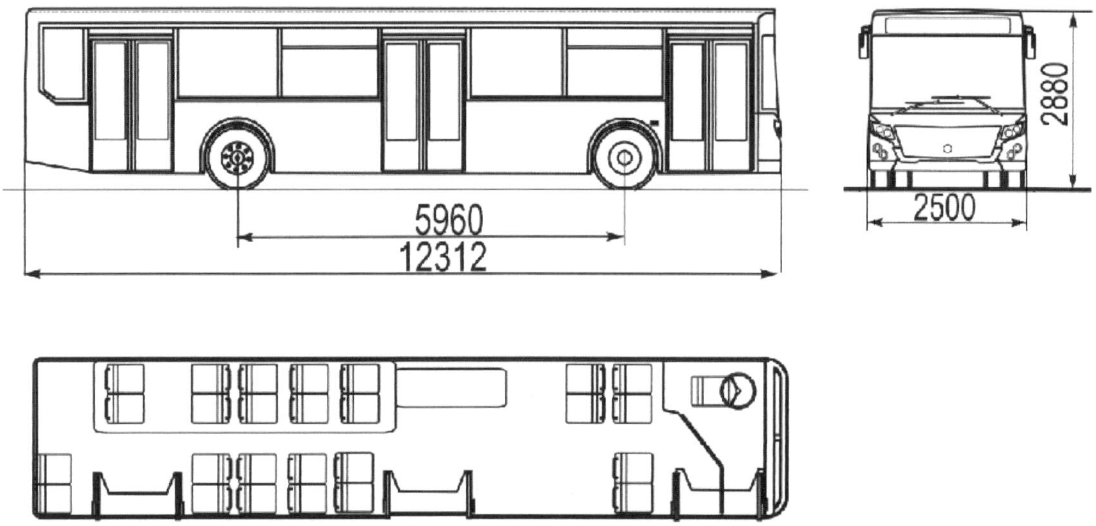

Figure 4. Overall dimensions of LiAZ 5292.30.

The electric bus movement parameters registration for further analysis on a PC was made using a CAN bus through the MATLAB Simulink software package [[3]]. The following parameters were recorded:

- time of motion t, s;
- voltage on the motor U, V;
- current on the motor I, A;
- speed of the electric bus v, km/h (calculated by the controller with the help of a standard rotor speed sensor).

The traveled distance s and the electrical energy J spent for acceleration and recovered during braking of the electric bus were calculated by numerical integration of the recorded data using the following formulas:

$$s = \int_{t_1}^{t_2} v dt$$

$$J = \int_{t_1}^{t_2} UIdt.$$

In the course of the tests, the driver was asked to travel a measuring section of 100 m in length during 20 s so that a minimum amount of energy was used for movement.

The calculated optimal phase trajectory for the section of the route under consideration is presented in figure 5. The calculation was carried out taking into account the dependence of the efficiency of the electric machine on the rotational speed of the rotor and the perceived load.

IOP Conf. Series: Materials Science and Engineering **315** (2018) 012014 doi:10.1088/1757-899X/315/1/012014

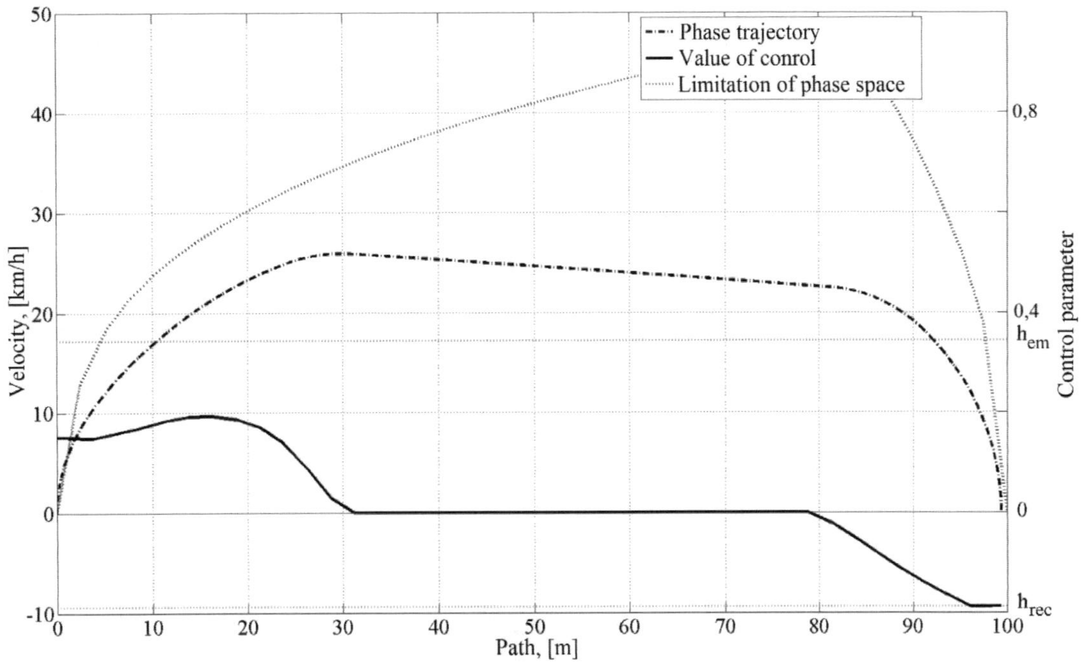

Figure 5. The calculated optimal phase trajectory.

The optimal control law on the estimated measuring site is:

1. Acceleration (the route segment from 0 to 31 m in figure Figure 5);
2. Movement in the coasting mode (the route segment from 31 m to 79 m in figure Figure 5);
3. Braking with recuperative brake (the route segment from 79 m to 100 m in figure Figure 5, in the end, the electric bus is stopped by the mechanical brakes).

Due to the shortness of the section of the route in question, there is no speed maintenance mode. The calculated value of the spent energy is 238 kJ.

For the collection of the statistics and reducing the impact of a possible slope of the road the driver had to move 5 times uniformly in the forward and reverse directions. The number of measurements was chosen on the basis of the requirement for the width of the confidence interval $\pm\sqrt{\widetilde{D}}$ with probability value $P = 95\%$. As a result, it was obtained that it is necessary to conduct at least 7 measurements [5].

Figure 6 presents the results of the road tests.

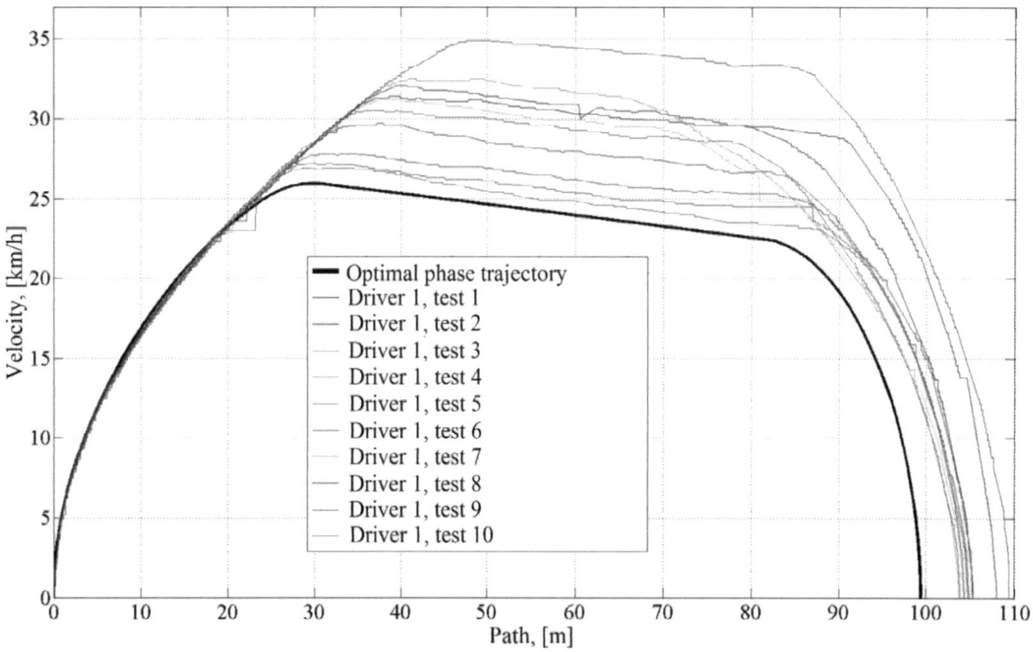

Figure 6. Road test results.

According to figure 6 it can be concluded that the driver can not independently maintain a uniform motion even on a simple section of the route, which increases the average energy consumption for traffic along the measurement site.

After the experiment, a comparison was made between the calculated data reflecting the energy consumption of the electric bus when driving according to the obtained energy-efficient control law, with the results of driving the test driver. The results of the road test data processing are given in the table 1.

Table 1. Results of the road test data processing.

Distance traveled s, m	Spent time t, s	Spent energy J, KJ
107,99	19,71	343,88
103,83	20,36	285,99
104,35	20,81	252,44
103,69	20,27	282,13
105,29	22,00	244,24
104,22	20,51	290,35
109,39	19,61	394,47
105,32	19,86	324,97
104,73	21,94	265,05
104,84	20,95	282,75

The result: $J = 296{,}63 \pm 32{,}71$ kJ

The results of the experiment confirmed that the proposed method makes it possible to store up to (296,63 kJ-238 kJ) / 296,63 kJ 100% = 19,7% of the energy when moving along the measuring site, due to the optimal choice of the timing of the switching of the driving regimes and providing a greater repeatability of the implementation of the law of motion.

IOP Conf. Series: Materials Science and Engineering **315** (2018) 012014 doi:10.1088/1757-899X/315/1/012014

The article is written based on the results of work carried out with the financial support of the Ministry of Education and Science of the Russian Federation under the agreement No. DN01 / 0001/216/15 between OOO LiAZ and BMSTU.

Conclusion

Comparison of the amount of energy spent on the movement obtained as a result of theoretical calculations with the energy expended when driving the electric bus by the test driver proved the possibility to save up to 19.7% of energy even on a simple part of the route. The received "gain" of energy is reached by an optimum choice of the moment of switching of modes of movement, and also due to the fact that the driver cannot maintain a uniform law of movement at repetition of the same site of a route. The obtained results prove the necessity of optimization of the laws of the automobile electric transport traffic between stops while driving along the city route, as well as the introduction of the traffic control system.

References

[1] Larin V V 2010 *The theory of motion of all-wheel drive vehicles*: a textbook (Moscow: publishing house BMSTU) p 391

[2] Bellman R 1960 *Dynamic programming* (Moscow: publishing house of foreign literature) p 400

[3] Butarovich D O, Kositsyn B B and Kotiev G O 2017 The method of developing an energy-efficient law governing the electric bus while driving along a city route // *Proceedings of NAMI*. **No. 2 (269)** pp 16-27

[4] Information on http://www.mathworks.com (reference date 01.08.2017)

[5] Spirin N A and Lavrov V V 2004 *Methods for planning and processing the results of an engineering experiment* Under the general editorship of Spirin N A (Ekaterinburg: GOU VPO UGTU-UPI) p 257

IOP Conf. Series: Materials Science and Engineering **315** (2018) 012015 doi:10.1088/1757-899X/315/1/012015

Isolated step-down DC -DC converter for electric vehicles

O V Kukovinets, K M Sidorov and V E Yutt

MADI University, Moscow, Russia

E-mail: olegakukovinets@gmail.com

Abstract. Modern motor-vehicle industrial sector is moving rapidly now towards the electricity-driving cars production, improving their range and efficiency of components, and in particular the step-down DC/DC converter to supply the onboard circuit 12/24V of electric vehicle from the high-voltage battery. The purpose of this article - to identify the best circuitry topology to design an advanced step-down DC/DC converters with the smallest mass, volume, highest efficiency and power. And this will have a positive effect on driving distance of electric vehicle (EV). On the basis of computational research of existing and implemented circuit topologies of step-down DC/DC converters (serial resonant converter, full bridge with phase-shifting converter, LLC resonant converter) a comprehensive analysis was carried out on the following characteristics: specific volume, specific weight, power, efficiency. The data obtained was the basis for the best technical option - LLC resonant converter. The results can serve as a guide material in the process of components design of the traction equipment for electric vehicles, providing for the best technical solutions in the design and manufacturing of converting equipment, self-contained power supply systems and advanced driver assistance systems.

1. Introduction

Communication systems sensitive devices and navigation systems of unmanned vehicles and cyber security of unmanned driving – it is impossible to imagine an unmanned electric transport vehicle without these essential components. Isolated step-down voltage converter is very important component too. Thanks to advanced technical solutions in the field of converting equipment, and in particular in the field of automotive converters, electric vehicle can fully meet all the requirements of safety, environmental and energy efficiency as well as automatic control.

Problems of energy saving and efficiency come to the fore in almost all applications, including electric vehicle: the higher efficiency of the conversion devices, the more the range distance of electric car, so this direction is important. One of the most important ways to improve the efficiency of the device is to increase the efficiency of DC-DC converters of the power supply. The losses in the Converter are divided into the switching (dynamic) that occur when power switches are working in the key mode and the losses in conductivity (resistive). The method to decrease the resistive losses is decreasing the amplitude of the ripple current. There are dynamic losses in this scheme. These losses can be minimized through circuit design and software.

Reducing losses not only increase the energy efficiency of the Converter and the electric car, but also can simplify the design of the Converter, as it will decrease the heating of its elements. [2]

IOP Conf. Series: Materials Science and Engineering **315** (2018) 012015 doi:10.1088/1757-899X/315/1/012015

Exploring the existing modern market of automobile DC/DC converters, it is possible to generate requirements for the design of energy-efficient and competitive Converter:

Table 1. Parameters of the modern converter.

The direction of the transformation:	From high voltage to low
Efficiency:	>92%
Input voltage:	220-410 V
Output voltage:	12-16 V
Power:	2,5 kW
The minimum power density:	1 W/cm3
Capacity (without radiator cooling)	2.5 liters

2. Analysis of topology

2.1. Full bridge phase-shift topology

Full bridge PWM Converter is shown in figure 1. The primary circuit is built on the principle of full bridge, secondary rectifier - with tap from mid point, and a series connected inductor required to operate this type of converters with PWM control.

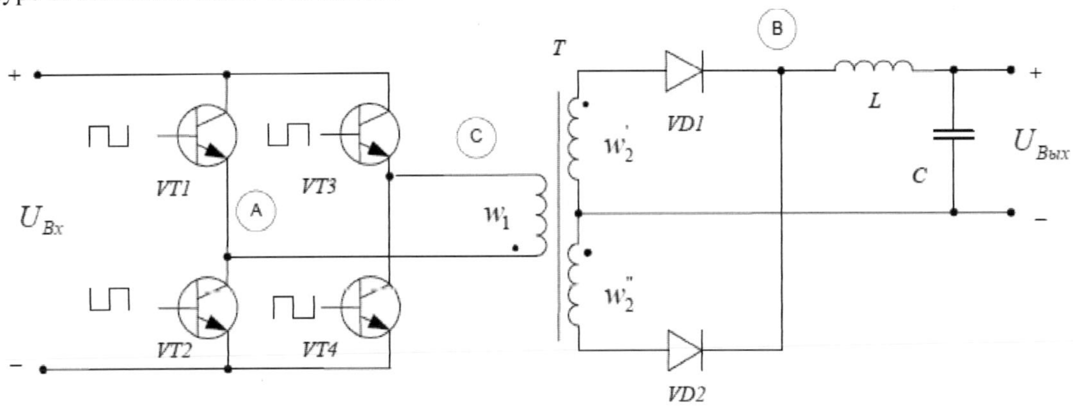

Figure 1. A simplified diagram of the full bridge Converter.

For better understanding of the operation principle of this Converter we should consider the circuit shown in figure 2. It is the basis of all step-down regulators, where the power circuit includes a series connected inductance.

IOP Conf. Series: Materials Science and Engineering **315** (2018) 012015 doi:10.1088/1757-899X/315/1/012015

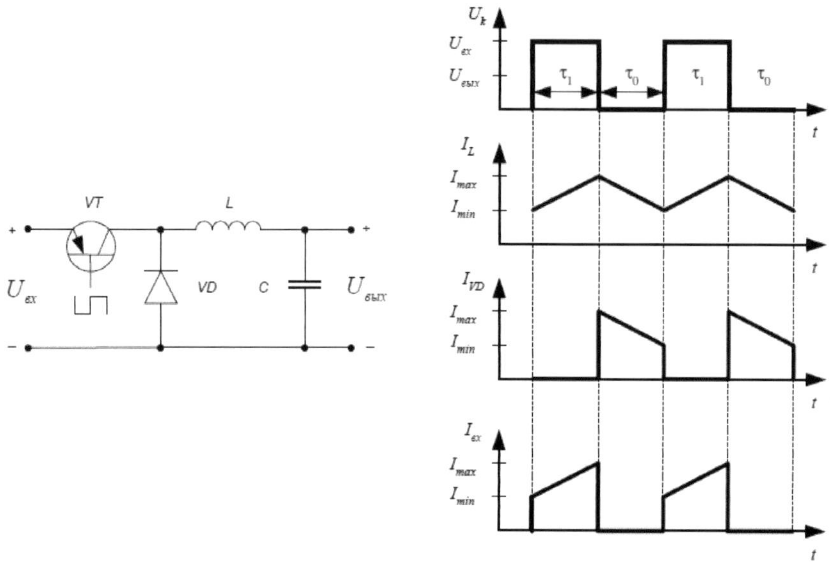

a) б)

Figure 2. Lowering scheme (a) and timing diagram (b).

We assume that the transistor and the diodes are perfect, and the switching frequency is so high that the voltage across the load not depend on switching time. In addition, we will consider the steady operation when the load current, the input and output voltages do not depend on switching time. [1]

Then the operation will be described by the equation:

$$Uout = Uin\,\frac{\tau_1}{\tau_0+\tau_1} = Uin\cdot\gamma \tag{1}$$

Where τ_1 – time of the closed state of the key, τ_0 – time of the open state of the key, γ is the duty ratio. [1]

There is a transformer (figure 1) with a transformation ratio k, then the equation will be as follows:

$$Uout = Uin\,\frac{\tau_1}{\tau_0+\tau_1}\,\frac{1}{k} = Uin\cdot\frac{\gamma}{k} \tag{2}$$

From this equation it is clear that the output voltage is not regulated in frequency. The output voltage is regulating by timing relationships of pulse and pause. Pulses to the primary winding of the transformer are supplied by closing the keys diagonally: in even half-cycles, the keys VT1 and VT4 are closed, and in odd ones, VT2 and VT3.

There are static and dynamic losses on the scheme shown in figure 1.

Dynamic losses in the transistor are caused by the considerable current flowing. Transistor for a short time is in the active mode, this is a cause of considerable current.

However, there is energy ten times greater than in the saturation state. It is clear that the released energy of switching operation, is proportional to the duration of the transient phenomena. So, to minimize released energy high speed transistors must be used. To reduce dynamic losses (figure 3) ZVS method is using (zero voltage switching), i.e., switching with zero voltage across the transistor. These overvoltages are due to resonant processes in the moments of switching. With the result that, considerable power produces in the time period of the active state of the key (the moment of transition from cutoff to saturation or Vice versa). ZVS can significantly reduce this power, which is clearly seen in figure 3 and figure 4 and turns “hard” switching to “soft”.

IOP Conf. Series: Materials Science and Engineering **315** (2018) 012015 doi:10.1088/1757-899X/315/1/012015

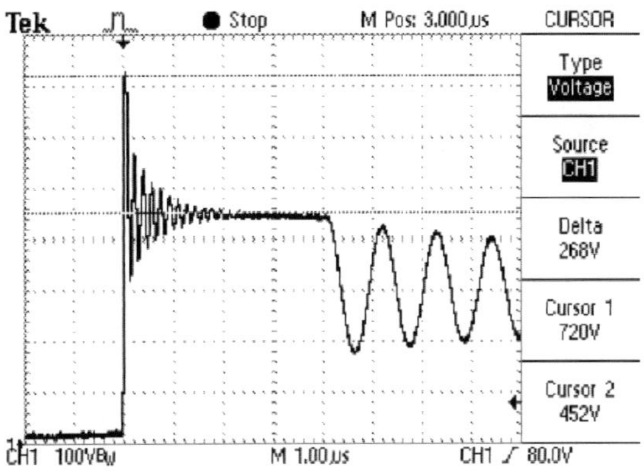

Figure 3. Overvoltages due to the switching of the transistor.

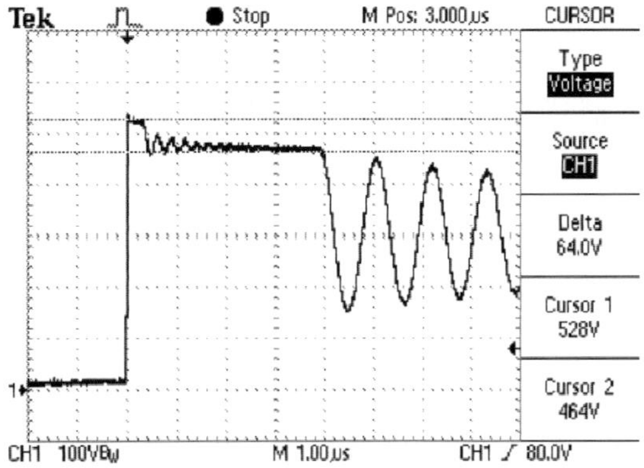

Figure 4. Normal switching of the transistor.

However, the mode switching at zero voltage can only be received under certain loads. This is due to the fact that despite the obvious symmetry of both half-bridges the topology and the control signals operation modes are significantly different. For the slave (second) half-bridge ZVS mode can only be received when the load is larger than a certain threshold. Decreasing this threshold is achieved only by increasing the leakage inductance of the transformer (or an inductor in series with the transformer), which inevitably leads to increased losses. The second important drawback: there are times when the primary winding is short-circuited at intermediate phase shifts (i.e., greater than 0° but less than 180°) through the two opened keys. At this time there is a sufficiently large current (circulating current), which cause the losses of course.

Static losses are due to the voltage drop on the circuit elements under the action of the current flowing during the periods of time between shifts. It is important to note that due to the skin effect high frequency current heats the wires. Frequency current is considerably greater than the permanent current of the same magnitude. The static losses in the switch are minimized by choosing transistors with low resistance in saturation mode, and Schottky diodes (with low direct voltage drop). Synchronous rectification is widely used method to create powerful converters with low output voltage. The main idea is to replace the diodes on the secondary side with the field effect transistors (MOSFET), which have a significantly smaller voltage drop in the open state. In this case the special scheme applies to synchronize the working secondary side with the main side.

There are surges in the hard switching of transistors because of difficult implementation of ZVS on the secondary side. To reduce the threat of the surge (dangerous, because it can damage the transistors and radiates noise) bulky snubbers and filter chokes were applied.

In 2011, the firm VALEO has designed the first prototype of such a Converter shown in figure 5.

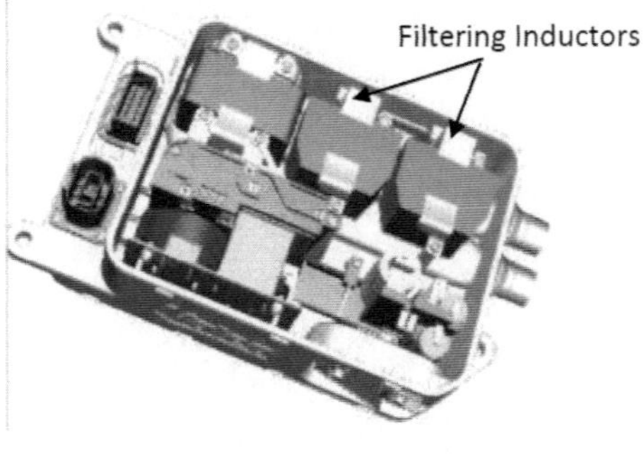

Figure 5. VALEO's Converter.

In the figure 5 one can notice two large chocks are designed to filter from noise. They take up very much space, so specific volume suffers greatly. The sizes of the inductors are determined by a large current in the power circuit.

Due to the hard switching of the transistors in the secondary circuit, the frequency is limited to 100 kHz, which limits the using of miniaturized magnetic components in this Converter. This Converter is based on full bridge circuit with phase shift and has an efficiency of 92%, with a volume of 4.8 liters and specific volume of 0.61 W/cm^3.

We can conclude that this topology is not suitable as parameters of the volume and efficiency do not meet modern requirements shown in the table 1.

2.2. The series resonant topology

Half-bridge series resonant Converter (shown in figure 6) is simple and well-known scheme for HV/LV conversion.

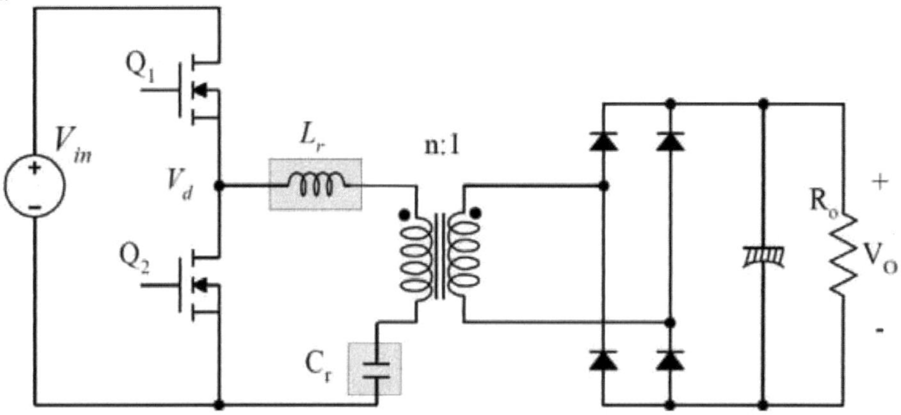

Figure 6. Series LC resonant Converter.

The most effective way to obtain high levels of specific power and specific volume is to increase the operating frequency so that the size of magnetic components has been significantly reduced. Resonant converters can operate at high frequencies due to the very small switching losses. In converters of this type the adjustment of the output voltage occurs by PFM (pulse-frequency) modulation. Half-bridge series resonant Converter is simple and well-known scheme for HV/LV conversion.

Resonance circuit consists of inductor Lr, resonant capacitor Cr and the transformer. The transient response represented by the following equation:

$$G = \frac{1}{1+j\frac{\pi^2}{8}Q\left[\frac{f_s}{f_r}-\frac{f_r}{f_s}\right]} \tag{3}$$

where $f_r = \frac{1}{2\pi\sqrt{L_r C_r}}$, f_s – switching frequency. Q- quality, $Q = \frac{\sqrt{L_r/C_r}}{R_{ac}} = \frac{Z_c}{R_{ac}}$; Rac – quality, $R_{ac} = R_{out}\frac{8}{n^2\pi^2}$.

The circuit of the load rectifier is connected in series with the LC resonant circuit, as shown in figure 6. In this configuration, resonant circuit and load operate as a voltage divider. The impedance of the resonant circuit is changing, when the frequency of the voltage Vd at the output of the half bridge is changing. The input voltage is divided between this impedance and the reflected load resistance. Thanks to voltage divider, the gain of the DC voltage LC serial resonant Converter is always <1. In the state of low load, its impedance is very large in comparison with the resistance of the resonant circuit; all the input voltage is on the load (it is clear from the graph in figure 7 for the curve Q = 1 (10 % load)).

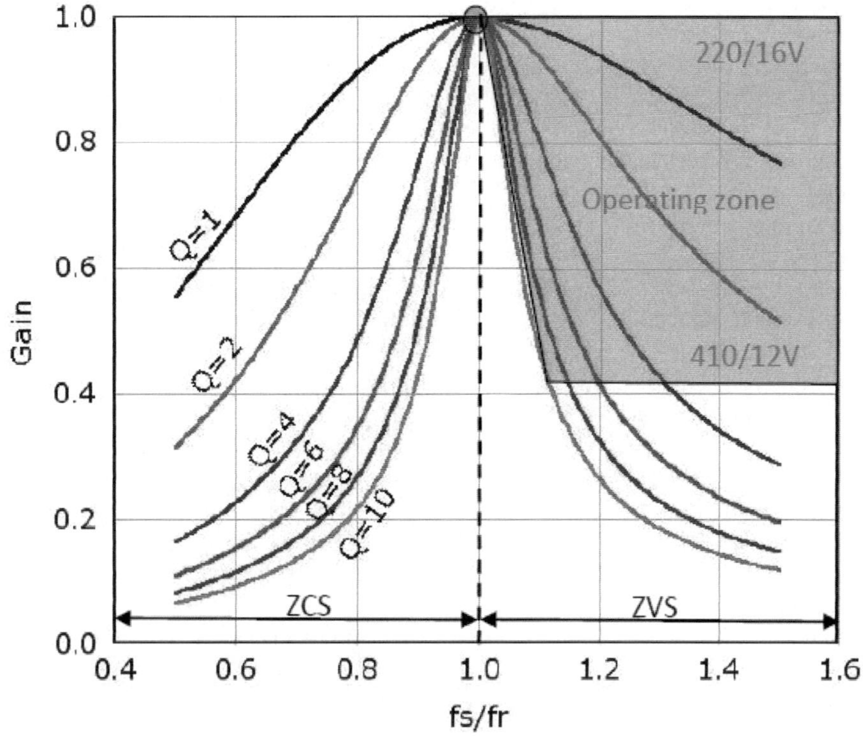

Figure 7. The transfer function of voltage to a series resonant LC Converter.

The output voltage adjustment is practically impossible in this mode. Theoretically, frequency should be infinite to regulate the output at idle. Some significant negative effects are appearing at low load due to the high switching frequency: skin effect and proximity effect. These effects are negate the whole idea of high-performance transformations. That's why this topology is also a poor choice in automotive application.

2.3. LLC resonance circuit

There is LLC resonance circuit in figure 8.

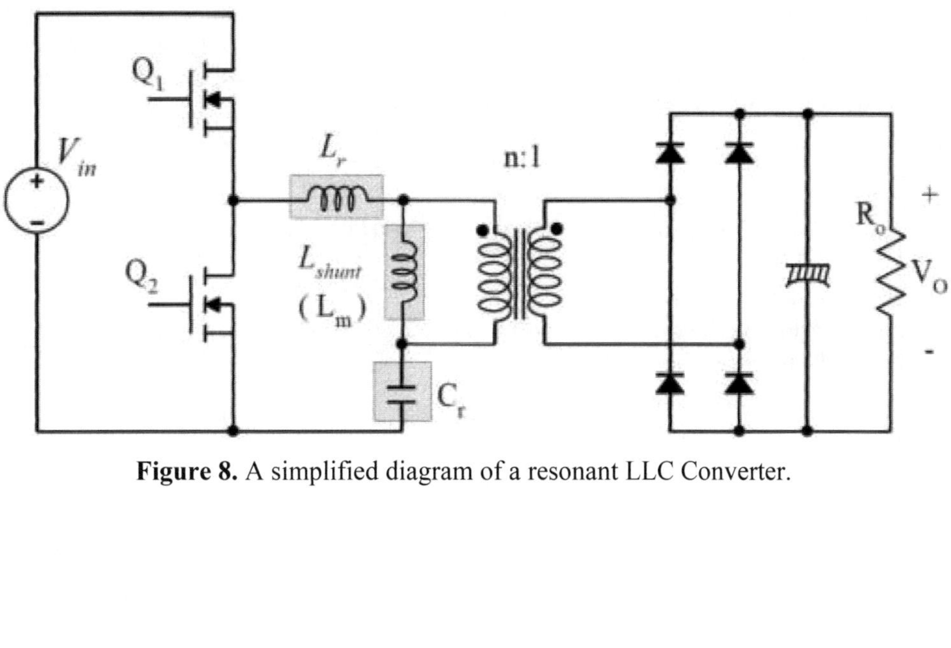

Figure 8. A simplified diagram of a resonant LLC Converter.

IOP Conf. Series: Materials Science and Engineering **315** (2018) 012015 doi:10.1088/1757-899X/315/1/012015

LLC resonant Converter have been proposed to overcome the limitation of series resonant converters. It is a modification of the serial LC resonant Converter implemented by placing a shunt inductor parallel to the primary winding of the transformer, as shown in figure 8. This topology has not received much attention when it was first presented. This is because of the circulating current increasing in the primary winding with a shunt inductor. It can not be beneficial to the circuit operation. However, it can be very effective in improving the efficiency for the use of high input voltage, where the switching losses dominate much more than the loss of conductivity. In the most practical design, this shunt inductor is realizing with the magnetizing inductance of the transformer. Diagram of the LLC resonant Converter looks the same as a serial resonant LLC Converter. The value of magnetizing inductance Lm is the only Difference from the other scheme. The series resonant Converter's magnetizing inductance is much larger than the resonant inductance (L_r) serial LC resonant circuit. But in the LLC resonant Converter scheme the inductance (Lr) is just 3 ~ 8 times larger, which is usually implemented by introducing an air gap in the transformer as leakage inductance. Minimization of magnetic components and their integration in the transformer is a huge advantage of an LLC Converter over other topologies.

Figure 10 illustrates the work of the LLC-Converter. The drain current I_{ds2} takes negative values, while it flows through the inverse diode of MOSFET and the voltage drop is practically zero (drop voltage on the open diode).

The drain current takes negative values, while it flows through the inverse diode of MOSFET and the voltage drop is practically zero (the voltage drop across the open diode).

At this moment the MOSFET opens and the switching losses are significantly reduced, so the efficiency of the Converter reaches 95%, which is a very good value. Besides, it stands to mention that form voltage is very close to the sine wave, thereby reducing the level of interference.

LLC topology is the best one for the implementation of advanced energy efficient voltage Converter.

3. Conclusions

Converters with sinusoidal current waveform have a number of advantages compared to converters with rectangular current. The lack of large values of di/dt and dv/dt simplifies the filtering and control of electromagnetic interference. Switching losses are lower, as current at the time of switching is zero. The recovery losses of rectifier are low, because of small magnitude di/dt. The load on all components is less, which gives improved reliability of the device.

The maximum operating frequency of the switching converters is mainly determined by the magnetic components. Achievement of the required performance is an easier task in the case of the resonant Converter, because the transformer passes only a sine wave at the fundamental frequency. Converters square wave needs to pass the harmonics and the fundamental frequency. Therefore resonant converter require a transformer with a wider range of frequencies to avoid unacceptable losses or distortions. [3]

Thus, resonant converters can achieve the advantages in the high-frequency operation without compromising cost, efficiency, reliability and electromagnetic compatibility. LLC scheme topology was considered as the best topology.

References

[1] Kondakov E V 2014 *Switching Voltage Converters* (Rostov-na-Donu: South federal University) pp 5 - 50
[2] Golishev V 2008 *Electronic components* pp 55-74
[3] Young S, Castino G 2006 *RESONANT CONVERTER POWER 500W FREQUENCY 100 KHZ MOSFET* pp 1-12

IOP Conf. Series: Materials Science and Engineering **315** (2018) 012016 doi:10.1088/1757-899X/315/1/012016

Development of driver's assistant system of additional visual information of blind areas for Gazelle Next

V Makarov[1], O Korelin[1], D Koblyakov[1], S Kostin[1,2] and A Komandirov[3]

[1] Nizhny Novgorod State Technical University n.a. R.E.Alekseev, Nizhny Novgorod, Russia
[2] LLC «Nizhpromplast», Nizhny Novgorod, Russia
[3] GAZ Group's United Engineering Center, Nizhny Novgorod, Russia

E-mail: makvl2010@gmail.com

Abstract. The article is devoted to the development of the Advanced Driver Assistance Systems (ADAS) for the GAZelle NEXT car. This project is aimed at developing a visual information system for the driver integrated into the windshield racks. The developed system implements the following functions: assistance in maneuvering and parking; Recognition of road signs; Warning the driver about the possibility of a frontal collision; Control of "blind" zones; "Transparent" vision in the windshield racks, widening the field of view, behind them; Visual and sound information about the traffic situation; Control and descent from the lane of the vehicle; Monitoring of the driver's condition; navigation system; All-round review. The scheme of action of sensors of the developed system of visual information of the driver is provided. The moments of systems on a prototype of a vehicle are considered. Possible changes in the interior and dashboard of the car are given. The results of the implementation are aimed at the implementation of the system - improved informing of the driver about the environment and the development of an ergonomic interior for this system within the new Functional Salon of the Gazelle Next vehicle equipped with a visual information system for the driver.

Advanced driver assistance systems (ADAS) serve to prevent accident situations, one of the reasons for which is the presence of so-called "blind" zones - zones around the car with reduced visibility for the driver. If you get into these areas of a nearby car or pedestrians, you cannot see them because of the special features of the vehicle. This can lead to an emergency situation when maneuvers are carried out. Partially "blind" zones can be reduced by installing spherical mirrors. But this is not enough. Therefore, worldwide development of systems for monitoring traffic conditions on the basis of active and passive sensors is underway. One of the first concepts with "transparent racks" was Nissan Pivo (figure 1). Company Jaguar Land Rover develops the concept of virtual vision Vision which includes in "transparent racks", "virtual windshields", "transparent hood", etc. (figure 2). Similar developments are being conducted in all leading automotive companies. On modern cars, systems are installed that monitor "blind" zones and help the driver to safely change lanes. These systems are based on the use of video cameras and radar. A system of circular vision is widely used that allows maneuvering in a cramped environment.

Figure 1. Nissan Pivo [1].

Figure 2. Jaguar [2].

The article deals with the developed system of visual information of the driver integrated into the windshield racks [3] carried out by the company "Nizhpromplast" together with specialists from the Nizhny Novgorod State Technical University n.a. R.E. Alexeev with the support of the Foundation for the Promotion of Innovation, the GAZelle NEXT car is used as the object of research.

As a result of the analytical review in the field of intelligent driver assistance systems - ADAS (Advanced Driver Assistance System) corresponding to the 1st and 2nd levels of the international classification of driver assistance systems SAE International, a concept for building a visual information system for the driver integrated into the windshield racks was developed. This is an extremely important task to improve traffic safety.

In general, the developed system of visual information will:

- Provide a warning to the driver about the possibility of a frontal collision; Control of blind areas; Warning of withdrawal from the band; Help with maneuvering; signs recognition;
- Allow the driver to receive information about the route; the presence of traffic jams; road- works; Speed limits; Travel time; The included transmission;
- Receive information: from front-end cameras of high resolution (road markings, interference - stationary and mobile, road signs, etc.); Video information from internal cameras to monitor the driver's condition; Input data from the vehicle's on-board network (data on speed, acceleration, engine operating mode, gearbox, brake system status, etc.); Navigation system data;
- Provide the driver with signals and visual information about the position of the vehicle relative to the lane; About finding a potential obstacle in the zone of visibility of cameras, incl. In the "blind" zones - Warning of dangerous proximity; Information on the location of the vehicle in the area of traffic signs (in conjunction with the navigation system); other information on traffic (traffic jams, road events, recommendations for choosing the optimal traffic regime, etc);
- Include a system of "transparent" vision in the areas of the lower part of the racks; Augmented reality system with information output to the windscreen; A system of visual and sound information about traffic events (presence of interference in the blind zone, assistance with reverse traffic, warning of dangerous rapprochement with other traffic participants, etc.); Sound and light indicators of road events.

The scheme of the sensors of the driver's visual information system is shown in Figure 3.

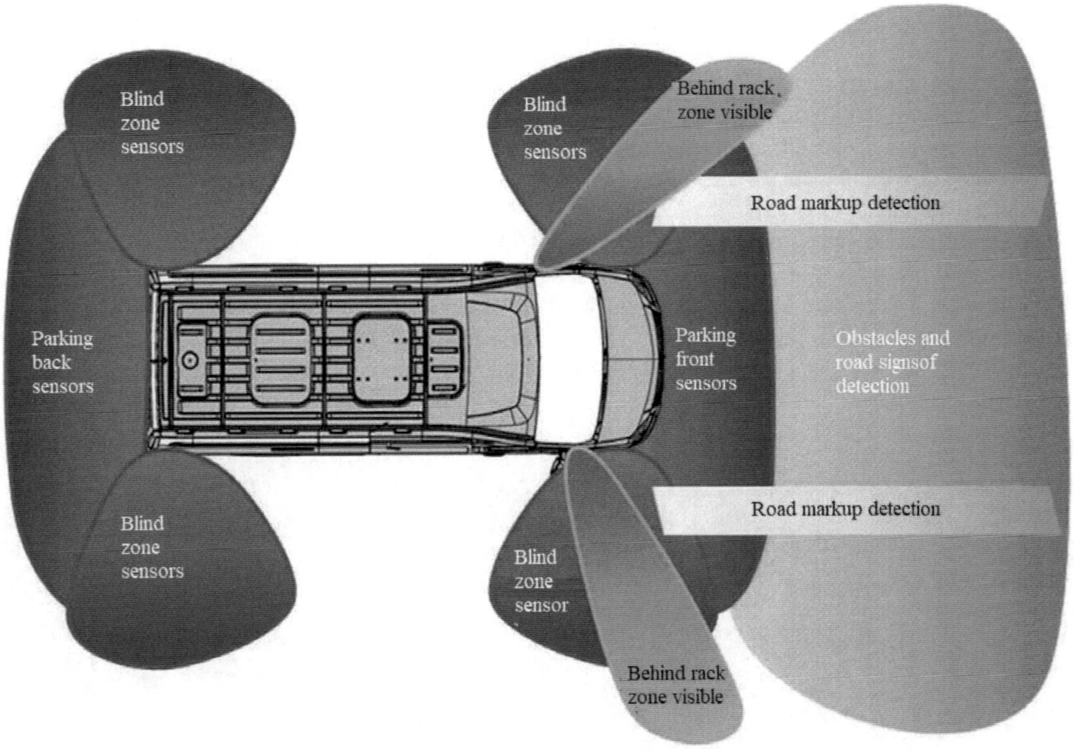

Figure 3. Action sensors of the developed system of visual information to the driver.

The composition of the visual and sound information system for the driver about the traffic situation includes:

1. Central computer for processing information and controlling the operation of the entire system.
2. Two displays – the main one, displaying video information from the cameras of the circular view, navigation information, technical characteristics of the vehicle (speed, engine modes, etc.), and a HUD display that outputs additional reality elements to the windshield of the vehicle.
3. Means of informing the driver, implementing the following functions:

- assistance in maneuvering and parking the driver (providing data on the presence of a potential obstacle in the field of view of the cameras);
- recognition of traffic signs, providing, in conjunction with the driver's navigation system, information on the location of the vehicle in the area of traffic signs;
- warning the driver about the possibility of a frontal collision with an obstacle;
- control of "blind" zones (providing the driver with data on the presence of a potential obstacle in blind areas));
- "transparent" vision in the windshield racks, extending the visibility zone, covered by the windshield racks;
- visual and audible information about the traffic situation, which implements information output to the main display, informs of dangerous road events with light indicators in the racks, reproduces warning sound signals about potential danger in road traffic;
- lane control (informing the driver about the vehicle's position relative to the lane and leaving the lane);
- monitoring the driver's condition (monitoring the driver's condition and actions for the subsequent analysis of the detection of errors in his actions in the event of emergencies);

- navigation to obtain information by the driver about the route, the presence of traffic jams, road works, speed limits, travel time;
- circular view, representing a system of additional reality, allowing the driver to see the position of his car at 3600 around.

According to the research conducted to date, there are no ready-made systems (both domestic and foreign) that fully implement all the items of the terms of reference for the project being developed. The complexity of the ADAS tasks in the framework of the project is currently conditioned by the usage of existing solutions implementing specific ADAS functions, integrating them into a single system, as well as further upgrading the system's intelligent solutions and adapting them to the specifics of the GAZelle Next platform. The market analysis made it possible to identify the most promising solutions in terms of development – Mobileye [4], Sparkauto [5], ParkMaster [6], Navitel [7], etc. The moments of installation of driver assistance systems for the prototype are shown in figure 4.

Figure 4. Fragments of installation, connection and calibration assistance systems for the driver on a prototype.

The complexity of their usage for project solutions is as these solutions are separate, independent systems, with their interfaces and input-output devices. Hence the conclusion is made about the necessity of using cross-platform equipment for integrating these solutions into a single system for visually informing the driver about the traffic situation, which requires selection and justification for the use of additional equipment, including various types of switches and signal converters.

After installing all systems on the car, full-scale tests were conducted to assess the performance. Recommendations and technical requirements for installing the visual information system for the driver were given. Proposals were prepared on the interior inside the new functional compartment (figure 5) and the installation of the elements (Figure 6) of driver information systems on the surrounding environment for the car GAZelle Next.

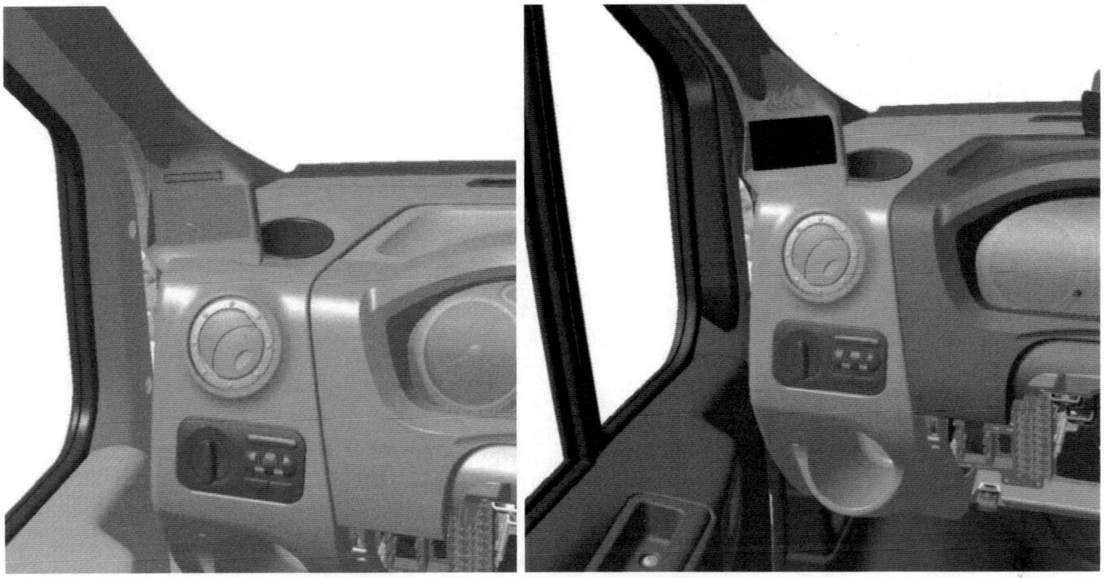

Figure 5. Possible changes in interior and dashboard elements.

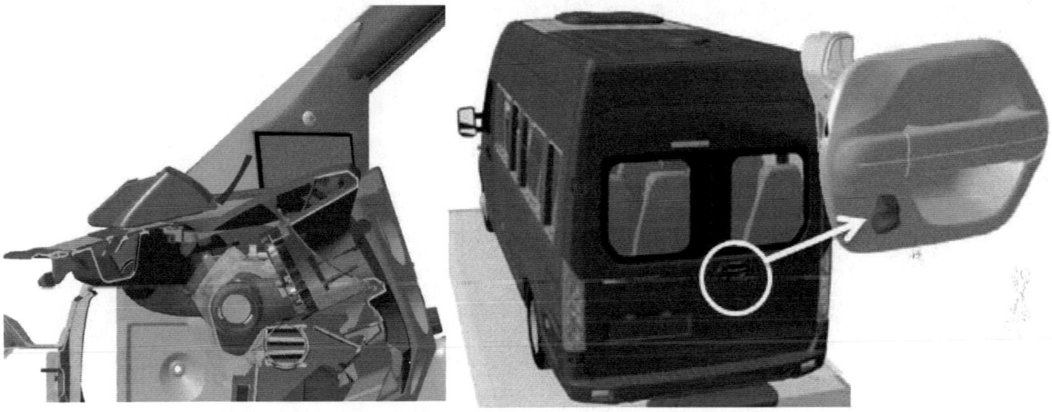

Figure 6. Installation of elements of driver information systems about the environment. Projector on the windscreen (on left), Rear View Camera (on right).

Practical implementation of the solutions described allows us to give recommendations and proceed to the creation of a modern competitive light commercial vehicle - the GAZelle NEXT car equipped with a visual information system for the driver.

References

[1] Electric car Nissan Pivo 3 at the Tokyo Motor Show 2011 available at http://www.electro-machines.ru/auto/elektromobil-nissan-pivo-3-na/

[2] Průhledné sloupky Jaguaru řeší zhoršující se výhled z vozů available at http://www.auto.cz/pruhledne-sloupky-land-roveru-resi-zhorsujici-se-vyhled-z-vozu-video-84660/

[3] Tumasov A V, Shiyan S A, Romanova A D, Shakolin V V and Tumanov A V 2017 Development of a visual information system for the driver, integrated into the windshield racks *Proceedings of the XVI International Youth Scientific and Technical Conference* (Nizhny Novgorod: NSTU n.a. R.E. Alekseev) p 248-249

[4] Mobilay Accident Prevention System available at http://www.mobile-eye.ru/
[5] Spark parking systems available at https://sparkavto.ru/
[6] ParkMaster available at https://www.parkmaster.ru/
[7] Navitel navigational program available at t http://navitel.ru/ru/

IOP Conf. Series: Materials Science and Engineering **315** (2018) 012017 doi:10.1088/1757-899X/315/1/012017

Development of a semitransparent ceramic heat-insulation for an eco-friendly combustion chamber of Low-Heat-Rejection diesel

V G Merzlikin[1,2], M O Gutierrez[3], A R Makarov[4], A A Bekaev[1], A V Bystrov[2] and F A Zagumennov[5]

[1] Department of Technology and Equipment of Mechanical Engineering, Moscow Polytechnic University, Bolshaya Semenovskaya, 38, Moscow, 107023, Russia
[2] Industrial Economics Department, Plekhanov Russian University of Economics, Stremyanny, 36, Moscow, 117997, Russia
[3] Tablet School Company, General Maldonado y Belisario Quevedo, Latacunga, 0501444, Ecuador
[4] Department of Mechanical Engineering and Instrument Making, RUDN University, Miklukho-Maklaya, 6, Moscow, 117198, Russia
[5] Department of Instrument Production Techniques, Bauman Moscow State Technical University, 2-ya Baumanskaya, 5/1, Moscow, 105005, Russia

E-mail: merzlikinv@mail.ru

Abstract. Efficiency of diesel has been studied using well-known types of the ceramic heat-insulating HICs- or thermal barrier TBCs-coatings. This problem is relevant for a high speed diesel combustion chamber in which intensive radiant component (near IR) reaches ~50% within total thermal flux. Therefore, in their papers the authors offered new concept of study these materials as semitransparent SHICs-, STBCs-coatings. On the Mie scattering theory the effect of selection of the specific structural composition and porosity of coatings on the variation of their optical parameters is considered. Conducted spectrophotometric modeling of the volume-absorbed radiant energy by the coating had determined their acceptable temperature field. For rig testings coated piston using selected SHIC (PSZ-ceramic ZrO_2+8%Y_2O_3) with a calculated optimum temperature gradient was chosen. A single cylinder experimental tractor diesel was used. At rotation frequency $n > 2800$ rpm the heat losses were no more than 0.2 MW/m^2. Executed testings showed ~2-3% lower specific fuel consumption in contrast the diesel with uncoated piston. Effective power and drive torque were ~2-5% greater. The authors have substantiated the growth the efficiency of this Low-Heat-Rejection (LHR) diesel due to the known effect of soot deposition gasification at high speed.Then unpolluted semitransparent ceramic thermal insulation forms the required thermoradiation fields and temperature profiles and can affect regulation of heat losses and reduction of primarily nitrogen dioxide generation.

1. Introduction

This paper examines already recognized method of efficiency increasing for Low-Heat-Rejection (LHR) diesel using well-known ceramic heat-insulating HICs- or thermal barrier TBCs-coatings which have been applied from the 1970s of last century in Russia [1-10] and abroad [13- 32]. But the authors suggest the reducing toxicity of exhaust gases can be achieved when using these ceramic coatings as semitransparent one taking into account their specific optical properties and formation a predetermined absorption regime of penetrating thermoradiation for control and management temperature of the exposed surface [7-10].

This problem is relevant for a diesel engine with combustion chamber (CC) in which intensive radiant (near-IR) component up to ~50% within total thermal flux [33-36]. Then these so called semitransparent coatings (SHICs or STBCs) can ensure the required heat rejection and controlled generation rate of exhaust gases (primarily nitrogen dioxide), caused by formation of specific temperature fields in their subsurface zones.

Since the end of the 20th century, extensive research of quasi-adiabatic engine with using of heat-insulating coatings has been suspended since the end of the 20th century [1-3, 14-17]. In our opinion, it was connected with unresolved problem of effect of radiant penetrating component (generated by red-hot soot particles) of total thermal flux falling on the heat-insulating coating, which can be semitransparent (see figure 1). Most well-known ceramic coatings based on oxides (Al_2O_3, SiO_2, Zr_2O_3) are semitransparent [7-10, 17, 36-38].

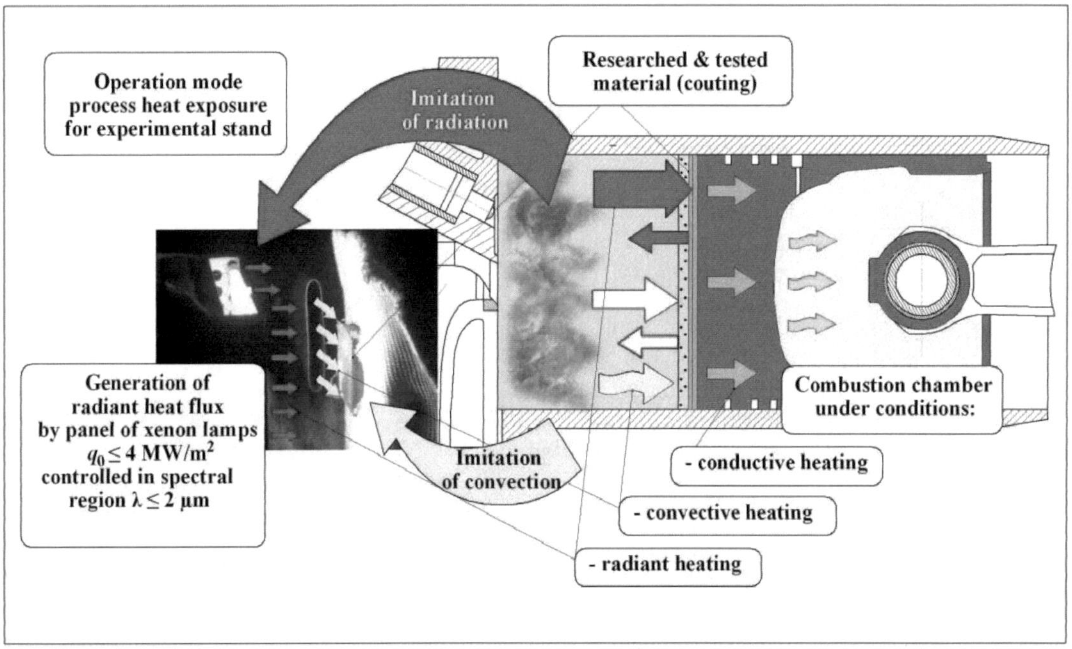

Figure 1. Block diagram of convective and radiant heat transfer inside diesel combustion chamber with an opportunity of physical modeling using radiative-and-convective cycling simulator [11, 12].

Earlier used traditional heat-insulating coatings were considered as opaque for IR radiation. This was true only for some coatings, for example, based on Si_3N_4, SiC [4, 20-22]. Otherwise the assertion will be erroneous and incorrect analysis of complex heat transfer in the coated combustion chamber (CC) would be conducted.

Even the latest developments in the field of thermal protection for LHR diesel engines still do not take into account the radiant component inside the combustion chamber and the optical properties of its ceramic coatings [4-6, 18-32].

But semitransparent coatings could be considered as opaque also because of the imperfection of the plasma-spraying technology, which causes the appearance of highly absorbent metallic particles of the plasmatron electrodes inside the coating.

All the opaque materials cause an increase of temperature on their irradiated surface due to high surface absorption of radiation within near IR. This effect resulted to impermissible overheating of the inner surface of the heat insulated CC' walls and intensive generation of the most toxic nitrogen oxides.

The temperature increase of the walls was noticed and it was indicated by the first publications [15-18], classical monographs [4, 18] and many recent researches. Eventual surface absorption of thermoradiation by highly absorbing coatings of CC walls is the primary reason for overheating which was not noticed.

Nowadays, numerous publications and special subject reviews show trend of increasingly growing interest to application and study of ceramic TBCs (HICs) during the last 10 years [4-6, 13, 19-32]. This was due to advance in deposition technologies with a predetermined porosity and structural composition.

The authors of one of the reviewes confirm commencement of TBCs intensive development of by global carmakers [25]: "Research to decrease costs and fuel consumption in internal combustion engines and technological innovation studies continue. Today, researchers work hard to improve engine efficiency via constructional modifications. For instance, alongside with development of advanced technology ceramics, ceramic coating applications in ICEs are taking place".

Usually, researchers are citing only the publications of the authors of this paper [29, 30]. Need to analyse "semitransparent characteristics of ceramics" - is emphasized, but their own methodology of radiant-conductive heat transfer has not been corrected.

The studies of ceramic coatings for ICEs in automotive industry are actually the result of earlier researches of characteristics of radiant and convective ablative thermal protection for aerospace aircrafts started in 1960-70s [39-43].

Numerous R&D of TBCs were carried out using classical apparatus of radiative transfer in semitransparent materials. But until now these developments have not been applied for research of non-destructible protective heat-insulating coatings in the auto industry [3-6, 13-32]. In aerospace industry applying semitransparent ceramic for turbine elements was a logical continuation of research of materials with a predetermined destruction threshold under continuous exposure of penetrating thermoradiation [44-47].

The term "Thermal Barrier Coating (TBC)" came to the automotive industry from the aerospace terminology. Such heat-insulating materials were used to design combustion chamber (CC) for adiabatic engines or their elements. At the same time, the first developers pointed out the need to study the role of radiation components in the combine exchange inside CC [1, 15-18]. Thus, in spite of the legendary developments of ceramic ICEs in previous decades many automotive engineers have not yet involved analysis of thermoradiation processes in their current researches [3-6, 13, 18-32].

2. Physical and mathematical simulation of radiant and heat conductivity transfer

The engine researches show that if the temperature of combustion chamber' inner walls becomes 2 times higher, the value of the heat transfer coefficient increases fivefold. Then the heat exchange in the boundary layer of the combustion chamber (CC) wall begins to intensify [1-3, 18]. This is because the flame (combustion source) at high surface temperatures is moving closer to the wall, causing a rise of the temperature gradient, since the distance from the source to the wall is decreasing.

As such, the studying heat exchange of semitransparent heat-insulating coatings (SHICs) focuses on research of the influence of its thermal regime on the temperature of the inner surface of combustion chamber walls and the heat transfer through these coatings. Methodology of analysis of radiant heat exchange and theoretical evolutions of temperature profiles based on calculated thermoradiation fields inside coating layer are presented in our papers [7-10]. In this paper, new boundary conditions and thermal characteristics obtained during conducted rig tests were used for theoretical calculations of temperature fields in semitransparent coatings.

One-dimensional optical two-layer model of semitransparent coating SHIC protecting metal substrate with surface reflection was used (see figure 1).

The diagram also presents the experimental radiative-and-convective cycling simulator. It allows to imitate a complex heat transfer inside the combustion chamber at an radiant flux up to ~5 MW/m^2 (with several simulators). Within wavelength range 0.3-2 μm in one-dimensional approximation for large areas of internal CC walls (up to hundreds of square centimeters) [11, 12]. In these papers authors have solved the main problem of creation of laboratory set up with broadband radiation source. This simulator' component includes high-intensity discharge xenon lamps, which is capable to generate the modelled impure radiation spectrum close to the continuous spectrum of radiation of red-hot soot particles in the typical combustion chamber (CC).

Partially yttria stabilized zirconia (PSZ) is the currently preferred coating for applications in CC diesel and turbines [4, 6-10, 13, 19-32, 36, 37, 44-47]. Among semitransparent ceramics with low thermal conductivity the coating has two import properties: the most high thermal expansion coefficient and a good erosion resistance.

A substances PSZ (ZrO_2+3-8%Y_2O_3) were used to produce flat tablets of various thickness intended for optical measurements of reflection (transmittance) coefficients for evaluation scattering and absorption indexes [7, 8]. It is a purified granular powder with white colour in the visible range. This substance has a transparency window up to 4 μm with a stable reflection coefficient 0.7-0.9 for thick layers [8, 38]. This material was used for plasma spraying and forming:

- homogeneous plane ceramic layer (thickness up to 0.5 mm) of semitransparent coating for spectrophotometric measurements [8] and
- same layer deposited on the piston surface of the tested tractor diesel.

So, alongside to heat conduction by lattice waves (phonons), inside semitransparent SHIC-coating heat is also transferred by a radiative component (photons). They become increasingly important at elevated temperatures. Thus, the total energy transfer through the coating exceeds the heat transfer caused by solid heat conduction alone.

Certainly, penetrating radiation reduces the efficiency of thermal barrier characteristics of semitransparent coatings and degrades SHIC insulating ability. But the effect of their own surface heat reradiation causes a decrease of surface temperature and allows displacing a temperature maximum from a surface semitransparent material into the depth of a SHIC-coating with the expansion of the subsurface zone of volumetric radiant heating.

Thus, the simulated optical properties of the coating would determine the temperature field. Experimental modeling of optical parameters is carried out using spectrophotometric measurements of control flat ceramic samples of different thicknesses.

Using the Mie scattering theory, the given optical parameters would be determined by the selection of the specific structural composition and the porosity of the coatings [8]. Technological structuring can ensure control and management of thermal conditions, form an acceptable temperature gradient, temperature regime of the coatings surface and gas in the combustion chamber [7-10, 44-47].

The opaque materials represent model in which radiation of red-hot soot particles inside combustion chamber does not penetrate through its exposed surface. This radiation is mainly absorbed by the combustion chamber (CC) walls surface (with the exception of insignificant surface scattering and a small surface reflection by the Fresnel law). In this paper the surface reflection coefficient of the irradiated opaque ceramic was assumed to be equal R_{op} = R_S = 20% for the heat-insulating HIC-coating [8, 10, 37, 38]. An opaque HIC-coating is seen as a material with the same known thermal physical characteristics as the semitransparent SHIC-coating.

3. Numerically simulated temporal temperature of heat-insulating coatings under the action of a radiant-convective monopulse

For rig testings, samples of semitransparent SHIC-coatings with best form of calculated temperature profiles due to volumetric subsurface overheating and reduced surface temperature determined specific optical properties of ceramic coatings were selected (see figure 2). Powder PSZ-ceramic based on ZrO_2+8%Y_2O_3 had the optimal optical characteristics experimental measurement with a serial spectrophotometer.

The ceramic powder was used to make check flat samples and to spray on the piston surface using plasma technology.

These coatings were semitransparent ceramics with insignificant absorption (κ = 14 m^{-1}) and a high scattering (σ = 2400 m^{-1}) in the near infrared region of the spectrum. The reflection coefficients are ~ 40% for thin (0.5 mm) and ~ 90% for thick layers.

Difference in the temperature profiles of opaque HIC- and SHIC-coatings analysis for the diesel combustion chamber (CC) was conducted under the following interaction conditions (close to the heat characteristics of bench tests): the model total thermal flux q_0 = 1.8 MW/m^2, fraction of radiant component ~ 50%. The middle temperature in the CC gas volume was constant $T_A(t)$ = 800 K. Turbulent heat exchange ratio is α_T =3000 MW/(m^2·K). Initial temperature T_0 = 500 K.

Figure 2 shows the calculated temperature profiles inside thick layers of polluted opaque (line 2) and rectified semitransparent (3) PSZ-ceramic coatings, soot deposition (1) and uncoated metal wall (4) under conditions of constant heat pulse of radiant and convective action on the internal walls of the diesel combustion chamber.

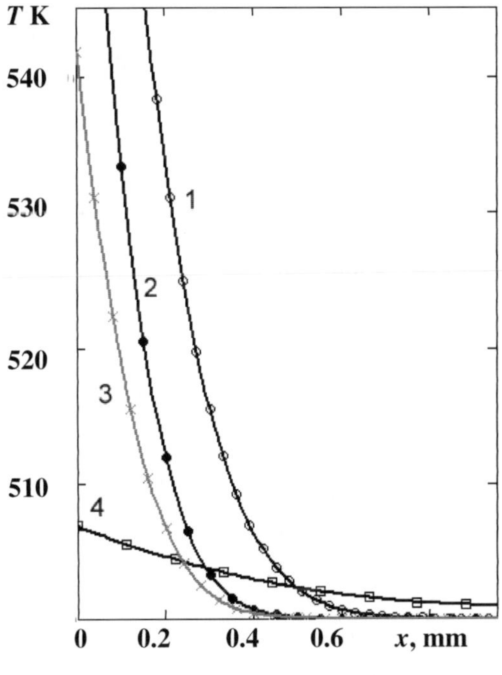

Figure 2. Numerically simulated temporal temperature in subsurface zone of heat insulation of combustion chamber exposed by radiant-convective monopulse during at t = 0.01 s of primary fuel combustion tact for the following layers:

- soot deposition (line 1);
- PSZ-ceramic coatings with equal thermophysical parameters and different absorption indexes
  - κ >> 14 m^{-1} (opaque ceramic, line 2) and
  - κ = 14 m^{-1} (semitransparent ceramic with scattering index σ =2400 m^{-1}, line 3).

Unprotected metal wall - line 4.
Back surfaces of coatings with thickness H = 1 mm is heat insulated.

The usage of opaque coatings and soot deposition increases the temperature of the piston head surface by 100-200 K (see figure 2, lines 1, 2) stimulating the generation of nitrogen oxides and often with an undesirable regime of convective-radiation heat transfer inside CC. Under these conditions, the ceramic coating can be destroyed due to a large forming subsurface temperature gradient.

Thus, the thermal regime of the semitransparent coating (see figure 2, line 3) is more suitable and can be controlled by changing only the optical parameters due to the specific structuring, for example, the selection the prevailing orientation for the scattering particles inside the ceramic layer [8, 40, 44-46]. In this case, coefficient of thermal conductivity seems to remain unchanged.

Proposed physical modeling of the optimal SHIC structure should contribute to the required thermal regulation of the CC' gas atmosphere, preventing its overheating and more active fuel self-ignition.

4. Numerically simulated temporal temperature of heat-insulating semitransparent SHIC- and opaque HIC-coatings at steady state

We use the standard conditions of ring testings for internal surface of the high-speed diesel combustion chamber (CC) wall. At harmonic characteristics simulation at rotational speed of $n = 3000$ rpm they are to be the following: total thermal flux ~ 1.8 MW/m^2 at the "hot" phase 10 ms ("cold" phase 30 ms) with fraction of radiant component ~ 50% which varies from 0 up to 0.9 MW/m^2 during 2 ms [2, 3] (in short wavelength diapason 0.8-2 μm); direct heat loss through coated piston $q_w(t)$ changes up to 0.20 MW/m^2 at the "hot" phase. Gas temperature in CC volume changes from 330K to 2000K; oscillation amplitude of heat turbulent transfer coefficient varies from 70 to 2000 W/(m^2·K); gas emissivity 0.2 - 0.6. Heated top layer of coating surface considered as black body with own radiation in long wavelength diapason 2-5 μm.

Duration of cyclic changes of values of the characteristics is achieved for a moving piston with SHIC-coating at complex radiant heat exchange inside CC of high-speed tractor engine TMZ-450D (see table 1, figure 1, 3).

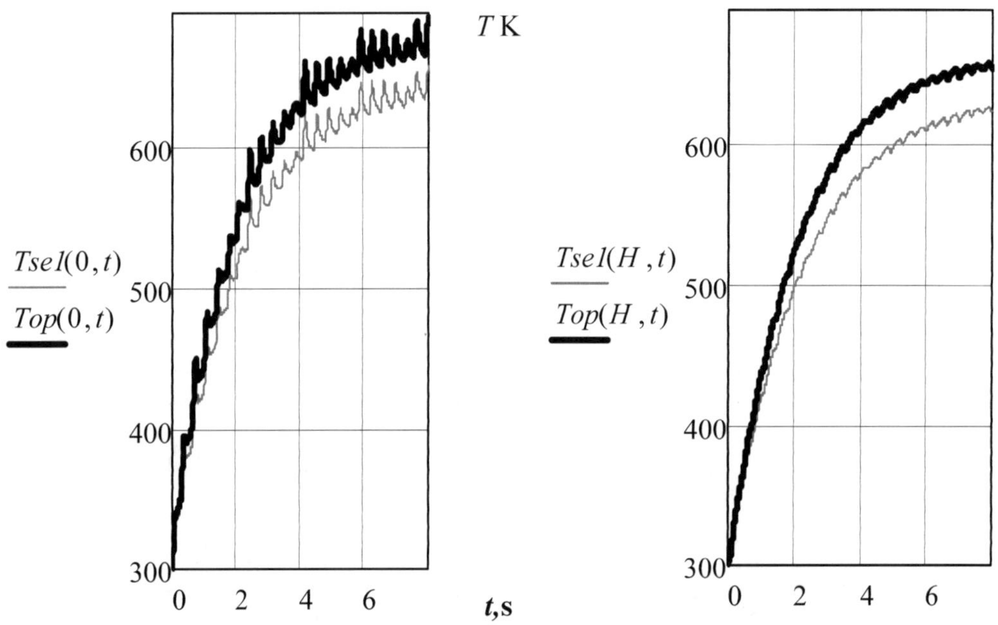

a b

Figure 3. Cyclic temporal temperature variations $T(x,t)$ for the front (exposed) (a) and back (b) surfaces of the 0.5 mm ceramic top layer for optical parameters of the following coatings:

- opaque HIC-coating (surface heating at $R_S = 20\%$) with temperature distributions $T_{op}(0,t)$, $T_{op}(H,t)$;
- semitransparent SHIC-coating (volumetric subsurface heating at absorption index $\kappa = 14$ m^{-1} and scattering one $\sigma = 2400$ m^{-1}) with temperature distributions - $T_{se1}(0,t)$, $T_{se1}(H,t)$.

A surface of oxidized metal substrate has reflection coefficient $R_{me} = 90\%$. Malty cycle of piston movement at n=3000 rpm.

The temperature distributions in semitransparent layers with different optical models are simulated for the boundary conditions of internal surface of CC wall.

For example, the highest temperature of front (exposed) surface of semitransparent coating could be reached $T_{se}(0, t) \sim 680$K with transmissivity coefficient $\tau_{se} \sim 79$ % (absorption $\kappa = 10$ m^{-1} and scattering $\sigma = 1000$ m^{-1} indexes) and substrate reflection $R_{me} = 10\%$.

But changing of SHIC-coating structure allows to decrees transmissivity τ_{se} up to 56 % (at more high scattering $\sigma = 3000$ m^{-1}) and this surface temperature becomes lower $T_{se}(0, t) \sim 660$K.

At increasing of substrate reflection R_{me} up to 90% there is the same result ~ 660K for SHIC-coating and its temperature (on the back side of the coating) can be lower T_{se} $(H,t) \sim$ 615K in comparison temperature $T_{op}(H,t) \sim 655$K for opaque coating.

When all radiant and heat fluxes are absorbed on the front surface of opaque HIC-coatings then there is a high temperature gradient for opaque coatings in comparison semitransparent SHIC-coatings.

The smaller absorption and more higher scattering of SHIC causes a decreasing of the surface temperature of these SHIC-coatings in comparison with opaque one, near 20–30 degree lower during one cycle (1st strike of combustion) of piston movement.

Table 1. The main performance characteristics, design and energy parameters of a single cylinder experimental diesel engine TMZ-450D (Russian production) [9].

Characteristics of the engine	Physical dimensions	Numerical value or description
Location of cylinders	-	Vertical
Piston distance / Cylinder diameter	mm	80/85
Piston-swept volume	dm^3	0.454
Compressive ratio	-	20
Rated speed	min^{-1}	3000
Indicated power	kW	8.0
Specific fuel consumption at rated	g/ (kW·h)	< 280

Optical models of such SHIC-coatings have the following characteristics: reflection of semi-infinite layer ~ 80÷90%, absorption $\kappa \sim 10$ m^{-1} and scattering $\sigma \sim 100\text{-}1000$ m^{-1} indexes at emittance coefficient ~ 0.98 in near IR (for coatings as black body in middle IR diapason ~2-5 μm).

Increasing of scattering of semitransparent coatings in 10 times leads to reducing temperatures on a surface and in subsurface area of SHIC about $\Delta T \sim 2$ -3 K for action of 1st heat monopulse.

These results agree with data on temperature regimes of ceramic thermal barrier TBC-coatings for aircraft engines [44, 45].

For steady state conditions the overheating of surface of opaque coatings also exceeds the corresponding value for semitransparent coatings.

Figure 3 show temporal temperature variation for the front (exposed) and back surfaces of the 0.5 mm top layer for opaque T_{op} and semitransparent T_{se1} coatings with optical model $\kappa = 14$ m^{-1}, $\sigma = 2400$ m^{-1} covering reflecting layer of metal substrate $R_{me} = 90\%$.

Peak temperatures of SHIC-coating keeps a value $T_S \sim 650$K.

For opaque coating the surface temperature by ~50 K is higher during steady state.

The steady state of thermal regime is reached for 6-8 s.

On a back surface of coatings temperature are stabilized at achievement 660 K for an opaque HIC-coating and 620 K for the semitransparent SHIC-coating.

It should be noted dependence of temperature maximum appearance on the back side of SHIC at decreasing of substrate reflection coefficient.

The simulation of temperature regimes shows essential dependence from radiating heat fluxes for semitransparent coatings.

It allows predicting optimal regimes of the operation for LHR diesel before development and production of coated combustion chamber using heat-insulating PSZ-ceramic.

5. Experimental results of rig tests of high-speed tractor engine with piston coated PSZ-ceramic

Let us consider practical application of semitransparent materials for heat insulation of CC piston using a single cylinder experimental diesel engine TMZ-450D (see figure 4, table 1).

The main objective of this research is standard rig testing and confirmation of the effectiveness of high-speed diesel engine combustion chamber (CC) with coated piston head using rectified PSZ-ceramic layer produced by a plasma-spraying technology.

It is well-known that the heat-insulating surface of the piston or other CC elements contributes to:

- growth of the intensity of combustion near the top dead centre, where the maximum pressure and temperature at the end of compression are reached;
- increase the brightness and temperature of the flame in the initial phase of combustion and shorten the time of the combustion process.

In this paper, the influence of selected piston' heat-insulating coating (PSZ-ceramic based on ZrO_2 + 8%Y2O3 with highly reflective and weakly absorbing in near IR) on heat losses reduction was evaluated through ring tests of high-speed tractor engine diesel combustion chamber.

The mechanical and fuel-energy characteristics of CC with a heat-insulated piston were determined in comparison with an unprotected piston (see figure 4).

a b

Figure 4. Picture of unprotected (a) piston and protected one (b) by semitransparent PSZ- ceramic.

The figure 5 shows usage of semitransparent heat insulation decreases the maximum value for heat rejection $q_w(\varphi)$ almost equidistantly by 10% depending upon the crankshaft angle within range from 20 to 80 degrees in comparison with the traditional diesel CC without heat insulating inserts.

The CC combustion intensity with coated piston can be regulated using semitransparent coatings with selected structure for volumetric absorption of penetrating radiant energy. This effect takes place at transmittance coefficient $\tau_{se} \sim 66\%$ for the coating with absorption $\kappa = 14$ m^{-1} and scattering $\sigma = 2400$ m^{-1} indexes which theoretically determined the temperature profile with the minimum temperature gradient.

As noted above, the modeled structure of the plasma coating was produced via required optical parameters using spectrophotometric control method.

Thus, the desired temperature profile for the CC heat-insulated walls with a predetermined subsurface temperature gradient will be selected by the forming thermoradiation and temperature field inside the appropriate structural composition of SHIC for coated piston or other CC elements.

Heat rejection from the working environment in the absence of semitransparent thermal insulation occurs in a wider range of CRA changing up to 80° at combustion process. But SHICs application allows showing the effect of the regeneracy of useful heat accumulated inside coated walls.

During the expansion stroke, the accumulated heat from the subsurface zone of this coating can be returned back to the diesel CC, i.e. it additionally turns into a useful work.

Some heat-energy and fuel characteristics of the unprotected piston and coated one of high-speed tractor engine TMZ-450D (Russian production) are shown on figures 5-7.

Rig test of this diesel with the use of semitransparent thermal insulation based on zirconium oxide with yttrium showed that the best results are obtained with high engine speeds from ~2800 to ~3400 rpm.

This is because, that combustion of fuel occurs almost completely near the top centre compression stroke, since the surface temperature of the coated piston is higher than for unprotected one in CC. It will promote more favourable conditions of spontaneous fuel ignition and reduction of its combustion time. Thus, efficiency of the given operating regime of typical diesel tractor engine can be managed and controlled at use SHIC with optimal selected optical parameters in a wide range of their changes at negligible varying of thermophysical parameters.

Conducted standard rig testing of the diesel allowed to estimate the heat losses up to 0.2 MW/m^2 through coated piston (see figure 5).

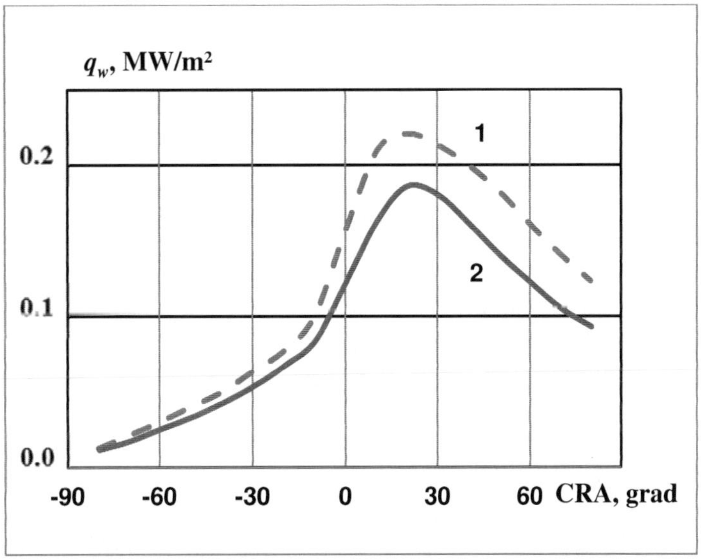

Figure 5. Heat losses q_w through the uncoated (line 1) piston head and coated (line 2) one with semitransparent heat-insulating ceramic coating vs crankshaft rotation angle (CRA) for high-speed tractor engine TMZ-450D (Russian production).

The hourly fuel consumption was determined by the weight method and was calculated based on measurements of time and expenditure of the fuel dose. Experimental test showed 2-3% lower specific fuel consumption (see figure 6); ~2-5% greater power and turning moment (see figure 7) on a single cylinder experimental diesel at $n > 2800$ rpm and it is due to the application of semitransparent ceramic coating.

The maximum pressure of the diesel cycle was increased by 3% at decreasing of the exhaust gases temperature up to 7%. This shows improvements in the diesel's working process, especially in the expansion cycle, where fuel energy turns into useful work.

The principle of the obtained results was confirmed in a number of works, including paper G. Woschni "Heat Insulation of Combustion Chamber Walls – a Measure to Decrease the Fuel Consumption of IC Engines" (1987) [15].

The data obtained from the tractor diesel testings with a coated piston by SHIC allowed confirm volumetric overheating qualitatively based on the predicted above calculated temperature distributions in this coating as well as provide a model version of the application of a completely opaque coating (based on a polluted semitransparent coating and soot deposition sample) (see figures 2, 3).

In the process of fuel combustion, the internal surface of CC is polluted with soot deposition (see figure 8, (b), (c)). In this case, the advantages of applying semitransparent coatings to opaque ones disappear. But with the increase in diesel engine speed, soot can begin to be gasified intensively [3] (see figure 8) [3].

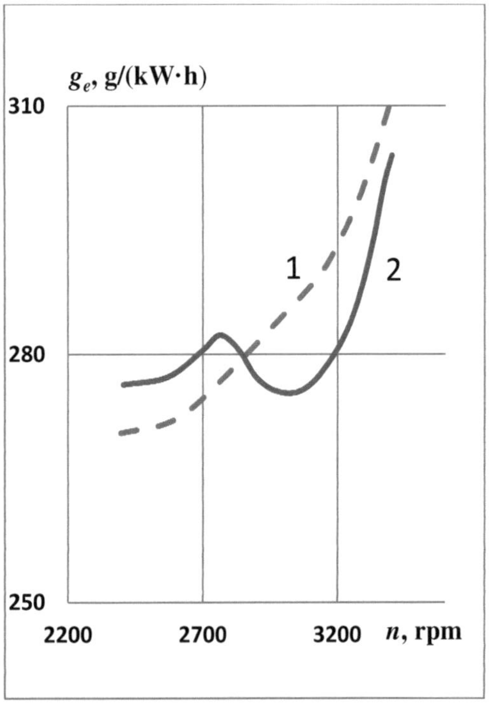

Figure 6. Experimentally measured values of the specific fuel consumption g_e depending on diesel speed n at rig testings (See explanations on figure 5).

In figure 8, the model of forming and removal of soot deposition is presented for a heat-insulated combustion chamber using semitransparent PSZ-ceramic (with selected structure and optical parameters) for operating modes of a single-cylinder experimental engine at various speeds of rotation: 0 rpm - initial moment (a); in the process of the speed increasing (b, c, d).

In the process of fuel combustion, the internal surface of CC is polluted with soot deposition (see figure 8, (b), (c)). In this case, the advantages of applying semitransparent SHIC-coatings in comparison with opaque HIC-coating disappear. But with the increase of diesel engine speed, soot can begin to be gasified intensively (see figure 8, (d)) [3].

The conducted ring testings confirmed the prospects of using semitransparent SHIC-coatings to increase the efficiency of the diesel and possibility of the surface temperature regulation for irradiated ceramic insulation at control the nitrogen dioxide generation.

6. Conclusion

Experimental researches of heat-insulated elements of the diesel combustion chamber (CC) and theoretical evolutions of thermoradiative and temperature fields allow to make a case for the advantages of the semitransparent heat-insulating coatings application which can ensure:

(1) volumetric absorption of penetrating thermal radiation in near IR (as fraction of total heat flux) inside subsurface zone of semitransparent heat-insulating SHICs- or thermal barrier STBCs-coatings;
(2) accumulation of penetrating and absorbed thermoradiation at a combustion process and it will promote the thermal regeneration effect during other strokes of piston moving;
(3) formation of subsurface temperature maximum and low temperature gradient inside exposed semitransparent heat-insulating (thermal barrier) materials and coatings at combine radiant-convective heat exchange;
(4) thermal cooling of a heat-insulated metallic wall (substrate) taking into account the conductive heat removal to the irradiated surface of coating, its own reradiation in the long wavelength (2-5 μm) range and volume reflection in the short wavelength (0.8-2 μm) range;
(5) thermoregulation of internal heat insulated surface of combustion chamber based on modeling optical properties (reflection – transmissivity or absorption - scattering) for SHIC-coatings;
(6) simulation of surface temperature for piston head with semitransparent SHIC-coatings (in contrast to the use of opaque HIC-coatings and uncoated metal surface of piston) causing the possibility of reduction NO_x generation which will determine improved management by temperatures of diesel combustion-exhaust system;

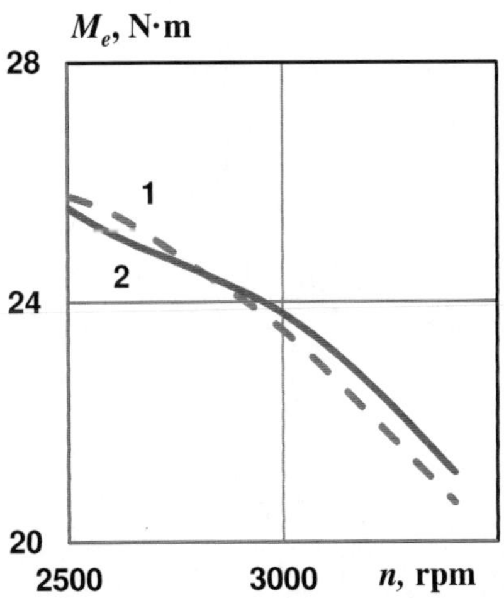

Figure 7. Turning moment M_e depending on diesel speed n at rig testings (See explanations on figure 5).

(7) controlling surface temperature of coated piston using semitransparent the SHIC-coatings with layer thickness ~ 0.5 mm and following optical characteristics: reflection coefficient ~ 70÷90% for semi-infinite layer, absorption κ~ 1-20 m^{-1} and scattering σ ~ 100-1000 m^{-1} indexes (in the short-wave region); emittance coefficient $\varepsilon \sim 0.98$ (in the long-wave region);
(8) regulated thermal stresses and the damage threshold based on the selection of optical parameters of coatings due to variation of the structure, changing distribution of absorbed thermal radiation and formation of desired temperature profile;
(9) standard rig testing confirmed of the effectiveness of combustion chamber of high-speed diesel engine with coated piston head using PSZ-ceramic (with selected composition ZrO_2+8%Y_2O_3) layer produced by a plasma-spraying technology
(10) heat losses do not exceed value ~ 0.2 MW/m^2 through coated piston;

(11) improving characteristics of the LHR diesel at n > 2800 rpm: ~2-3% lower specific fuel consumption; ~2-5% greater effective power and drive torque in comparison with uncoated piston

(12) physical modeling of the optimal SHIC-coating' structure should contribute to the required thermal regulation gas atmosphere of the combustion chamber, preventing its overheating and better self-ignition of the fuel.

The authors made the case for rise in the efficiency of LHR diesel with increasing rotational speed due to of soot gasification. This brought about the changes of the coating' transparency and possibility of wide range of variation scattering and absorption values and structure modeling.

The opportunity of reconstructing the structure of the coating as "smart material" to monitor thermoradiative and temperature fields inside ceramic heat insulation for control heat losses and formation exhaust gasses is shown.

Application of the innovative ceramic with adjustable structure as semitransparent heat-insulating SHICs-coatings (thermal barrier STBCs-coatings) for new generation of diesel will be promote increase efficiency engines with control of nitrogen dioxide generation.

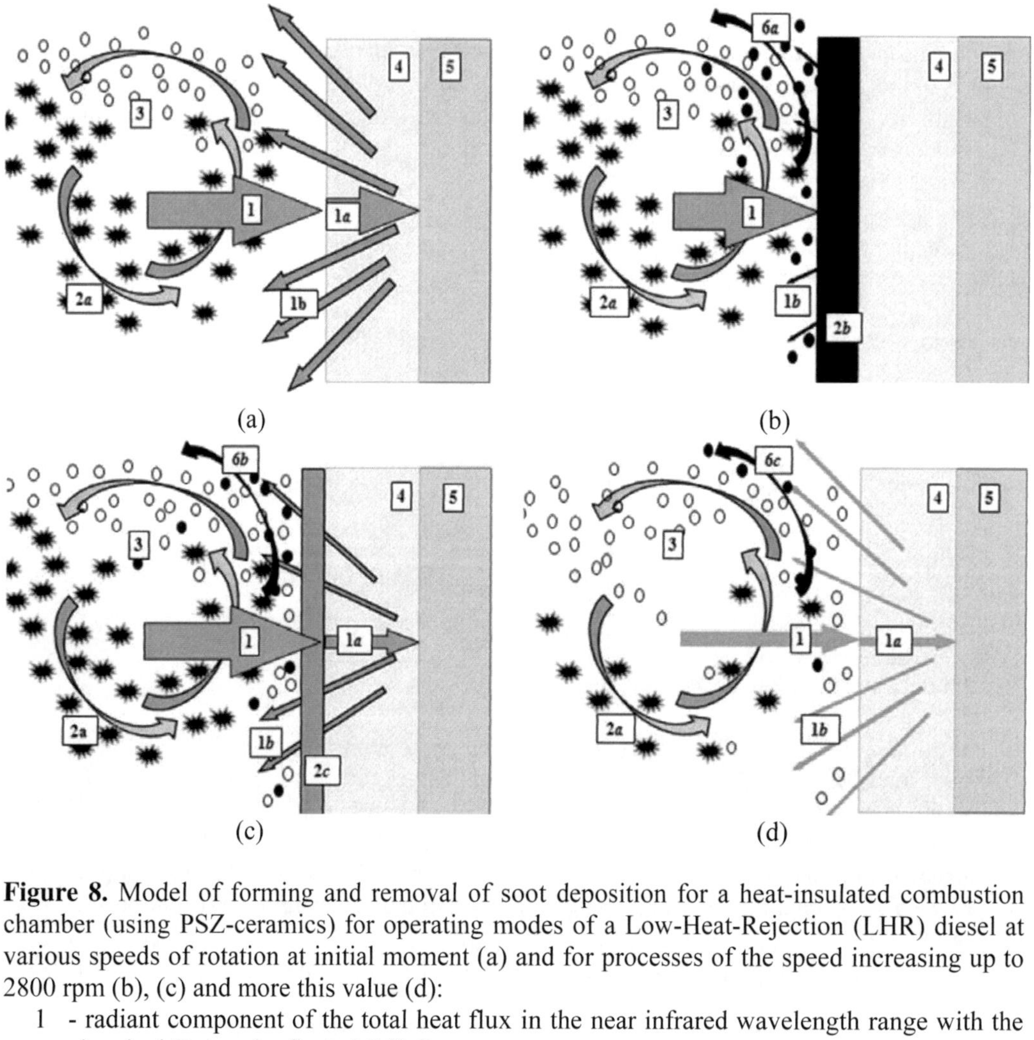

Figure 8. Model of forming and removal of soot deposition for a heat-insulated combustion chamber (using PSZ-ceramics) for operating modes of a Low-Heat-Rejection (LHR) diesel at various speeds of rotation at initial moment (a) and for processes of the speed increasing up to 2800 rpm (b), (c) and more this value (d):

1 - radiant component of the total heat flux in the near infrared wavelength range with the absorbed (1a) and reflected (1b) fluxes;

2 - flow of individual red-hot particles (2a) and a soot deposition (2b, 2c) above ceramic

coating;
3 - flow of gasified carbon molecules;
4 - heat-insulating semitransparent coating;
5 - metal substrate;
6 - generation of nitrogen oxides molecules with a maximum (6*a,* b) and a negligible (6*b*, c; 6*c*, d) concentrations at high and low temperatures of heated surface respectively for opaque (2*b*, b) and semitransparent (2*c*, c) soot deposition or its absence (d) above SHIC-coating.

References

[1] Kostin A K 1979 *Thermal Stress of Internal Combustion Engines* (Moscow: Russ Publ Mech Eng) p 231

[2] Bazhaykin A N 1992 Characteristics of ignition and combustion of fuel jets in heat-insulated combustion chamber *Journal of Engine Building* **4** pp 23-8

[3] Kavtaradze R Z 2001 *Local Heat Transfer in the Piston Engines* (Moscow: Bauman Moscow State Tecnical University) p 592

[4] Kavtaradze R Z, Onishchenko D O, Zelentsov A A, Kadyrov S M and Arypzhanov M M 2011 Calculus-experimental study of influence of thermal insulation piston and cylinder liner for generation nitric oxides in combustion products high-speed diesel *Herald of the Bauman MSTU., Ser. Mech. Eng.* **4** pp 83–102

[5] Dudareva N Yu, Kal'shchikov R V, Dombrovskii O P and Butusov I A 2015 Experimentally studied thermal piston-head state of the ICE with a thermal layer formed by micro-arc oxidation method *Nauka i Obrazovanie. MGTU im. N.E. Baumana* (*Electronic Materials*) **05** pp115–125

[6] Onishchenko D O, Pankratov S A and Smirnov A Yu 2016 Effect of the partial heat insulation of the d. e. combustion chamber on heat transfer into the cooling system *Herald of the Bauman MSTU., ser. Mech. Eng.* **3** pp 81–89

[7] Merzlikin V, Timonin V, Gutierrez O M and Sidorov O 2007 New selectively absorbing and scattering heat-insulating coatings of the combustion chamber for LHR Diesel *SAE Technical Paper* 2007-01-1755

[8] Merzlikin V, Sidorov O, Cheranev S and Antonakopoulos N 2011 Optimal spectral optical and thermo radiating characteristics of semitransparent heat-insulating coatings for Low-Heat-Rejection diesel engines *Proc. 11th Int. Conference on Engines and Vehicle ICE2011 (Capri – Naples)* Booк of abstracts 41

[9] Gutierrez O M 2007 *Reduction of heat losses and thermal stress of diesels using semitransparent ceramic coatings* (Moscow: Extended abstract of PhD dissertation, Moscow State Technical University "MAMI") p 28

[10] Merzlikin V G, Parshina S A, Garnova V Yu, Bystrov A V, Makarov A R and Khudyakov S V 2017 Rig test of diesel combustion chamber with piston coated optically simulated semitransparent PSZ-ceramic *Proc. 13th Int. Conference on Engines and Vehicle ICE2017* (*Capri - Naples*) 2017-24-0129

[11] Merzlikin V G, Tovstonog V A, Eliseev V N 2014 Thermal Energy Radiator *RF Patent* 2,529,894

[12] Eliseev V N and Tovstonog V A 2013 Evaluation of possibility of using tubular gas-discharge radiation sources for simulating thermal regimes of large-size space structures *Herald of the Bauman Moscow State Tech. Uni., Ser. Mech. Eng.* **2** pp 111-16

[13] Uchida N and Osada H 2017 A new insulation concept for heavy-duty diesel engines to reduce heat loss from the wall *Proc. 13th Int. Conference on Engines and Vehicle ICE2017* (*Capri - Naples*) 2017-24-0161

[14] Annand W J D 1963 Heat transfer in the cylinders of the reciprocating Internal Combustion Engines *Proc. Inst. Mech. Engin* **177** 973-990
[15] Woschni G, Spindler W and Kolesa K 1987 Heat insulation of combustion chamber walls – a measure to decrease the fuel consumption of IC Engines *SAE International Paper* 870339
[16] Morel T and et al. 1989 Heat transfer in a cooled and an insulated diesel engine *SAE Technical Paper* 890572
[17] Ogury T and Inaba S 1972 Radiant heat transfer in diesel engines *SAE Technical Paper* 720023
[18] Rakopoulos C D and Giakoumis E G 2009 *Diesel engine transient operation. principles of operation and simulation analysis* (Springer-Verlag London Limited) p 390
[19] Azadi M A 2013 A review of tbc effects on diesel engine performance and components lifetime *Int. J. of Autom. Eng.* **3** pp 305-317
[20] Das D, Majumdar G, Sen R S and Ghosh B B 2013 Evaluation of combustion and emission characteristics on diesel engine with varying thickness of PSZ coated piston crown *Int. J. of Innovative Research in Sci., Eng. and Tech.* **2** (10)
[21] Ramu P and Saravanan C G 2009 Effects of ZrO_2-Al_2O_3 and SiC coating on diesel engine to study the combustion and emission characteristics *SAE International Paper* 2009-01-1435
[22] Sankar V 2014 Thermal barrier coatings material selection, method of preparation and applications – a review *Int. J. Mech. Eng. & Rob. Res.* 510-517
[23] Domakonda V K and Puli R K 2012 Application of thermal barrier coatings in diesel engines: a review *Energy and Power* **2**(1) pp 9-17
[24] Kosaka H, Wakisaka Y, Nomura Y, Hotta Y, Koike M, Nakakita K and Kawaguchi A 2013 Concept of "Temperature Swing Heat Insulation" in combustion chamber walls, and appropriate thermo-physical properties for heat insulation coat *SAE Int. J. Engines* **6** (1) 142-149
[25] Ciniviz M, Salman M S, Canl E, Köse H and Solmaz Ö 2012 Ceramic coating applications and research fields for ICEs", *Ceramic Coatings – App. in Engineering* 195-234
[26] Bovo M, 2014 *Principles of heat transfer in Internal Combustion Engines from a modeling standpoint* (Gothenburg: PhD Diss., Dep. of Appl. Mech., Chalmers Uni. of Tech) p 37
[27] Blomqvist Ch 2014 *Thermal Barrier Coatings for Diesel Engine Exhaust Application* (Karlstad: Master's degree in Eng. Sc. & Mech. Eng., Karlstads Uni.) p 73
[28] Mohan B 2015 *Performance and emission optimization of CIDI Engine through various fuel injection strategies* (Singapore: PhD Diss., National Uni. of Singapor) p 250
[29] Hoffman M A, Lawler B J, Filipi Z S, Güralp O A and Paul M N 2014 Development of a device for the nondestructive thermal diffusivity determination of combustion chamber deposits and thin coatings *J. Heat Transfer* **136** 7
[30] Dubouil R 2012 *Simulation study of heat transfers in a motor electric hybrid powertrain thermics* (Nantes: Ecole Centrale de Nantes) p 266
[31] Carmona D D V 2014 *Thermal barrier coatings for efficient combustion* (Stockholm: Master's Degree in Materials Sc. & Eng., School of Ind. Eng. & Manag., KTN Campus) p 96
[32] Güralp O A 2008 *The effect of combustion chamber deposits on heat transfer and combustion in a homogeneous charge compression ignition engine* (Michigan: PhD Diss., Uni. of Michigan, Dep. of Mech. Eng.) p 107
[33] Chirkov A A 1962 About the level of scientific research of heat transfer in internal combustion engines *Bulletin of Mechanical Engineering* (Yaroslavl: Yaroslavl Ins. of Technology) **6** pp112-124
[34] Ryabov D I and Sviridov U B 1958 Research of some features of burning sprayed fuels. *News of Academy of Sciences of the USSR OTN*
[35] Pflaum W 1961 Die Warmeubertragung bei dieselmaschinen mit unci ohne auflagung *Motor Technische Zeitung* **3** pp 570-574
[36] Dannecker R, Noll B, Hase M, Krebs W, Schildmacher K.-U, Koch R. and Aigner M 2007

Impact of radiation on the wall heat load at a test bench gas turbine combustion chamber: measurements and CFD simulation Am. Soc. of Mech. Eng. (ASME), *Proc. of Turbo Expo. Power for Land, Sea, and Air* (*Montreal*) vol 4 parts A & B **GT2007-27148** 1311-1321
[37] Makino T, Kunitomo T, Sakai I and H. Kinoshita 1984 *Heat Transfer. Jpn. Res.* **13** 33-36
[38] Novitski L A 1980 *Optical properties of materials in low temperatures* (Moscow: Mashinostroyenye) p 243
[39] Avduevskiy V S 1972 *Fundamentals of the flight theory of space vehicles* (Moscow: Publ. "Mashinostroenie") p 345
[40] Tovstonog V A 1993 The evaluation of fireproof properties of light-scattering coatings *High Temp.* **31**(4) 202–208
[41] Boeringer J C and Spindler R J 1963 Radiant heating of semitransparent materials *AAIA Journal* **1**(1) pp 84-88
[42] Viskanta R and Grosh R J 1961 Heat transfer in a thermal radiation, in absorbing and scattering medium *Int. Development in Heat Transfer* (New York: ASME) **part IV** pp 820-828
[43] Howe J T, Green M J and Weston K C 1973 Thermal protection by subliming volumetric reflective materials in convective and intensive radiant environments, *AAIA J.* **11**(7)
[44] Siegel R 1996 Internal radiation effects in zirconia thermal barrier coatings *AIAA J. Thermophysics Heat Trans.* **10**(4) pp 707-709
[45] Manara J, Arduini-Schuster M, Rätzer-Scheibe H-J and Schulz U 2009 Infrared-optical properties and heat transfer coefficients of semitransparent thermal barrier coatings *Surface and Coatings Technology* **203**(8) pp 1059-1068
[46] Wang L, Eldridge J I and Guo S M 2014 Comparison of different models for the determination of the absorption and scattering coefficients of TBCs *Acta Materialia* **64** pp 402-410
[47] Karaoglanli A C, Ogawa K, Türk A and Ozdemir I 2013 Thermal shock and cycling behavior of thermal barrier coatings (TBCs) used in gas turbines *Progress in Gas Turbine Performance*, ed E Benini chapter 10 p 268

IOP Conf. Series: Materials Science and Engineering **315** (2018) 012018 doi:10.1088/1757-899X/315/1/012018

Improvement of ecological characteristics of the hydrogen diesel engine

T Natriashvili, R Kavtaradze and M Glonti

Raphael Dvali Institute of Machine Mechanics, Tbilisi, Georgia

E-mail: t_natriashvili@yahoo.com

Abstract. In the article are considered the questions of influence of a swirl intensity of the shot and injector design on the ecological indices of the hydrogen diesel, little-investigated till now. The necessity of solution of these problems arises at conversion of the serial diesel engine into the hydrogen diesel. The mathematical model consists of the three-dimensional non-stationary equations of transfer and also models of turbulence and combustion. The numerical experiments have been carried out with the use of program code FIRE. The optimal values of parameters of the working process, ensuring improvement of the effective and ecological indices of the hydrogen diesel are determined.

1. Introduction

The problems of preservation of the environment and providing the transport with power are the basic problems for the modern civilization. These problems are directly connected with piston engines - the basic users of natural fuels, playing an essential role in the atmospheric pollution.

It is necessary to underline that soot is practically absent in combustion residues of the hydrogen diesel engine that against toughening acts on ecology is the main advantage in comparison with the diesel engine conversion into the gas-diesel [1]. In this connection, improvement of the ecological indices of the hydrogen diesel, first of all, means decrease of nitrogen oxides concentration.

The aim of this work is to study such little-investigated questions solution of which is necessary at conversion of the serial diesel engine into the hydrogen diesel: Influence of the injector construction on the ecological indices of the hydrogen diesel engine and determination of the air swirl motion optimal intensity.

2. Mathematical model of the hydrogen diesel engine working process

The model of the operating process proceeding in the cylinder of a hydrogen diesel engine is constructed taking into account the turbulent ignition processes and the chemical kinetics of formation of the nitrogen oxides [2, 3, 4]. The model is based on the generalized transfer equation, the vector and tensor forms of which are given in the table 1. In this equation Φ is one of the dependent variables, Γ_Φ - generalized transfer coefficient (of diffusion, viscosity, thermal, and temperature conductivity), S_Φ - the source term. At substitution of the relevant expressions from the table into the generalized equation a system of three-dimensional nonstationary transfer equations is obtained: equations of momentum (Navier-Stokes), equations of energy (Fourier-Kirchhoff), and equations of continuity and diffusion (Fick). The following notation is assumed in the equations from the table 1: $D/D\tau$ - substantial derivative;

W - vector of the gas velocity (m/s); ρ - density of the gas (kg/m^3); p - pressure (N/m^2); G_i - projection of the vector of density of volume forces (N/m^3) on axis Ox_i of the Cartesian coordinate system; C - concentration (kg/m^3); H - total specific energy (J/kg); $V_\mu = \frac{1}{3}\mu \cdot grad\left(div\vec{W}\right)$ - term representing volume deformation; μ - dynamic viscosity (kg/(m s)); c_p - thermal capacity at constant pressure (J/(kg K)), w_r - velocity of the chemical reaction per unit volume (kg/(s m^3)); Qr - amount of heat educed per unit of mass (J/kg); λ - heat conductivity (W/(m K)); δ_{ij} - Kronecker symbol; D_c - diffusion coefficient (m^2/s); $\dot{m}$ - intensity of the mass source, the rate of the change of mass of the chemical component per unit volume (kg/(s m^3)). The radiation heat flux $\frac{\partial q_{Rj}}{\partial x_j}$ from the radiation source plays a substantial role only in the heterogeneous combustion of the liquid diesel fuel, which is accompanied by the intensive formation of the solid microparticles of soot - main generators of radiation. In case of the hydrogen burning the water steam is the basic source of thermal radiation whose selective radiation is negligibly small [5, 6, 7, 8].

As a result of averaging according to the Favre method [9] the original system of equations of three-dimensional nonstationary transfer is transformed into the system of Reynolds equations containing new unknown quantities with turbulent fluctuations of the parameters (see table 1). A relatively new k–ζ–f model of turbulence is used to close the Reynolds system [10].

Table 1. The basic equations of mathematical model of the working process of a hydrogen diesel engine.

EQUATIONS OF LAWS OF CONSERVATION AS PARTICULAR CASES OF THE GENERALIZED DIFFERENTIAL EQUATION

$$\frac{\partial}{\partial\tau}(\rho\Phi) + div\left(\rho\vec{W}\Phi\right) = div\left(\Gamma_\Phi grad\Phi\right) + S_\Phi \Leftrightarrow \frac{\partial}{\partial\tau}(\rho\Phi) + \frac{\partial}{\partial x_j}\left(\rho W_j\Phi\right) = \frac{\partial}{\partial x_j}\left(\Gamma_\Phi\frac{\partial\Phi}{\partial x_j}\right) + S_\Phi$$

$\Phi = W_i$, $\Gamma_\Phi = \mu$, $S_\Phi = G_i - \frac{\partial p}{\partial x_i} + V_\mu$	Initial form	$\rho\frac{DW_i}{D\tau} = G_i - \frac{\partial p}{\partial x_i} + \frac{\partial}{\partial x_j}\left[\mu\left(\frac{\partial W_i}{\partial x_j} + \frac{\partial W_j}{\partial x_i} - \frac{2}{3}\delta_{ij}\frac{\partial W_k}{\partial x_k}\right)\right]$
	Reynolds form	$\bar{\rho}\frac{D\overline{W}_i}{D\tau} = \overline{G}_i - \frac{\partial\bar{p}}{\partial x_i} + \frac{\partial}{\partial x_j}\left[\mu\left(\frac{\partial\overline{W}_i}{\partial x_j} + \frac{\partial\overline{W}_j}{\partial x_i} - \frac{2}{3}\delta_{ij}\frac{\partial\overline{W}_k}{\partial x_k}\right)\right] - \overline{\rho W_i' W_j'}$
Equation of conservation of the momentum (Navier–Stokes equation)	New unknown quantities	$\overline{\rho W_i' W_j'}$ - tensor of the Reynolds turbulent stresses determined by the oscillatory components of velocity; $\overline{\tau_{ij}} = \mu\left(\frac{\partial\overline{W}_i}{\partial x_j} + \frac{\partial\overline{W}_j}{\partial x_i} - \frac{2}{3}\delta_{ij}\frac{\partial\overline{W}_k}{\partial x_k}\right)$ - tensor of the viscous (turbulent) stresses determined by the averaged values of the velocity components.

Table 1. (Continued)

$\Phi = H$, $\Gamma_\Phi = \frac{\lambda}{c_p}$, $S_\Phi = \frac{\partial p}{\partial \tau} + \frac{\partial}{\partial x_i}(\tau_{ij}W_j) + G_jW_j + w_rQ_r + \frac{\partial q_{R_j}}{\partial x_j}$ Energy conservation equation	Initial form	$\rho\frac{DH}{D\tau} = G_jW_j + \frac{\partial p}{\partial \tau} + \frac{\partial}{\partial x_i}(\tau_{ij}W_j) + \frac{\partial}{\partial x_j}\left(\lambda\frac{\partial T}{\partial x_j}\right) + w_rQ_r + \frac{\partial q_{Rj}}{\partial x_j}$
	Reynolds form	$\bar{\rho}\frac{D\bar{H}}{D\tau} = \overline{G_j}\overline{W_j} + \frac{\partial \bar{p}}{\partial \tau} + \frac{\partial}{\partial x_j}(\overline{\tau_{ij}}\overline{W_j}) + \frac{\partial}{\partial x_j}\left(\lambda\frac{\partial \bar{T}}{\partial x_j} - c_p\overline{\bar{\rho}T'W_j'}\right) + \overline{w_r}\overline{Q_r} + \frac{\partial \overline{q}_{R_j}}{\partial x_j}$
	New unknown quantities	$c_p\overline{\bar{\rho}T'W_j'}$ - is turbulent enthalpy transfer $\bar{\rho}c_pT'$ by means of velocity fluctuation W_j';
$\Phi = 1$, $\Gamma_\Phi = 0$, $S_\Phi = 0$ Mass (continuity) conservation equation	Initial form	$\frac{\partial p}{\partial \tau} + \frac{\partial}{\partial x_i}(\rho W_j) = 0$
	Reynolds for	$\frac{\partial \bar{p}}{\partial \tau} + \frac{\partial}{\partial x_i}(\bar{\rho}\overline{W}_j) = 0$
$\Phi = \frac{C_l}{\rho}$, $\Gamma_\Phi = D_l\rho$, $S_\Phi = \dot{m}_l$. Diffusion (concentration) equation	Initial form	$\frac{DC}{D\tau} = \frac{\partial}{\partial x_j}\left(D\frac{\partial C}{\partial x_j}\right) + \dot{m}$.
	Reynolds form	$\frac{D\bar{C}}{D\tau} = \frac{\partial}{\partial x_j}\left(D\frac{\partial \bar{C}}{\partial x_j} - \overline{C'W_j'}\right) + \overline{\dot{m}}$,
	New unknown quantities	$\overline{C'W_j'}$ - turbulent diffusion transfer of concentration C_l of component ℓ by means of the velocity fluctuation W_j'.

In the equations is used, the rule of summation according to the twice-repeating index $i, j, k = 1,2,3$ *and averaging is realized according to Favre; i.e ,the density is used as a weight function.*

The wall functions are used to describe the flow of the working medium and heat exchange in the wall zone [11]. The process of turbulent combustion is simulated on the base of the well-known model of Magnussen–Hjertager [4, 12], which has also been well tested in the theory of piston engines. Its advantage is that there are no turbulent fluctuations of parameters in it. The model assumes that combustion of the fuel and air mixture occurs instantly and consequently the average combustion rate is determined by the time of turbulent mixing of the reagents and is written as follows:

$$\overline{w}_r = \frac{A}{\tau_t}\bar{\rho} min\left(\overline{m}_T, \frac{\overline{m}_{O_2}}{L_0}, \quad B\frac{\overline{m}_{comb.pr.}}{1 + L_0}\right), \tag{1}$$

where L_0 is the bulk stoichiometric quantity of air; $\tau_t = \bar{k}/\bar{\varepsilon}$ - time scale of turbulent mixing; A and B - empirical coefficients that take into account the influence of turbulence and fuel parameters on the rate of the chemical reaction. It is evident that for the concrete conditions they should be specified according to the experimental data. We note that the experimental indicator diagrams of the hydrogen diesel engine under study were used for their determination.

The local formations of the nitrogen oxides in the combustion chamber of the hydrogen diesel engine are simulated with the use of the extended mechanism of Zeldovich [4, 13], developed for NO whose portion in the total concentration of all nitrogen oxides [NOx] in the piston engines is 90% and more [4]. In this connection it is assumed that [NO_x] ≡ [NO].

The model was numerically realized on the base of the 3D-CFD code FIRE developed by the AVL List GmbH [14], designed for numerical solution of specific problems of the theory of piston engines. The core of the FIRE is based on the numerical method of control volumes with the use of the improved algorithm SIMPLE described in detail in [2, 14].

3. Model verification

The research objects were the experimental, four-stroke, single-cylinder supercharged diesel engine with direct injection of the gaseous hydrogen and its basic variant with injection of the diesel fuel. The experimental diesel engines were single-cylinder sections of the series diesel engines MAN with a ratio of piston stroke/cylinder diameter S/D = 30/24 cm/cm and a crankshaft rotation rate of n = 800 min^{-1}. The injection pressures were pi_{nj} = 350 bar and p_{inj} = 300 bar, and the degrees of compression were ε= 13,5 and ε=17 for the basic and hydrogen diesel engines, respectively [15, 16].

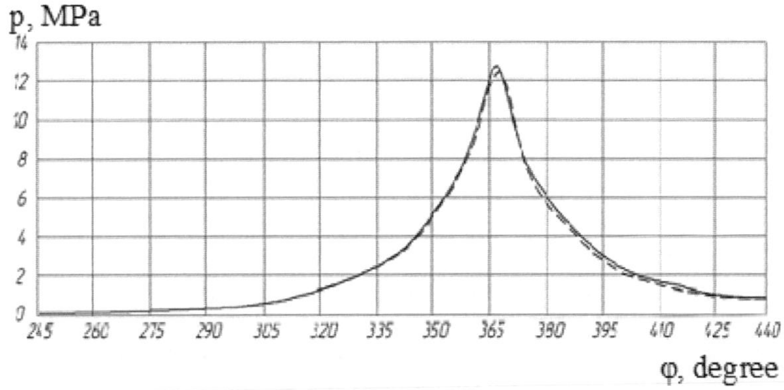

Figure 1. Indicator diagrams of the hydrogen diesel engine: experiment (- - -) and 3D modelling (–). Factors in the model of combustion (1) A=10, B=1. φ – angle of rotation of the crankshaft (ARC).

The model of operating process of the hydrogen diesel engine is verified by comparing the results of the numerical calculation with the experimental data (figure 1). The values of the empirical coefficients A = 10 and B = 1 were determined in the Magnussen-Hjertager combustion model (1), which closely correlate the experimental indicator diagrams with the calculated data (the difference in the maximum pressure of the cycle did not exceed 2 %). It should be noted that a comparative analysis of the operating processes of the basic and hydrogen diesel engines was carried out at approximate equality of the effective horse powers.

Some studies on the selection of certain structural (structure of the injection nozzle, swirl intensity of the charge) parameters having influence on the local formation of the nitrogen oxide in the combustion chamber were carried out to improve the ecological indices of the hydrogen diesel engine.

4. Influence of the injector design on formation of the nitrogen oxides

A series of numerical experiments carried out for variants of the injection nozzle has shown that the maximum and minimum values of the total mass portions of nitrogen oxides per cycle in terms of the cylinder volume are obtained with the use of nozzles 6×0,7 mm and 18×0,5 mm (z × d_c _ number of nozzle holes z and the nozzle hole diameter d_c in mm) and amount to $[NO_x]_\Sigma$= 0,00158 and

$[NO_x]_\Sigma$=0,00027 respectively. The remaining nozzles give the intermediate results (figure 2). It is seen that with increase of the number of injection nozzle holes, hydrogen is more uniformly distributed over the volume of the combustion chamber, the hydrogen-air mixture is homogenized and the level of local temperatures in the combustion areas decreases that leads to decrease in concentration of NO*x*.

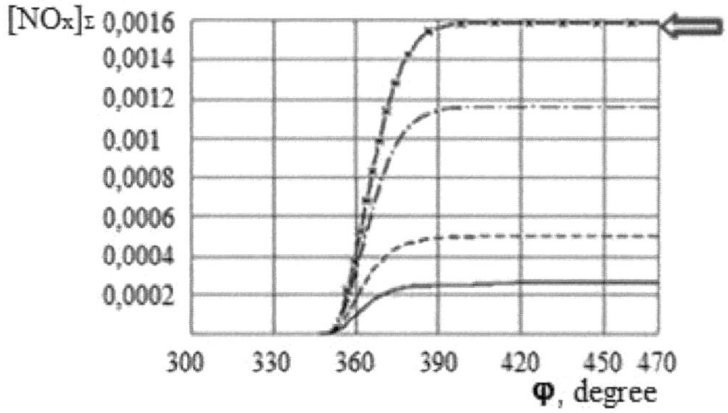

Figure 2. The total concentrations of the nitrogen oxides for different structures of the spray unit: 1. (— x —) 6×0,7 mm; 2. (— • —) 4×0,5 mm + 4×0,7 mm; 3. (- - - -) 12×0,5 mm; 4. (—) 18×0,5 mm. The arrow shows the experimental value for the nozzle 6×0,7 mm. φ – angle of rotation of the crankshaft (ARC).

5. Influence of intensity of the air whirl on formation of the nitrogen oxides

Usually, the whirling motion of the charge, together with the characteristic of fuel injection and the nozzle structure, ensures a high-quality proceeding of the carburetion and ignition processes, and first of all decrease of the ignition delay. It is possible to achieve the low specific fuel consumption at the optimal intensity $D_n=n_{air}/n$ (n_{air} - frequency of the air whirling motion, n - frequency of the crankshaft rotation) of the whirling motion. On the other hand, it leads to increase of the NO_x emission. The influence of the whirling motion on formation of the nitrogen oxides in the hydrogen diesel engine has a completely different character (figure 3). The numerical experiments showed that a significant increase in the air whirl at the end of compression contributes to fast diffusion with the hydrogen. Already at the onset of supply, the hydrogen is diluted by air, so that its amount in the local areas turns out to be lower than the concentration ignition limits (less than 4% by volume). Self-ignition starts later after the hydrogen concentration increases. It should be noted that at the same time the ignition delay increases [17, 18, 19], the combustion process is delayed, the maximum velocity of heat generation moves away from the upper dead point (UDP), the local temperatures of the working medium decrease, and, as a result, the nitrogen oxide emission decreases (figure 3). In addition, the delayed energy conversion process leads to decreased rate of increase of the pressure and reduction of noise. However, such a change in the heat generation velocity leads to decrease in the efficiency of the cycle. Thus, increase of the charge whirling motion intensity in the hydrogen diesel engine leads to worsening of its efficiency indices.

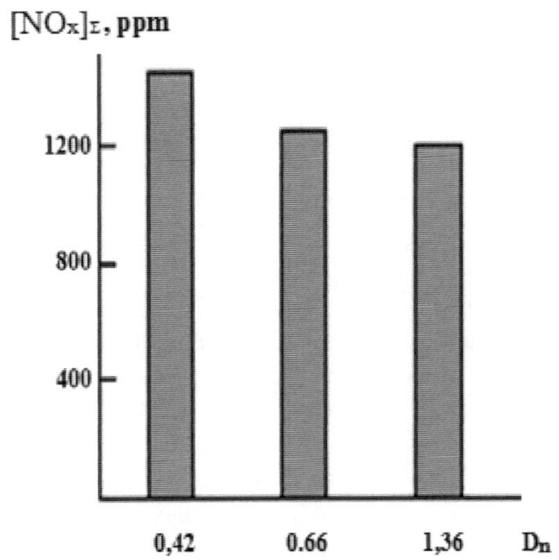

Figure 3. Influence of intensity of the charge whirling motion on the concentration of the nitrogen oxides in the products of combustion of the hydrogen diesel. The average effective pressure p_e=8,5 bar; frequency of rotation of the crankshaft n=800 min^{-1}; the air excess factor $\lambda_{air.}$=2,4_2,5; angle of outstripping of the hydrogen injection $\varphi_{inj.}$=15^0; the air temperature and pressure at the inlet T_k=296 K и p_k=1,29 bar; nozzle №1 (6×0,7 mm).

5. Conclusions

Out of the possible methods of converting the serial diesel engines into the hydrogen ones, the concept of the hydrogen diesel engine operating by self-ignition of the gaseous hydrogen is promising. The main problem of improving the ecological indices of the hydrogen diesel engine is minimization of the nitrogen oxide emission, which can be realized by changing the structural and regulating parameters.

The optimal (from the standpoint of efficiency and environmental compatibility) combination of the operating process parameters for the stationary hydrogen diesel engine under study is: spray 18×0,5 mm for the hydrogen supply; cyclic supply of the hydrogen m_c=0,49 g/cycle; the air excess factor λ_{air}=2,5. In this case, the concentration of the nitrogen oxides in the products of combustion is $[NO_x]_\Sigma$=950 ppm, and the indicator efficiency of the hydrogen diesel engine η_i=0,48 that is quite admissible for this type of the diesel engine.

Acknowledgement

The mentioned work is being realized under financial support of Shota Rustaveli National Scientific Foundation (Grant FR/241/3-170/14). Any opinion expressed in the present publication belongs to the authors and it may not express opinion of Shota Rustaveli National Scientific Foundation.

References

[1] Kavtaradze R Z 2011 *Thermal-Physical Processes in Diesel Motors Converted to Natural Gas and Hydrogen* (In Russian) (Moscow: Bauman State Technical University Publishing House) p 238

[2] Andersen D Tannehil G and Pletcher R 1984 *Computational fluid mechanics and heat transfer* (Hemisphere Publishing Corporation: New York, Cambridge, Sydney) p 776

[3] Merker G, Schwarz Ch, Stiesch G et al, 2016 *Verbrennungsmotoren. Simulation der Verbrennubg und Schadstoffbildung* 3 Auflage (Teubner-Verlag: Stuuugart, Leipzig, Wiesbaden) p 820

[4] Kavtaradze R Z 2016 *Theory of Piston Engines. Special chapters*. Second edition (In Russian) (Moscow: Bauman State Technical University Publishing House) p 590

[5] Kavtaradze R Z 2016 *Local heat exchange in piston engines*. Third edition (In Russian) (Moscow: Bauman State Technical University Publishing House) p 515
[6] Kavtaradze R Z, Gaivoronskii A I and Fedorov V A et al 2007 Calculation of Radiative-Convective Heat Transfer in the Combustor of Diesel Engine. *High Temperature* vol 45 No 5 pp 673-80
[7] Kavtaradze R Z, Onishchenk D O and Zelentsov A A et al 2009 The influence of rotational charge motion intensity on nitric oxide formation in gas-engine cylinder *International Journal of Heat and Mass Transfer* 52 pp 4308–16
[8] Kavtaradze R, Zelentsov A, Gladyshev S et al 2012 Heat Insulating Effect of Soot Deposit on Local Transient Heat Transfer in Diesel Engine Combustion Chamber *SAE International Paper* p 12
[9] Favre A 1965 Equations des Gaz Turbulents Compressibles *1. Formes Generales. J. Mecanique.* vol 4 pp 361-90
[10] Hanjalič K, Popovač M Hadziabdič M et al 2004 Near -Wall Elliptic-Relaxation Eddy-Viscosity Turbulence Model for CFD *Int. J. Heat Fluid Flow* No 25 p 897
[11] Popovač M and Hanjalič K 2005 Compound Wall Treatment for RANS Computation of Complex Turbulent Flow. *Proc. 3rd Int. Conf. on Comput. Fluid and Solid Mechanics* pp 1-28
[12] Magnussen B F and Hjertager B H 1976 On Mathematical Models of Turbulent Combustion with Special Emphasis of Soot Formation and Combustion *16th Int. Symp. on Combustion. Cambridge*. pp 719-29
[13] Warnatz J, Maas U and Dibble R 2001 *Verbrennung: Physikalisch-Chemische Grundlagen, Modellierung und Simulation, Experimente, Schadstoffentstehung* 3. Auflage (Berlin: Heidelberg: Springer-Verlag) p 326
[14] 2016 FIRE *Users Manual Version 2016* (Graz, Austria: AVL List GmbH) pp 1-9
[15] Rottengruber H, Wiebicke U, Woschni G et al 1999 Investigation of a Direct Injecting Hydrogen Diesel-Engine *World Hydrogen Conf. XII. Buenos Aires* pp 1255–64
[16] Rottengruber H, Wiebicke U, Woschni G et al 2000 Wasserstoff-Dieselmotor mit Direkteinspritzung, hoher Leistungsdichte und geringer Abgasemission. Teil 3: *Versuche und Berechnungen am Motor* /*MTZ* No 2 pp 122–8
[17] Kavtaradze R, Zeilinger K and Zitzler G 2005 Ignition Delay in a Diesel Engine Utilizing Different Fuels. *High Temperature* vol 43 No 6 pp 951-60
[18] Kavtaradze R Z, Natriashvili T M and Zelentsov A A 2015 Ignition Delay and Emission of the Noxious Substances in Double-Fuel Engines Working on the natural Gas and Syngases Chapter 15 (p. 109-20) in the Book: *Innovative Methods for Improvement of Technical, Economic and Ecological Efficiency of Motor Cars* (New-York: NOVA-Publishers) p 138
[19] Kavtaradze R Z 2016 Improving the Ecological Indices of a Hydrogen Diesel Engine with Direct Gaseous Hydrogen Injection *Journal of Machinery Manufacture and Reliability* vol 45 No 4 pp 307–15

IOP Conf. Series: Materials Science and Engineering **315** (2018) 012019 doi:10.1088/1757-899X/315/1/012019

Navigation of the autonomous vehicle reverse movement

M Rachkov and S Petukhov

Moscow Polytech, Moscow, 107023, Russia

E-mail: michyur@gmail.com

Abstract. The paper presents a mathematical formulation of the vehicle reverse motion along a multi-link polygonal trajectory consisting of rectilinear segments interconnected by nodal points. Relevance of the problem is caused by the need to solve a number of tasks: to save the vehicle in the event of a communication break by returning along the trajectory already passed, to avoid a turn on the ground in constrained obstacles or dangerous conditions, or a partial return stroke for the subsequent bypass of the obstacle and continuation of the forward movement. The method of navigation with direct movement assumes that the reverse path is elaborated by using landmarks. To measure landmarks on board, a block of cameras is placed on a vehicle controlled by the operator through the radio channel. Errors in estimating deviation from the nominal trajectory of motion are determined using the multidimensional correlation analysis apparatus based on the dynamics of a lateral deviation error and a vehicle speed error. The result of the experiment showed a relatively high accuracy in determining the state vector that provides the vehicle reverse motion relative to the reference trajectory with a practically acceptable error while returning to the start point.

1. Introduction

The development of systems for the vehicle automatic return based on the trajectories measured and stored in the onboard computer corresponds to the general trend of the development of autonomous and remotely operated vehicles. Among such may be planet-walkers, mobile robots, robotic combat vehicles, cars, etc. It is assumed that the data about landmarks are obtained by means of a stereo camera system on board the vehicle. The solution of the problem is connected with the use of adaptive control and intellectual information processing.

Relevance of the problem is caused by the need to solve a number of tasks: to save the vehicle in the event of a communication break by returning along the trajectory already passed, to avoid a turn on the ground in constrained obstacles or dangerous conditions (for example, while demining), or a partial return stroke for the subsequent bypass of the obstacle and the continuation of the forward movement.

Determination of vehicle trajectory parameters and a location of the vehicle from the navigation measurements, calculation of the trajectory corrections and elaboration of control system settings for the upcoming maneuvers constitute the essence of the navigational task, which is similar to the navigation of space vehicles [1]. The vision usage for navigating mobile systems has been studied, in particular, in [2-5]. The aim of the work is development of these studies in the direction of solving the problem of reverse motion optimizing by using the passed trajectory information.

IOP Conf. Series: Materials Science and Engineering **315** (2018) 012019 doi:10.1088/1757-899X/315/1/012019

2. Formulation of the problem

The task of automatic return of the vehicle on memorized landmarks provides two stages of vehicle movement:

- movement along the working area in an autonomous or remote control mode with saving trajectory parameters and a layout of the landmarks *P*, the length of the traversed path *l*, the angle φ of trajectory rotation and the rotation angle of the stereo block β;
- automatic return by reverse motion with the help of the stored point plans and newly obtained point plans for corrections of the trajectory.

It is assumed that the vehicle moves along a multilink polygonal trajectory consisting of rectilinear lengths l_k connected by nodal points $S_k, k = 1, 2, ..., N-1$. Such a trajectory requires a minimum number of parameters and the amount of memory to store in the on-board computer. At some nodal points, called reference points, landmark measurements are made (figure 1).

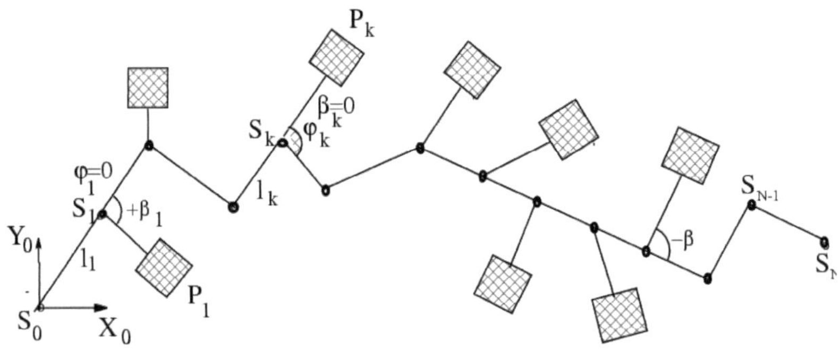

Figure 1. Diagram of binding the nominal trajectory to the plans of the neighborhoods P_k.

The vector of navigation measurements M_k in the *k*-th reference point can generally be written in the following form:

$$M_k = col(l_k, \varphi_k, \beta_k, P_k). \tag{1}$$

If the measurement is not made at the node then the measurement vector is incomplete and $M_k = col(l_k, \varphi_k, 0, 0)$. The trajectory passed and stored in the process of tele-controlled forward movement into the working area is nominal for automatic return of the vehicle.

The method of navigation with forward movement assumes that the route of movement is predetermined, and the path is refined provided that it passes near the landmarks. To measure landmarks on board the vehicle, a stereo block of cameras is placed on a vehicle controlled by the operator by means of the radio channel. The formation of point plans takes place in the following sequence.

1) During the stop, the operator searches for landmarks by turning the pointing device of the stereo unit on the angle β_k.

2) Stereo images are inserted into the on-board computer and their processing is performed.

3) A compressed representation of landmarks on the reference surface of motion is formed by projections of vertical lines in the form of point plans P_k in a coordinate system associated with the vehicle.

4) The parameters $(l_k, \varphi_k, \beta_k)$ and the resulting point plans P_k are saved for the automatic return of the vehicle.

5) The next section of the trajectory at each node S_k is set by the settings (φ_k, l_{k+1}), which are transmitted by the operator to the board.

The method for navigating an autonomous vehicle while moving to a work area requires parameters ($\tilde{l}_k, \tilde{\varphi}_k, \tilde{\beta}_k$), where k = 1, .., N-1. A pre-compiled orientation map stored in the on-board computer memory in the form of a sequence of point plans $\tilde{P}_k$ is used for navigation at reference points. Wherein:

- landmarks are captured in the field of view of cameras at the point S_k by turning the stereo block to an angle $\tilde{\beta}_k$;
- the trajectory correction is performed on the basis of the coordination results of the plans $\tilde{P}_k$ and the newly obtained point plans P_k that are stored for automatic return;
- measured values l_k, φ_k, β_k, and point plans P_k of landmarks are used for automatic return.

The principle of autonomous navigation in automatic reverse motion is based on the use of the operational machine relative map (MRM), automatically formed in the process of direct movement of the vehicle at the nodal and reference points. Such a card is represented in the memory of the on-board computer in the form of a stack. Each "sheet" of the MRM corresponding to the reference point k contains parameters $(l_k, \varphi_k, \beta_k)$ along with a point plan P_k. If there are no reference points in the nodal points, then the corresponding "sheet" of the MRM is represented only by parameters φ_k, l_k. In the reverse motion, the MRM parameters are extracted from the stack computer memory in the reverse order, in accordance with the stop point number of the vehicle. In the MRM, the point plans are linked through the parameters of the trajectory and form a single information field.

The parameters φ_k, l_k of the trajectory are worked out by the control system with errors, so there is a deviation of the motion from the planned point S_{k-1} to the point S_k' at which the stereoscopic survey is performed and the point plan P_k' is obtained. As a result of the alignment of the plan P_k' with the plan P_k, the deviations from the trajectory ($\Delta\bar{r}_k, \Delta\varphi_k$), $\Delta\varphi_k = \varphi_k' - \varphi_k$ and the control parameters ($\hat{r}_k, \hat{\varphi}_k$) that ensure the vehicle motion to the point S_{k-1} are determined (figure 2).

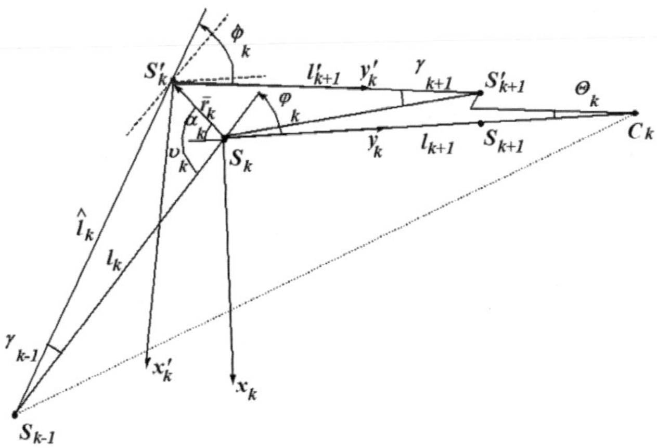

Figure 2. Diagram for determining path length $\hat{l}_k$ and heading angle $\hat{\varphi}_k$.

The central problem is determination of the benchmark plans for calculating the navigation parameters $(\bar{r}_k, \Theta_k)$ shown in Figure 2, and the control parameters $(\hat{l}_k, \hat{\varphi}_k)$ to continue motion to the next node. Diagram of the reverse movement trajectory is shown in figure 3.

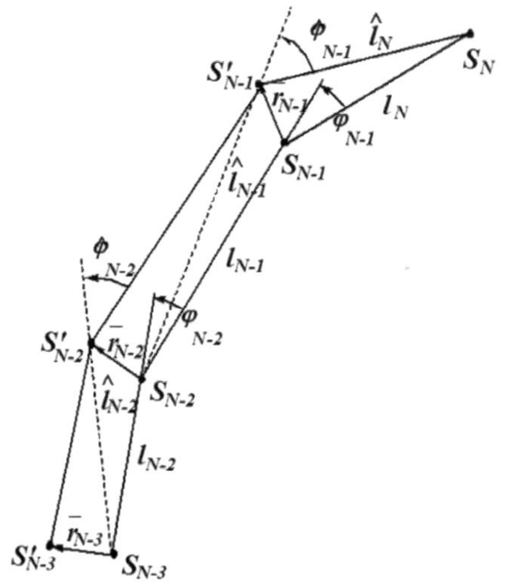

Figure 3. Diagram of the reverse movement trajectory.

Mathematical statement of navigation and control tasks with automatic return is the same for remote control and autonomous movement. The definition of the parameters l_k, φ_k, β_k and P_k has the random nature, therefore the nominal trajectory is probabilistic, as well as the return path.

The motion model in the vicinity of the stored trajectory is based on the representation of the disturbed motion relative to the planned trajectory by a system of linear differential equations. It is assumed that the vehicle moves rectilinearly between the nodal points at a constant speed. The main error component is caused by the constantly increasing lateral deviation due to an error in the course, which leads to deviations from the target point $\delta x_k \approx V \cdot T_{k+1} \cdot \delta\varphi_k$, where T_{k+1} is duration of the movement from point S_{k+1} to point S_k. Both quantities δl and $\delta\varphi$ are random and determine the deviation from the point, which is assumed to be subordinate to the two-dimensional normal distribution law.

Errors in estimating deviations from the nominal trajectory of motion are determined using the multidimensional correlation analysis apparatus based on the dynamics of the error variation of the lateral deviation ρ_1 and the vehicle speed ρ_2 of deviation from the nominal trajectory. The quantities ρ_1 and ρ_2 are components of the state vector $\bar{\rho}$. Differential equations of perturbed motion in matrix form are as follows:

$$\frac{d\bar{\rho}}{dt} = A\bar{\rho}, \qquad \bar{\rho} = \begin{vmatrix} \rho_1 \\ \rho_2 \end{vmatrix}, \qquad A = \begin{vmatrix} 0 & 1 \\ 0 & 0 \end{vmatrix}, \tag{2}$$

where A is the matrix of a linear differential equations system,

$$\rho_1 = \delta x_k, \qquad \rho_2 = d\rho_1/dt = V \cdot \delta\varphi_k, \tag{3}$$

under the initial conditions: $\rho_1(0) \in (0, \delta r), \rho_2(0) \in (0, V \cdot \delta\theta)$ with a correlation matrix of initial deviations

$$\Sigma_\rho = \begin{vmatrix} \sigma_1^2 & K_{12} \\ K_{21} & \sigma_2^2 \end{vmatrix}, \qquad K_{12} = K_{21} = K\sigma_1\sigma_2, \tag{4}$$

where σ_1 and σ_2 are the variance of the errors in the determination of ρ_1 and ρ_2 ; K is the correlation coefficient.

The state of the system $\overline{\rho}(T)$ at the final instant of time T is found by means of a transition matrix $\Phi(T,t), t \in (0,T)$, which is determined from the solution of the conjugated system [6]. With the help of this transition matrix, the state vector of the system is determined at the final instant of time T according to the state at the current time:

$$\overline{\rho}(T) = \Phi(T,t)\overline{\rho}(t); \rho_1(T) = \Phi_1(T,0)\overline{\rho}(0), \quad (5)$$

where $\Phi_1(T,0)$ is the first row of the transition matrix.

The maximum possible deviation from the node S_k due to a position error δr and an error in the implementation of the rotation angle $\delta\varphi$ is obtained by substituting the value $T = \hat{T}$ of the predicted motion time from point S_{k+1} to point S_k and the maximum possible deviations in the components of the state vector $\overline{\rho}(0)$ at the initial moment:

$$\rho_1(\hat{T}_i) = \begin{vmatrix} 1 & \hat{T} \end{vmatrix} \begin{vmatrix} \delta r \\ 3V\sigma_\varphi \end{vmatrix} = \delta r + 3V\hat{T}\sigma_\varphi. \quad (6)$$

The predicted value of the variance of deviation from the calculated trajectory at the instant $\hat{T}$ relatively the initial instant $(t = 0)$ is:

$$\sigma_x^2 = \Phi_1(\hat{T},0)\Sigma_\rho\Phi_1^*(\hat{T},0), \quad (7)$$

where $(*)$ is the transpose, K is the correlation coefficient, which determines the orientation of the dispersion ellipse [6].

The purpose of the correction is to limit the deviation r_0 from the starting point S_0, but at the same time the current deviations r_k from the nominal trajectory, controlled at the reference points, also should not exceed the permissible value: $r_k \le R_d$, where $k = 1,2,...,N-1$, and R_d is the maximum permissible deviation. This condition, in addition to safety considerations while returning, involves taking camera pictures from closely spaced points to the nominal trajectory and minimizing the differences in the plan of landmarks P_k' from the plan P_k. The beam width of the possible return paths is determined at the node points relative to the nominal trajectory by scattering ellipses R_k, $k = N-1, ..., 0$.

3. Experimental results

As an experimental example, an interval control was tested, the necessary condition of which is the movement of the vehicle from each current node S_k to the next node S_{k-1} and the control parameters are selected from the approximation conditions near this point.

The hardware included an on-board computer, and the passed distance was measured by the rotation of the driving wheels. Rotation of the vehicle was performed by rotating left and right wheels in opposite directions. The software consisted of a set of programs for automatic analysis of stereo images, the formation and coordination of point plans, calculation of the guidance parameters ($\hat{r}_k, \hat{\varphi}_k$) in the vicinity of the k-th reference point. A variant of the vehicle motion trajectory in the room along 6 reference points and 4 points of rotation and the trajectory of reverse motion to the initial point are shown in figure 4.

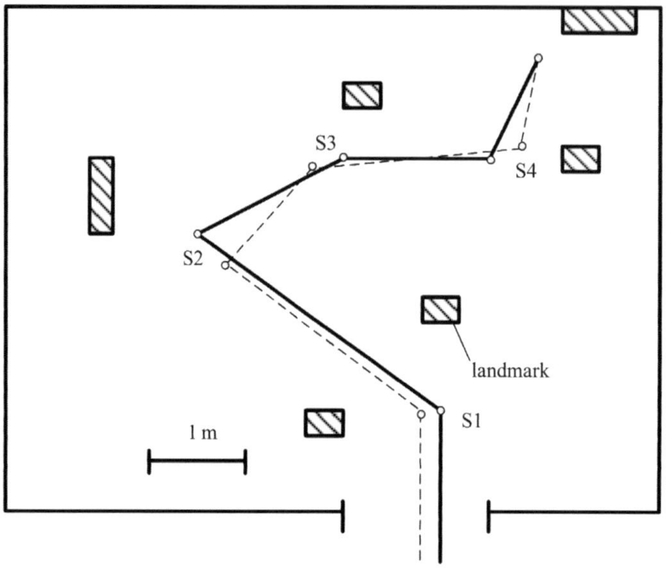

Figure 4. Plan of the room and the trajectory of motion
(solid line - direct movement, dashed line - reverse movement).

The result of the experiment showed sufficient accuracy of determining the state vector $(r_{N-k}, \Delta\hat{l}_{N-k}, \hat{\varphi}_{N-k})$ according to the results of the point plans matching, providing reverse movement relative to the reference trajectory with an acceptable error about 0.2 m at the start point.

4. Conclusion

A mathematical formulation of the vehicle reverse motion along a multi-link polygonal trajectory consisting of rectilinear segments interconnected by nodal points is determined. The method of navigation with direct movement assumes that the route of movement is measured, and the reverse path is corrected by using landmarks. To measure landmarks on board, a block of cameras is placed on a vehicle controlled by the operator through the radio channel.

Errors in deviation estimating from the nominal trajectory of motion are determined using the multidimensional correlation analysis apparatus based on the dynamics of the lateral deviation error and the vehicle speed error.

The principle of autonomous navigation in automatic reverse motion is based on the use of an operational machine relative map automatically formed during the forward movement of the vehicle at the node and reference points. The vehicle motion model in the vicinity of the stored trajectory is based on the representation of the disturbed motion relative to the planned trajectory by a system of linear differential equations. It is assumed that the robot moves rectilinearly at a constant speed between the nodal points.

The result of the experiment showed a relatively high accuracy in determining the state vector that provides a reverse motion relative to the reference trajectory with a practically acceptable error while vehicle returning to the start point.

References

[1] Alekseev K and Bebenin G 1974 Space vehicles control *Mashinostroenie* p 340 (in Russian)
[2] DeSouza G and Kak A 2002 Vision for Mobile Robot Navigation *IEEE Trans. on PAMI* №24 pp 237-267

[3] Petukhov S and Ivaniugin V 1999 Interaction with external environment of robot by man-machine interface *2nd Inter. Conf. on Climbing and Walking Robots* (Portsmouth: UK) pp 445-452

[4] Petuhov A, Rachkov M and Petuhov S 2007 Application of compression post-processing algorithms for remote control of mobile robots *Mechatronics, automation, control* **1** pp 17-24 (in Russian)

[5] Gradeckij V, Rachkov M and Petuhov S 1999 Measuring systems of a robotic complex for operation in an atomic reactor (preprint № 656) *IPM RAN* p 46 (in Russian)

[6] Petukhov S and Rachkov M 2009 Navigation Method of Autonomous Robot Backward Motion by Remembered Landmarks *12th Int. Conf. on Climbing and Walking Robots and the Support Technologies for Mobile Machines* (Istanbul: Turkey) pp19-25

IOP Conf. Series: Materials Science and Engineering **315** (2018) 012020 doi:10.1088/1757-899X/315/1/012020

Methods of monitoring the technical condition of the braking system of an autonomous vehicle during operation

A Revin, V Dygalo, G Boyko, M Lyaschenko and L Dygalo

Volgograd State Technical University, Volgograd, Russia

E-mail: dygalo@vstu.ru

Abstract. Possibilities of diagnosing of a technical condition of braking system of the autonomous vehicles with automated modules while in service are considered. The concept of sharing of onboard means and stands for diagnosing is presented.

1. Introduction

The braking system is an integral part of autonomous vehicles. It is the units of the braking system (in addition powertrain) which are the most important in the implementation of the algorithms for the operation of autonomous vehicles. It is obvious that a lot depends on quality control of its technical condition. Saturation of autonomous vehicles with automation means certainly contributes to increased autonomy and traffic safety. In the process of operation, the operating parameters of these systems deviate from the nominal ones, as a result of wearing out and other factors [5, 6, 7]. It is known to be effective to use any systems in operation only if the methods of monitoring of their technical condition are successfully implemented. Especially this applies to the use of automated systems, because otherwise you can get an even more dangerous situation than without it. However control of the technical condition of the braking system' in an automated autonomous vehicles (even automatic systems) is not always suitable. What are the "pitfalls" in the implementation of the existing problem? They are as follows:

2. Problem

First of all, this is the absence on autonomous vehicles with automated and automatic systems of traditional signs of serviceability. When the vehicle is stopping traces left on the surface of asphalt by braked wheels is one of the signs. These traces make it possible to judge in the first approximation the efficiency of the brake drive and, first of all, whether the braking torque developed at the wheels of the car reaches the maximum possible moment of adhesion in these road conditions. At the same time, the wheel wear on dry asphalt concrete indirectly indicates sufficient braking power, which follows from the dependence shown below

$$M_{br} \geq R_z r_d \varphi (S_x) \qquad (1)$$

M_{br} – braking torque on the wheel, R_z - normal load, r_d - dynamic radius of the wheel, $\varphi (S_x)$ – coefficient of adhesion to the road.

By these tracks one can indirectly deduce another important indicator - the unevenness of the work of the braking mechanisms, which arises for various operational reasons. The non-simultaneous occurrence of traces of noise is the motive for getting the vehicle bench tested.

In the presence of automated elements in the braking system of the autonomous vehicles, the situation changes radically. The absence of tracks on the surface of the road can be regarded in two ways. On the one hand, this may indicate the effective operation of the system and, on the other hand, the inadequate efficiency of the functioning of the elements of the brake drive and its inability to provide in principle the magnitude of the maximum coupling torque (see the formula). The latter facilitates the braking of the wheel in the subcritical region of the φ (S_x) diagram. But in the first case, not everything is so unambiguous. The situation is also complicated by the fact that during the operation of the vehicle, the stand inspection of the technical condition is carried out periodically at the next maintenance through a certain mileage or once a year with the TRP. However, it is known that the sudden failure of the elements of the braking system of the autonomous vehicles, especially automated, entails the aggravation of the severity of the consequences. So, according to the evidence of the traffic police of European countries, the sudden failure of the automated system led to more serious consequences from the road accident.

3. Solutions

It is known that manufacturers equip their products with a self-monitoring system aimed at checking the integrity of electrical circuits and the signal level. Thus, opening of the supply circuit of the modulator or sensor will immediately turn off the units and modules of the system. Obviously, this approach does not exclude all possible failures of its elements. So, you can add a change in the cross-section of the modulator channels due to their clogging and delay when the valves operate, an increase in the angular compliance of the stator of the wheel sensor due to weakening of its fastening, etc. Such faults do not control the self-diagnosis system and do not give a signal of failure.

Therefore, there is an urgent need to improve on-board diagnostics of the technical state of the elements of the braking system, capable of obtaining objective information for the driver about the quality of the work process and the condition of the braking system as a whole during the intercontrolled operation period. In our opinion, during the first stage the diagnosis should be carried out by on-board diagnostics and based primarily on the parameters of efficiency, i.e. give a general "integral" assessment, thereby, making up for the driver lack of objective visual criteria. Consequently, the basis of such on-board diagnostics should be realized braking dynamism and its correspondence to the set value. At the same time, it is advisable to further improve on-board diagnostics of the braking system of autonomous vehicles in the direction of improving the quality of control based on the developed structural and investigative schemes. However, in doing so, it is necessary to remember the inherent operation of the autonomous vehicles, a significant variation in the characteristics of external conditions and, first of all, the fluctuation of the coefficient of adhesion in both the longitudinal and transverse directions.

As is known, it is possible to obtain information about the current state of the braking system of autonomous vehicles in two ways: airborne control and by means of bench diagnostics. When solving the task, each of them has its own pros and cons. Improving airborne diagnostic tools is certainly appealing, since it is possible to obtain information in the on-line mode, i.e. in the intercontrol period. However, in this case, the autonomous vehicles design requires the introduction of a number of additional sensors (working medium pressure sensors in service brake cylinders, brake pad wear sensors, deceleration sensors, etc.), and an additional diagnostic unit for recording and processing current information. The situation is further complicated by the fact that the fluctuation of the coefficient of adhesion does not allow unambiguously to set the standard normal values of the signals of all additional sensors. This will require an analysis based on the statistical data obtained in the series of inhibitions.

4. "Advantages and disadvantages" of stand diagnostics

In bench diagnostics, the phenomenon of fluctuations in the coefficient of adhesion along the way is reduced to a minimum, and control over the technical state of the elements of the braking system is carried out in fixed and therefore completely reproducible conditions, which is an undoubted "plus".

All the existing stands, which can diagnose the braking system of the autonomous vehicles, can be divided into two large groups: roller power and drum-type inertial. Stands of the area type are omitted

IOP Conf. Series: Materials Science and Engineering **315** (2018) 012020 doi:10.1088/1757-899X/315/1/012020

from consideration due to the need to provide relatively high speeds if there are automated modules on the autonomous vehicles.

When assessing the quality of functioning of an automated braking system in laboratory conditions on inertial stands, an important condition is the maximum possible reconstruction of the actual conditions of interaction of the tire with the road surface. The latter is achieved, on the one hand, due to the adequate normal distribution of the normal loads in the contact spot of the tire, and on the other hand, the reconstruction of the law of variation of the coefficient of adhesion from slippage. The first condition determines at least a one and a half to twofold ratio of the diameters of the running drum and the wheel of the autonomous vehicles, and the second is the need to introduce a cleaning system for the drum in the contact spot from the tire wear products, for example, by applying and then removing the kaolin powder from the spot. All that has been said above significantly complicates the construction and increases the dimensions of the inertial-type stand, transforming it into a research category. This excludes their mass application in the dialer stations and in the Transport Company's.

In the wide-spread roller power braking stand on rollers, longitudinal riffling is applied to eliminate slippage of the tire, which practically excludes the possibility of reproducing the tires close to real road conditions. However, this function is not required in the power stand, since the main task is to assess the maximum possible realization of the braking torques on the wheels and the difference in braking forces for compliance with permissible standards. Therefore, this design does not allow diagnosing an automated system. Rollers virtually eliminate the possibility of wheel slippage.

Let's try to turn the noted disadvantage into an advantage, eliminating the unstable link "wheel-rollers", which can be achieved by introducing special simulators or electronic control circuits to the moment of resistance of the brake rollers. The way out of this situation can by means of diagnosis using the simulation of movement and brakes of the autonomous vehicles. In the first way, the system can be triggered on the stand by two methods: by using a special mode dialer that simulates the signals of the wheel speed sensor or software loaded into the control unit, which imitates signals at the program level in the diagnostic mode.

5. Diagnostic method

At the first approach, several well-known autonomous vehicles of the same brand, model and model year are stripped of the main characteristics of the wheel speed sensor of each wheel, as well as from the vehicle speed sensor to determine the average values. The signals with the received characteristics are supplied to the input of the system control unit with the aid of a master. In this case, a pulse generator capable of providing emulation of the signals from the wheel rotation sensors and the vehicle speed sensor can be used as the setpoint generator. The generator is controlled dynamically from a personal computer, so it can work according to a predetermined law.

The procedure for checking the effectiveness of the system is as follows. The four wheel speed sensors and the vehicle speed sensor are disconnected, and the outputs from the driver are connected in their place, after which the autonomous vehicles is mounted on the roller power stand. When the wheels rotate, the driver presses down on the brake pedal with the required force. At the same time, the operator activates the wheel rotation condition sensor, which generates signals in accordance with the previously obtained characteristic, thereby imitating the periodic occurrence of each of the wheels. The conclusion about the technical condition of the braking system of the autonomous vehicles is carried out on the basis of the analysis of the oscillogram by the method of "comparing the states" of the tested vehicle and taken for a standard of a known serviceable autonomous vehicles.

The second approach is aimed at simulating the movement of the autonomous vehicles "inside the control unit". The program code of the control unit incorporates an additional special program responsible for diagnosing the system on a power roller stand. The test procedure is similar to the first case, except for the connection of the diagnostic equipment to the OBD-II connector.

The second approach is more promising, since the main drawback of the method using an external signal setter [2, 3, 4] is the connection of additional cables, which takes some time, since the control unit on some autonomous vehicles can be located in hard-to-reach places. So, for example, on a Volvo

V40 CC model in 2014, the battery must be removed before access to the ABS unit and it is advisable to take into account the need for access to the control unit during the development stage [1]. The figure shows, as an example, the appearance of the assembled signal emulator board [8, 9, 10].

The software method is more preferable, because it is more technological. In this case, there is no need to interfere with the electric circuit of the autonomous vehicles and, therefore, the testing time is shortened, which allows increasing the efficiency of the brake stand, reducing training costs for the operator and reducing the probability of error. It must be remembered that only the manufacturer is entitled to interfere with the software of the automated module unit.

Figure 1. Appearance of the assembled emulator board.

Therefore, for the application of the software method, it is necessary to develop a basic universal algorithm for the operation of the diagnostic program, which can be embedded in the code of the ABS control unit with minimum costs, i.e. it is necessary to work together with the specialists of the system manufacturer. If we consider the workshop or the official dealer center as an organization specializing in certain brands of autonomous vehicles, then the manufacturer of ABS should be interested in improving the quality of service.

6. Conclusions

Taking into account the features of the functioning of the braking system of autonomous vehicles with automated diagnostic elements, it should be carried out on a complex basis, including both on-board diagnostics during the intercontrol period and on the most common power roller stand provided with the mode dial.

Further development of on-board diagnostic tools for autonomous vehicles with automated modules should be carried out on the basis of developing new diagnostic methods, using for this multi-level structural and investigation schemes, primarily focused on performance parameters.

Advanced element-wise diagnostics should be carried out in bench conditions, with the maximum possible exception of random factors affecting the process, which is possible with the use of widely used bench equipment with a relatively small cost and overall dimensions, for example, power roller stands, with appropriate modification.

It is advisable to refine the power roller stands in the direction of creating special mode controllers that allow using the standard equipment of the stand to estimate the delay time of the modulator operation, the rate of change of the brake moments, the thresholds for setting up the logical unit of the systems, and so on. A perspective direction can be considered the inclusion of a self-diagnostic module in the control unit, with recorded test modes for testing on power roller stands.

References

[1] Revin A A Formation of the main operational properties of vehicles in braking mode *Automobile industry* 2014 **11** pp 3-5

[2] Dygalo V G Technologies for testing active safety systems for vehicles *Mechanical Engineering* 2012 p 387

[3] Dygalo V G Methodology of virtual physical tests of automated braking systems of wheeled vehicles dis. Dr.Sci.Tech (Place of protection: Volgograd, Volgograd. state. tech. un-t) 2015 p 348

[4] Dygalo V G Development of the device for simulating the signals of the wheel sensors of the active safety system *Izv. VolgGTU* Series "Terrestrial transport systems." Issue. 5 interuniversity. Sat. sci. Art. VolgGTU (Volgograd) 2012. **2** p 32-35

[5] Kurbanov I Sh, Smorchkov A E, Dygalo V G, Boyko G V, Kotov V V and Dygalo L V Influence of active safety systems on changing the technical condition of the car during operation (Electronic resource) Science of the XXI century: the experience of the past - a glimpse into the future: the mother. II Intern. scientific-practical. Conf. (Omsk) 2016 pp 186-192 Access mode: ttps://yadi.sk/mail/?hash=conz63HC0NLNyrMw6dYxovtd1xLQSYTGgvvxcas1Fk0%3D.

[6] Dygalo V G, Kotov V V, Boyko G V, Dygalo L V and Revin A A Complex assessment of the influence of instability of the characteristics of the brake wheel mechanism elements of the car arising during operation *Izvestiya VolgGTU* Ser. Ground transportation systems (Volgograd) 2015 **6** pp 16-19

[7] Revin A A, Dygalo V G, Yudina An A and Dygalo L V On the problem of the estimation of the influence of the instability of the characteristics of the elements of the "modulator-ABS braking mechanism" chain on the quality of the control commands *Izvestiya VolgGTU* Ser. Ground transportation systems. Issue. 12 (Volgograd) 2015 **6** p 42-43

[8] Melnikov G O, Dygalo V G, Revin A A and Larin E S Diagnostic complex for checking ABS and ESP (Electronic resource) *SCI-ARTICLE.RU* electronic periodical scientific journal 2013 **3** p 17-22 Access mode: http://sciarticle.ru/stat.php?i=diagnosticheskiy_kompleks_dlya_proverki_ABS_i_ESP

[9] Melnikov G O, Larin E S, Dygalo V G and Revin A A Complex for diagnostics of active safety systems of cars *Izvestiya VolgGTU* Series "Ground-based transport systems" Issue. 6 (Volgograd) 2013 **10** pp 102-105

[10] Melnikov G O, Larin E S, Revin A A and Dygalo V G A computer method of diagnostics of systems of active safety of the car. Diagnosis of ABS and ESP Aktual directions of fundamental and applied research Topical areas of fundamental and applied research: mater. Intern. scientific-practical. Conf. (Moscow) 2013 **2** *Academic* North Charleston (SC, USA) 2013 pp 81-85

IOP Conf. Series: Materials Science and Engineering **315** (2018) 012021 doi:10.1088/1757-899X/315/1/012021

Mathematical modelling of active safety system functions as tools for development of driverless vehicles

V Ryazantsev[1], N Mezentsev[2] and A Zakharov[1]

[1] Active Safety Sector, FSUE "NAMI", Moscow, 125438, Russia
[2] Intelligent Vehicles Department, FSUE "NAMI", Moscow, 125438, Russia

Abstract. This paper is dedicated to a solution of the issue of synthesis of the vehicle longitudinal dynamics control functions (acceleration and deceleration control) based on the element base of the vehicle active safety system (ESP) – driverless vehicle development tool. This strategy helps to reduce time and complexity of integration of autonomous motion control systems (AMCS) into the vehicle architecture and allows direct control of actuators ensuring the longitudinal dynamics control, as well as reduction of time for calibration works. The "vehicle+wheel+road" longitudinal dynamics control is complicated due to the absence of the required prior information about the control object. Therefore, the control loop becomes an adaptive system, i.e. a self-adjusting monitoring system. Another difficulty is the driver's perception of the longitudinal dynamics control process in terms of comfort. Traditionally, one doesn't pay a lot of attention to this issue within active safety systems, and retention of vehicle steerability, controllability and stability in emergency situations are considered to be the quality criteria. This is mainly connected to its operational limits, since it is activated only in critical situations. However, implementation of the longitudinal dynamics control in the AMCS poses another challenge for the developers – providing the driver with comfortable vehicle movement during acceleration and deceleration – while the possible highest safety level in terms of the road grip is provided by the active safety system (ESP). The results of this research are: universal active safety system – AMCS interaction interface; block diagram for the vehicle longitudinal acceleration and deceleration control as one of the active safety system's integrated functions; ideology of adaptive longitudinal dynamics control, which enables to realize the deceleration and acceleration requested by the AMCS; algorithms synthesised; analytical experiments proving the efficiency and practicability of the chosen concept.

1. Introduction

According to article [7], active safety is "a capability of a vehicle to prevent a road accident or to reduce its probability." The main functional aim of the active safety systems is to prevent critical situations and maintain steerability and stability of the vehicle movement – transport task fulfilment. At the very beginning, implementation of active safety systems caused a reduction of number of road accidents and severity of their consequences. However, the feeling of being safe changed the driver's mind and behavior, which caused stagnation and subsequent growth of number of road accidents. This unpleasant situation made engineers move to the next level of the road safety improvement – advanced driver assist systems (ADAS), semi-autonomous motion control systems (SAMCS) and tendency towards introduction of pilotless or driverless vehicles. In regulatory documents [15], there are six stages of driverless vehicle development requiring automatic vehicle acceleration and deceleration control.

However, active safety systems are and will remain the crucial element in the vehicle, because the maneuvers calculated by the AMCS cannot guarantee stable "vehicle+wheel+road" object motion.

There are papers [4, 5] dedicated to definition of limits or borders of sustainable movement in relation to the performance of the actuators. Nevertheless, the issues of autonomous driving and possibility to achieve the desired trajectories under the conditions of steerability and stability of the vehicle are still open.

2. Interface description

The authors proposed a set of rules, which provide interaction between the active safety systems and AMCS – hereinafter referred to as the universal interface – to control the longitudinal dynamics of the vehicle. The functional division of the tasks between these two systems dictates the necessity of such an interface. The AMCS task is to calculate the desired motion trajectory based on the data from computer vision and navigation systems, without taking into account the information about dynamic parameters of the vehicle, and the task of the active safety system is to provide the most efficient and safest driving along this trajectory.

Figure 1 shows the proposed interface for the acceleration control, where the primary control signal is the "required (target) acceleration" (m/s^2). According to the conditions of safe driving within the traffic, the AMCS calculates upper and lower limits of acceleration, which represent the acceptable error in the acceleration adjustment. Since, in the first approximation, the "vehicle+wheel+road" model is an inertial filter and features a transition process, its quality is set by the AMCS using a higher-order derivative – "a jerk (tug)" (m/s^3). This approach allows obtaining of the controllable transition process, which is comfortable for the user.

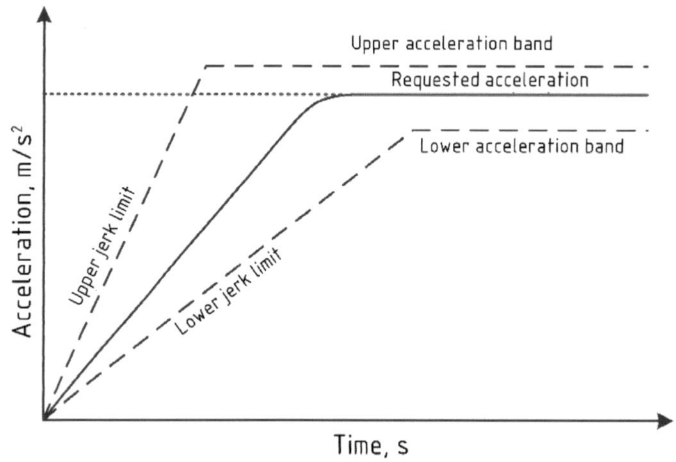

Figure 1. Universal acceleration control interface illustration.

3. Function description

Speed control, as well as realization of the acceleration or deceleration requested by the AMCS are preformed by active safety systems through the management of the engine torque or actuating fluid pressure in the brake mechanisms. Assessment of the "vehicle+wheel+road" object behavior is performed via the same information channel.

The management of the "vehicle+wheel+road" object dynamics is complicated because of the lack of prior information. Absence of initial data about the object causes the need to combine, in some ways, monitoring of the object and its management. In this case, the control actions are of a dual nature – they serve as a means of object monitoring and investigation and as a means to achieve the desired (i.e. optimal) condition – this is called "a dual control" [8-11].

Optimality of the longitudinal dynamics control can be formulated based on the primary aim of this control – ensure the vector change according to the requested values. Then, when there is no exact information, the optimum will be the minimum error between the requested and the actual realized characteristic of the phase vector change. This result can be obtained if the longitudinal dynamics control system is implemented within the class of adaptive systems [1]. The control part of the adaptive loop (figure 2) consists of two parts or elements: consistently-correcting contour and parallel-correcting contour. The first element includes adjustable parameters, the second one – constants.

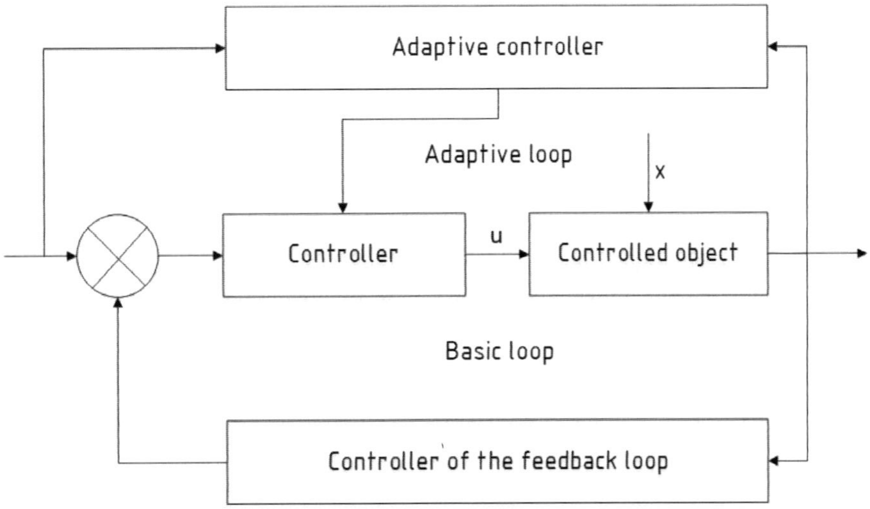

Figure 2. Structure diagram of adaptive loop.

The parallel-correcting loop is made as a program controller [2] using some basic models of the "vehicle+wheel+road" object [6, 13], which were defined by the condition of the equality of the forces applied to the vehicle.

$$m_a \frac{dV_v}{dt} = \sum_{i=1}^{n} R_{x_i} + W_L + W_H \tag{1}$$

where: m_a – vehicle weight;
V_v – linear speed;
R_x – tangential force applied to the tire-road contact area;
W_H – uphill movement resistance;
W_L – other movement resistances.

The consistently-correcting loop is a self-adjusting loop, which takes the minimum error of the phase vector as a qualitative criterion. The optimality based on a control error needs invariability of the parameter used. The works on the design of active safety systems [12] show that vehicle linear speed can be used as a parameter for the self-adjusting loop to analyse the vehicle motion. Then, the qualitative criterion takes the following form:

$$e = V_{st-st} - V_v \rightarrow min \tag{2}$$

where: V_{st-st}– desired vehicle speed;
V_v – vehicle speed.

IOP Conf. Series: Materials Science and Engineering **315** (2018) 012021 doi:10.1088/1757-899X/315/1/012021

The developed interface allows generating the signal of the desired speed (V_{st-st}) in the form of a difference equation (3). The need of a difference equation is caused by the problem of obtaining the second-order derivative in a technical system – this problem represents a separate task.

$$V_{st-st}(i+1) = \alpha\left(\widehat{V}_e(i) + \zeta j_{req}(i)\Delta t\right) + (1-\alpha)V_{sens}(i) \quad (3)$$

where: $\widehat{V}_e$– vehicle speed estimation (recommended speed);
j_{req}– requested target acceleration;
Δt – time interval;
α & ζ – correction coefficients;
V_{sens}– vehicle speed based on the wheel speed sensor data.

The correction coefficient α is introduced to allow usage of information from the wheel speed sensors in case of steady-state motion. The coefficient ζ is introduced based on the "jerk" parameters and is defined as follows:

$$\zeta = \begin{cases} 0, & if\ j_{st-st} - j_{req} \leq \varepsilon \\ \dfrac{j_{st-st}}{j_{req}}, & if\ j_{st-st} - j_{req} > \varepsilon \end{cases} \quad (4)$$

where: j_{st-st} – steady-state acceleration based on the "jerk";
ε – acceptable error.

4. Results

The results of the mathematical modelling (figure 3) of the function of the vehicle longitudinal dynamics control, which was performed in the MatLAB Simulink software system, show feasibility of the proposed interface. Figure 3b shows that the error between the requested and actual acceleration is 3%. Also, the vehicle speed estimation process (figure 3a) has a high convergence level.

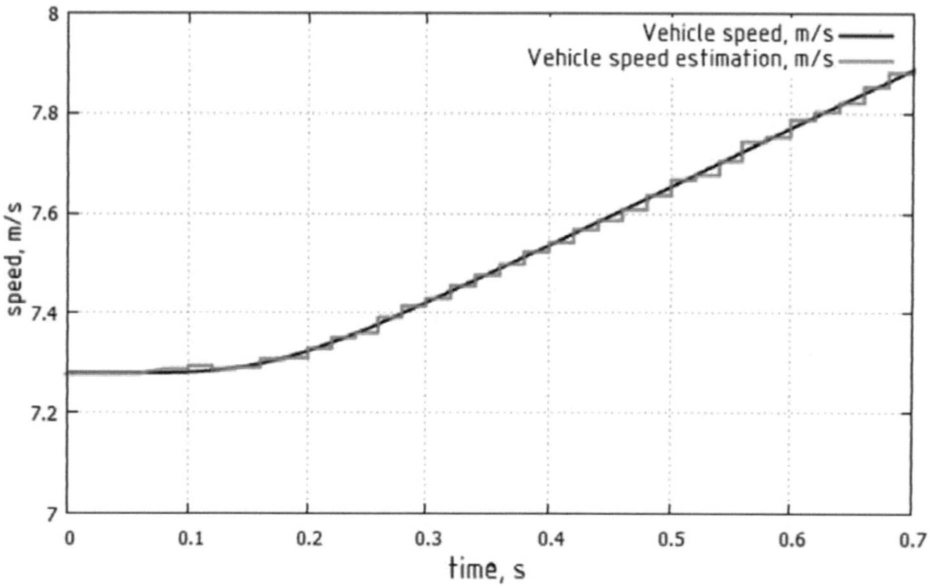

a)

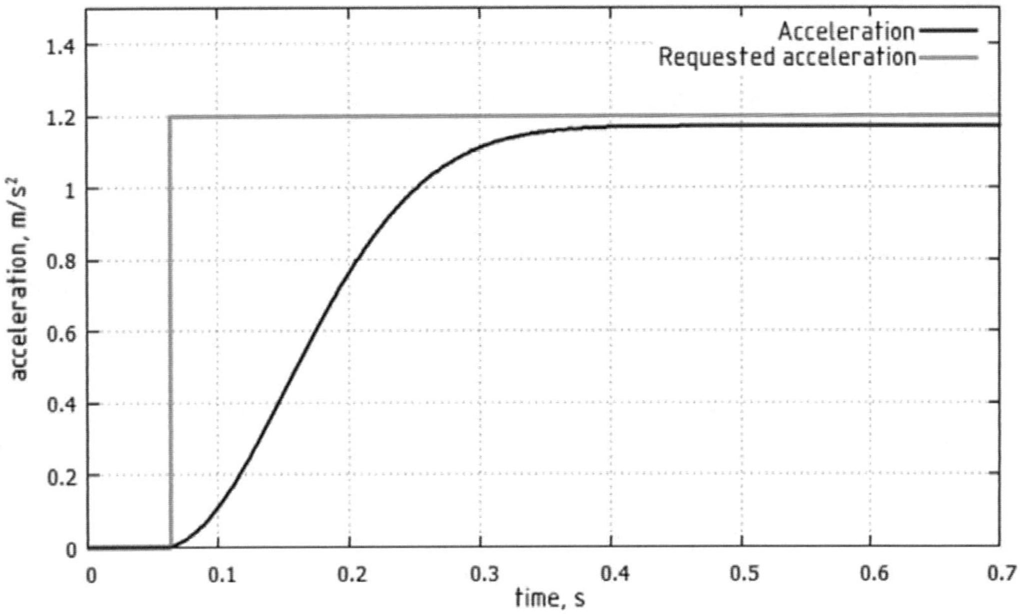

b)

Figure 3. Results of mathematical modelling of function of vehicle longitudinal dynamics control: a) vehicle speed dynamics; b) acceleration dynamics.

5. Summary

The proposed interface allows the AMCS to use physical values with a high-order derivative – acceleration/deceleration and acceleration/deceleration change rate ("jerk") – as the desired motion parameters, which allows the system to focus on its functional task. Conversion of the desired parameters on the part of the AMCS into the control actions for the vehicle actuators (actuating fluid pressure in the brake system and engine torque) is performed within the active safety system with ensuring the condition of vehicle stable motion based on the condition of the tire-road friction coefficient in the longitudinal direction. Technical implementation within the active safety systems requires conversion of the requested parameters into the speed vector of the vehicle and usage of adaptive control to manage the actuators. Usage of the "jerk" allows ensuring of controllable transition process, which is comfortable for the user, and preventive estimation of feasibility of the AMCS request.

References

[1] Aleksandrov A G 1989 *Optimal and Adaptive Systems* (Moscow: Vyssh. shk.) p 263
[2] Besekerskij V A and Popov E P 1972 *Theory of Automatic Control Systems* (Moscow: Nauka) p 768
[3] GOST 15971-90 Information processing systems Terms and definitions
[4] Debelov V V, Endachev D V, Evgrafov V V and Bahmutov S V 2017 Ways of implementation of vehicle driving mode control systems based on development of autonomous driving technologies *Proceedings of 99th International Scientific and Technical Conference* Irkutsk National Research Technical University pp 504-12
[5] Endachev D V 2016 Forecasting of curvilinear motion characteristics of driverless vehicle Candidate of Engineering Science dissertation: 05.05.03 (Moscow) p 183
[6] Smirnov G A 1990 *Wheeled vehicle moving theory*/ Smirnov G A *Mashinostroenie* (Moscow) p 352

[7] Sockov D A 1988 Improvement of vehicle active safety when braking Doctor of Engineering Science dissertation 05.05.03 (Vladimir) p 547
[8] Feldbaum A A 1960 *Dual control theory I Avtomatika i telemehanika* **21** №9
[9] Feldbaum A A 1960 *Dual control theory II Avtomatika i telemehanika* **21** №11
[10] Feldbaum A A 1961 *Dual control theory III Avtomatika i telemehanika* **22** №1
[11] Feldbaum A A 1961 *Dual control theory IV Avtomatika i telemehanika* **22** №3
[12] Nefedyev Ya N 1987 Theory, development and research of unified system for automatic control of anlilock braking of cargo vehicles Doctor of Engineering Science dissertation 05.05.03 (Moscow) p 307
[13] Edited by G. Mastinu and M. Ploechl 2014 *Road and off-Road Vehicle System dynamics Handbook* by Taylor & Francis Group LLC
[14] Hart P M 2003 Review of heavy vehicle braking systems Requirements (PBS requirements) draft report
[15] SAE3016 Taxonomy and definitions for terms related to driving automation systems for on-road motor vehicles (http://standards.sae.org/j3016_201609)

IOP Conf. Series: Materials Science and Engineering **315** (2018) 012022 doi:10.1088/1757-899X/315/1/012022

Distributed radar-based monitoring system for intelligent vehicles

A G Ryndyk, A V Myakinkov and S V Shishanov

Information Radiosystems Department, Nizhny Novgorod State Technical University n.a. R.E. Alekseev, 603950, N.Novgorod, Minin str., 24

E-mail: rs@nntu.ru

Abstract. The following article introduces a model of a distributed radar-based monitoring system for intelligent vehicles which is based on surround view monitoring and measuring the dimension of objects. Our model employs ultrawideband (UWB) transceiver modules with nearly omnidirectional antennas. We also suggest that the coordinates of the detected objects are to be measured using the information provided by several transceiver modules. The article shows that this system allows measuring the coordinates and dimensions of the objects detected around the vehicle with accuracy of 10 cm. The suggested model of the radar system will help provide an accurate and detailed estimation of the traffic conditions and increase the traffic safety.

1. Introduction

The end of the XX century witnessed emerging of the automotive radar systems market. Today car manufacturers use the radar systems as a part of the driver assistance systems which include mostly long range radars (up to 300 m) with the coverage of only up to 30 degrees [1]. Such radars alone cannot meet the safety requirements specified for the modern vehicles due to their coverage, as well as range and azimuth resolution limits. Therefore today we are witnessing the development of the systems based on the UWB short range radars. To insure the proper development, in 2004 European Telecommunications Standards Institute (ETSI) published standards for the automotive short range radars (up to 30m) operating in the 77-81 GHz band. Measuring the coordinates and the dimensions of the detected objects is essential both in city traffic and on cross-country roads. Any obstacles such as stones, fallen trees, sinkholes on cross-country roads, or road safety barriers, speed bumps, pedestrians or other vehicles in the city must be detected and accurately measured by the radar systems. For these purposes radar modules and antennas with 120 degree radiation pattern are usually employed as UWB transceiver modules [2].

Distributed radar-based monitoring system. This monitoring system can be used as a part of the driver assistance system, while the coordinates and dimensions measuring can serve the following purposes:

- to keep the vehicle within the driving lane;
- to ensure a safe vehicle maneuver, including safe passage of uncontrolled road intersections or pulling out of minor roads;
- to autonomously ensure collision avoidance;

- to detect objects (a pedestrian, a cyclist, or an animal);
- to assist parking.

Figure 1 shows the general structure of our model.

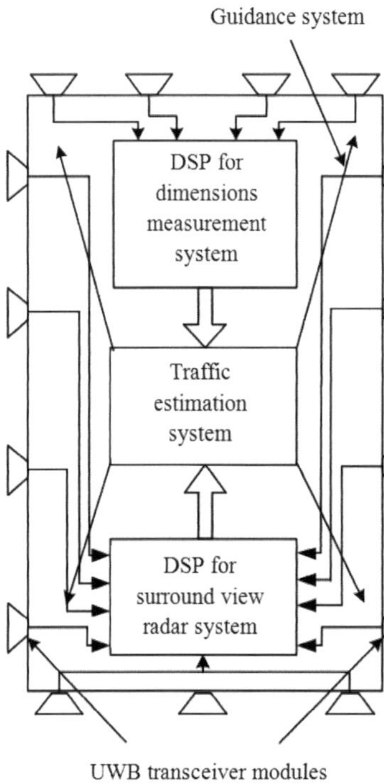

Figure 1. Block diagram of the radar-based monitoring system.

UWB antennas which provide the dimension measurements of the detected obstacles are found in the front part of the car. UWB short range surround view antennas are mounted all around the vehicle at regular intervals.

Data signal processing system includes DSP for dimensions measurement system as well as DSP for side view radar system. All the information on the detected objects, their coordinates and motion rates is processed in the traffic estimation system, which in an emergency will trigger the automatic braking system, lane keeping assist system, dynamic stability control, and anti-lock braking systems [3].

2. Surround view system

To ensure a 360° surround view the UWB transceiver modules must be mounted all around the car so that the blind spot, which is an area around the vehicle that cannot be directly observed by any of the modules, is eliminated as much as possible. Due to the specific design features of the car body, the number of the spots to locate UWB transceivers is very limited. Usually they are found under the front and rear bumpers, in the door handles, outside mirrors and fenders.

Coordinate measurements can be carried out using the well-known methods of calculating the intersection points of the position lines of the objects [4 – 8]. If there is only one object detected, its coordinates can be measured using the distance information provided by the two opposite-located transceivers. But when dealing with multiple targets, at least three transceivers are required to provide the distance measurements; otherwise the data is inconclusive. Figure 2 provides the transceiver location pattern. Numbers 1 and 2 represent the blind spot and the area which allows only the distance measurements. Accurate coordinate measurement of the objects detected in area, represented by number 2, cannot be carried out due to the object not being seen by at least three transceivers [7, 8].

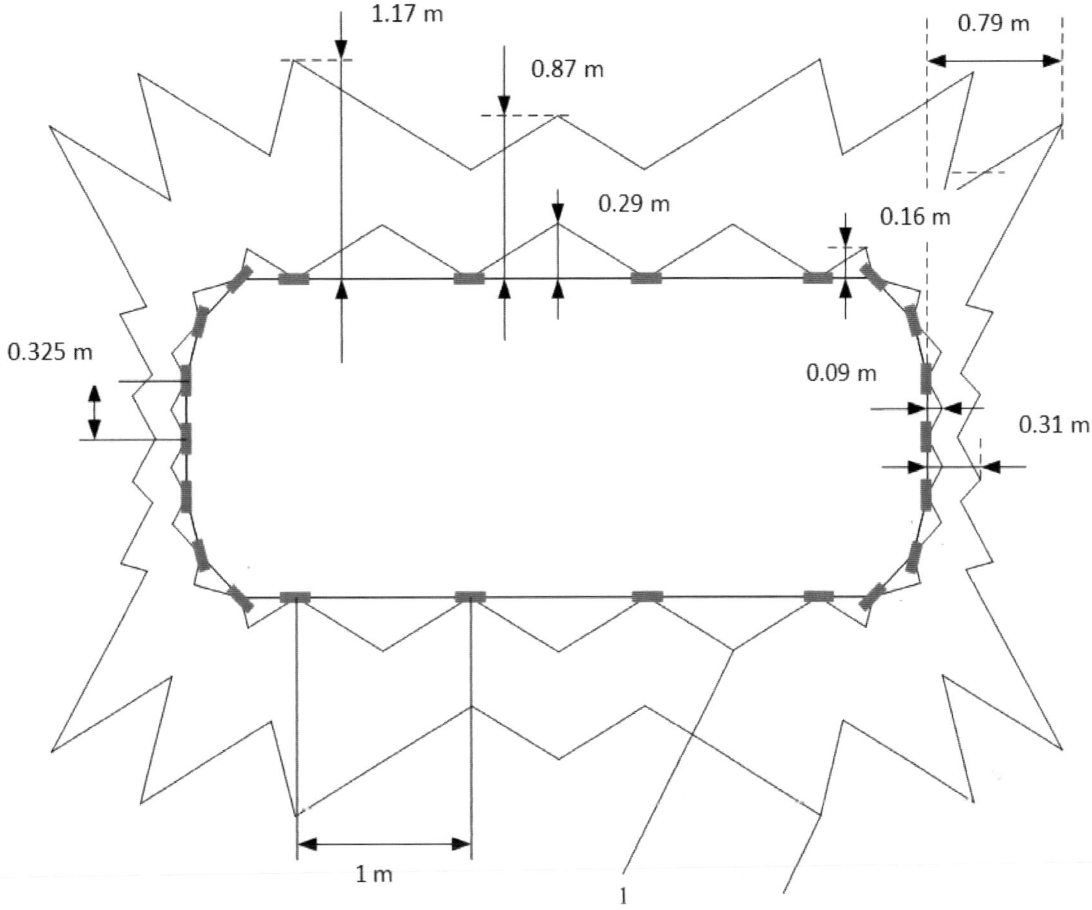

Figure 2. Blind spot and the area allowing distance measurements only.

On the front and rear bumper the UWB transceivers are mounted every 31.5 cm, while on either sides the interval is 1 m. The operating range of the system depends on the characteristics of the transceiver modules and normally stands at about 30 m [2].

Figure 3 provides the RMS errors in coordinate (x (a) и y (b)) measurement when using the distance measurement method with the x coordinate in case of a constant y = 20 m [8]. Numbers 1 and 2 represent the variations of the root standard deviation of the coordinate measurement error when using the long range method in case the interval between the modules is 2 m and 63 cm respectively.

Figure 3 shows that with the UWB transceiver modules mounted all around the car at regular intervals the RMS error in coordinate estimate is about 10 cm.

3. Dimensions measurement system in the front of the vehicle

Most of the detected objects are three-dimensional. Such objects can be presented as a number of independent point scatterers, generally referred to as highlights [9]. The range resolution of our model allows resolving the highlights of the particular objects such as road users, road infrastructure objects, or other obstacles.

Figure 4 (a) provides a geometric representation of a three-dimensional radar target. Generally the targets are not radar-transparent; therefore practically it's only possible to measure the height h and the lateral size a of an object. To make it easier to visualize we chose an objects with the highlights located along the two orthogonally related intersecting lines, which are represented by z and y axes in figure 4 (b).

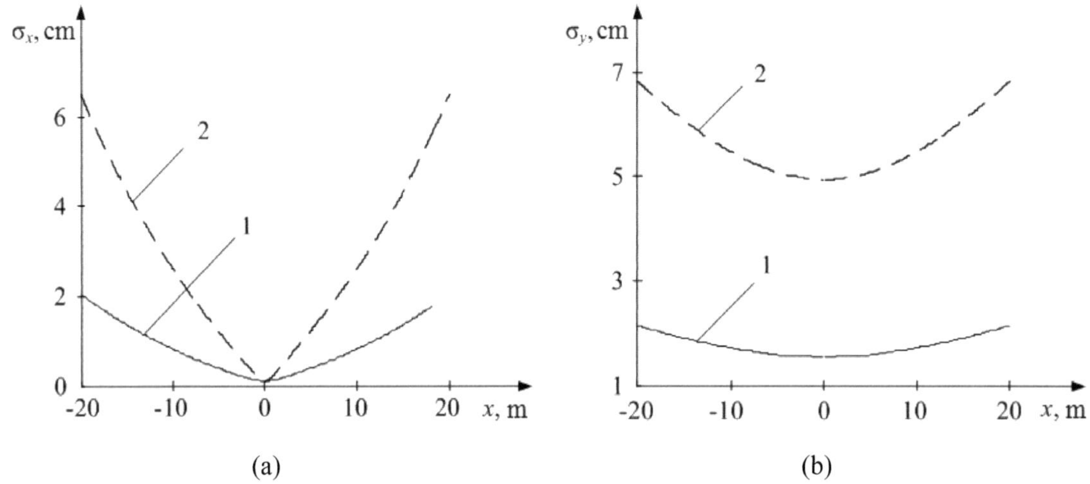

Figure 3. RMS errors in coordinate measurement when using the distance measurement method with the x coordinate in case of a constant y.

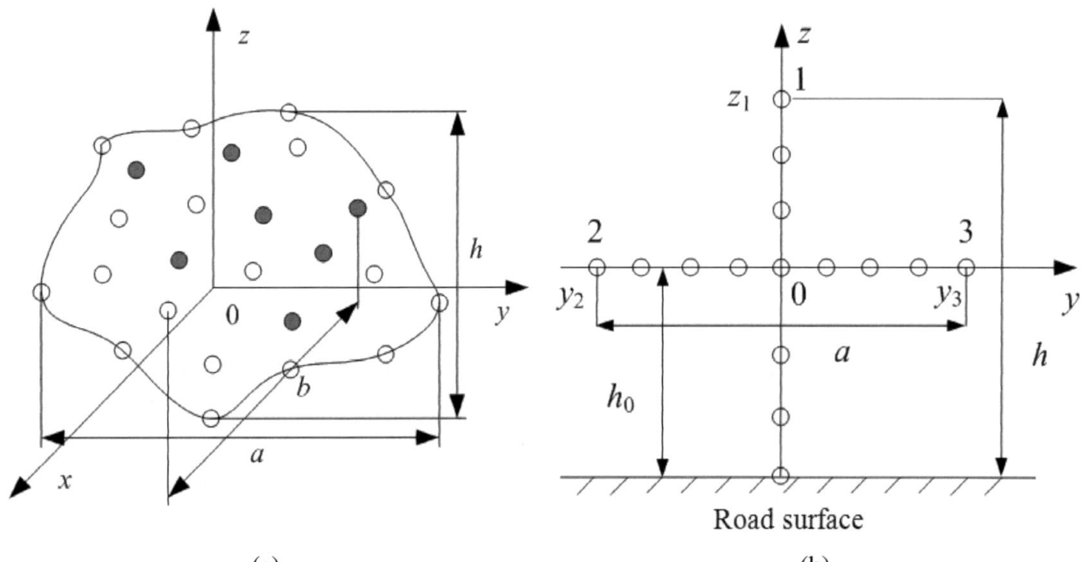

Figure 4. A model of a three-dimensional object.

h – height of the object;
h_0 – the distance between the road surface and the center of the coordinate system;
a – lateral size of the object;
b – longitudinal size;
z_1 – z coordinate of highlight No.1;
y_2 and y_3 – y coordinates of highlights No. 2 and 3 respectively.

Estimation of the highlights' coordinates allows us to estimate the dimensions of the detected objects. Normally the height of the target is measured with respect to some level, for example, the road surface for a vehicle radar system. Thus the height h of the target is equal to the z coordinate of the maximum value with respect to the road surface ($h = z_1 + h_0$). The lateral size of the object is equal to the difference between the y coordinates of the maximum and minimum values ($a = y_3 - y_2$) (figure 4 (b)).

Estimation of the coordinates of the highlights can be done by estimating the distance measurements by at least three transceiver modules. Figure 5 provides an image of a passenger car with UWB transceiver modules which are found in its front, particularly in its outside mirrors, the rearview mirror and the front bumper.

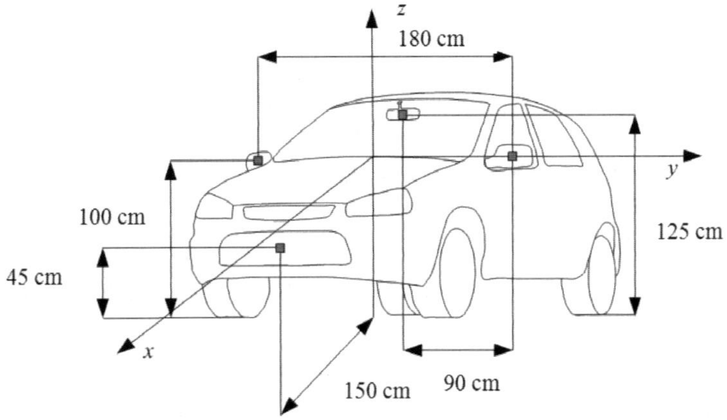

Figure 5. A passenger car with UWB transceiver modules.

The coordinates of the object can be estimated based on the distance measurements provided by three transceivers, position surfaces being spheres with radiuses R_1, R_2, and R_3 respectively. But the distance measuring method tends to give uncertain results in the situation with multiple targets. To avoid the uncertainty we use the information from four transceivers.

Figure 6 provides a system scheme, where 1, 2, 3 and 4 stand for receiving positions, T – for the target (or detected object), x_T, y_T, z_T – target coordinates, R_1, R_2, R_3 and R_4 – distance between the target and the receiving positions.

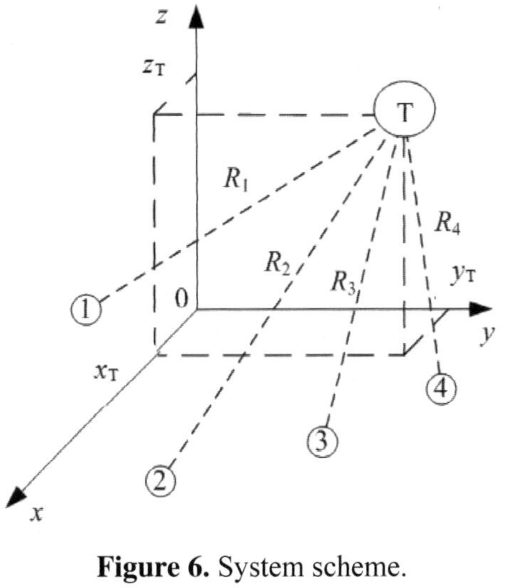

Figure 6. System scheme.

IOP Conf. Series: Materials Science and Engineering **315** (2018) 012022 doi:10.1088/1757-899X/315/1/012022

The coordinates of the target using the distance measuring method can be calculated by solving simultaneous equations:

$$\begin{cases} A_1 \cdot x^2 + A_2 \cdot x + A_3 \cdot y^2 + A_4 \cdot y + A_5 \cdot z^2 + A_6 \cdot z + A_7 = 0, \\ B_1 \cdot x^2 + B_2 \cdot x + B_3 \cdot y^2 + B_4 \cdot y + B_5 \cdot z^2 + B_6 \cdot z + B_7 = 0, \\ C_1 \cdot x^2 + C_2 \cdot x + C_3 \cdot y^2 + C_4 \cdot y + C_5 \cdot z^2 + C_6 \cdot z + C_7 = 0, \end{cases} \tag{1}$$

where coefficients A_1, A_2, A_3, A_4, A_5, A_6, A_7, B_1, B_2, B_3, B_4, B_5, B_6, B_7, C_1, C_2, C_3, C_4, C_5, C_6, C_7 are defined depending on the type of position surfaces which are actually second-order surfaces. The system of equations (1) can be solved in 8 different ways.

4. Accuracy of dimension measurements

Height h of the target is equal to the z coordinate of the maximum value with respect to the road surface (figure 4 (b)), while the lateral size of the object a accounts to the difference between the coordinates of the two highlights $y_3 - y_2$. Thus the accuracy of the height measurement depends upon the accuracy of the z coordinate estimation, while the accuracy in lateral size estimation depends upon the accuracy of the y coordinate estimation.

Assume that distance measurements for the highlights are not defined by any other conditions. This assumption is valid if in all the transceiver modules the highlights are resolved by distance estimates. Hence the estimates of y_2 and y_3 coordinates (figure 4 (b)) are independent either, and lateral size measurement error can be calculated as follows

$$\sigma_a = \sqrt{\sigma_{y_2}{}^2 + \sigma_{y3}{}^2}, \tag{2}$$

where σ_{y_2} and σ_{y_3} stand for the RMS errors in the y_2 and y_3 coordinate measurement.

Let us next resolve the RMS errors in the y and z coordinate measurement assuming that distance measurement errors are independent zero-mean Gaussian random variables.

Let us expand the coordinates estimation expressions, resolved from the simultaneous equations (1), into Taylor-series in the neighborhood of their true value. Using only those terms which have the first order derivatives, an expression for the RMS errors in coordinate measurement is as follows

$$\begin{aligned} \sigma_y &= \sqrt{\left(\frac{\partial y}{\partial \xi_1}\right)^2 \sigma_{\xi_1}^2 + \left(\frac{\partial y}{\partial \xi_2}\right)^2 \sigma_{\xi_2}^2 + \left(\frac{\partial y}{\partial \xi_3}\right)^2 \sigma_{\xi_3}^2}, \\ \sigma_z &= \sqrt{\left(\frac{\partial z}{\partial \xi_1}\right)^2 \sigma_{\xi_1}^2 + \left(\frac{\partial z}{\partial \xi_2}\right)^2 \sigma_{\xi_2}^2 + \left(\frac{\partial z}{\partial \xi_3}\right)^2 \sigma_{\xi_3}^2}, \end{aligned} \tag{3}$$

where $\partial y/\partial \xi_1$, $\partial y/\partial \xi_2$, $\partial y/\partial \xi_3$, $\partial z/\partial \xi_1$, $\partial z/\partial \xi_2$, $\partial z/\partial \xi_3$ represent partial derivatives of the y and z coordinates with primary measured parameters ξ_1, ξ_2 and ξ_3, while $\sigma_{\xi 1}, \sigma_{\xi 2}$ and $\sigma_{\xi 3}$ represent the RMS errors in in defining the primary parameters ξ_1, ξ_2 and ξ_3.

For the distance measuring method of coordinates estimate, the primary measured parameters are the distances to the object highlights: $\xi_1 = R_1$, $\xi_2 = R_2$, $\xi_3 = R_3$. Hence $\sigma_{\xi_1} = \sigma_{\xi_2} = \sigma_{\xi_3} = \sigma_R = c\sigma_\tau$, where c stands for the speed of light, $\sigma_\tau = 1/(q_{out} \cdot \Delta\omega_{eff.})$ – for the potential accuracy in signal delay measurement, $q_{out} = \sqrt{2E/N_0}$ is the maximum potential of signal to noise ratio at the outcome of the signal drive, where the signal is being stored coherently, E represents the received signal energy, N_0 – noise power spectral density, $\Delta\omega_{eff.}$ – the effective signal bandwidth.

To estimate the efficiency of the dimensions measuring we analyzed the UWB transceiver modules system with the modules mounted all around the vehicle (figure 6), assuming, the effective UWB signal bandwidth $\Delta f_{\text{eff.}} = 4$ GHz, proper detection expectation $D = 0.9$ and false alarm expectation $F = 10^{-6}$, signal to noise ratio at the outcome of the signal drive upon the given D and F values $q_{\text{out}} = 11.39$.

Figure 7 provides the RMS errors in estimating the height of the object using distance measuring method as a function of x coordinate and the height of the object h. For this we assumed that: 1) $x = 30$ m, $y = 0$ m (figure 7 (a)), 2) $y = 0$ m, $h = 1.8$ m (figure 7 (b)).

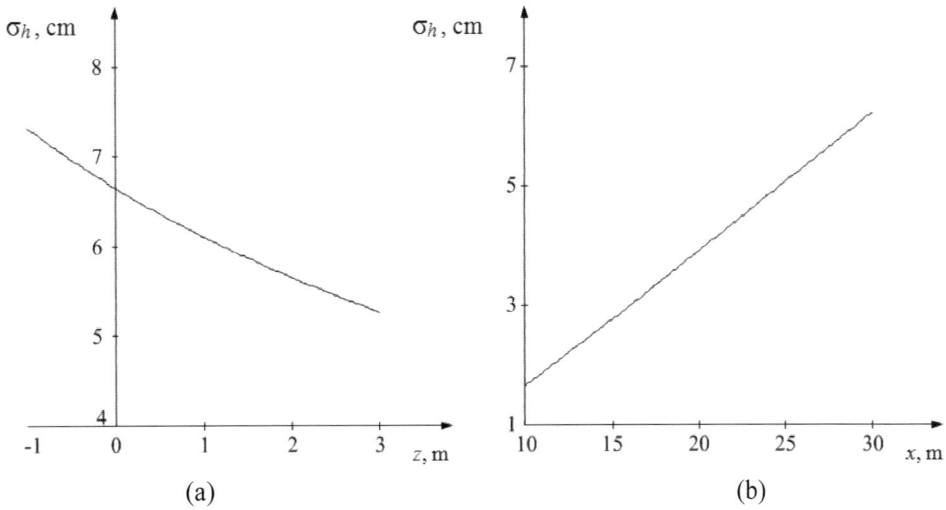

Figure 7. RMS errors in estimating the height of the object using distance measuring method as a function of x coordinate and the height of the object h.

Figure 7 shows that the less is the height of the object with x coordinate being constant, or the higher x coordinate value with the height being constant, RMS errors in estimating the height is always increasing, as the angle between the osculating planes to the position surfaces, drawn through their intersection point, is reduced (α and β angles in figure 8).

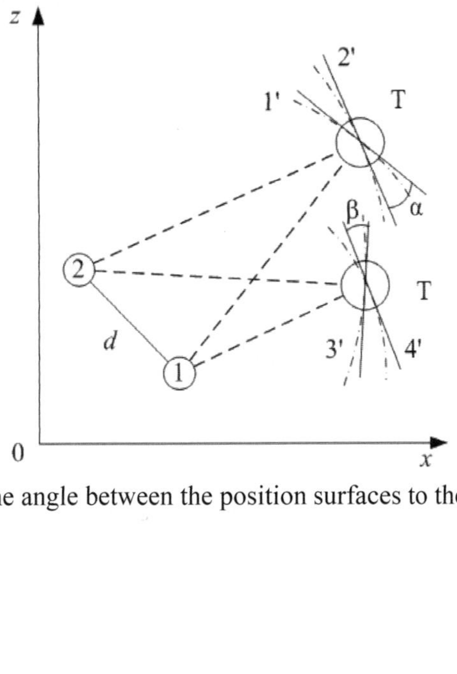

Figure 8. Ratio of the angle between the position surfaces to the height of the target.

1, 2 – transceiver modules;
T – target;
α and β – angles between 1` and 2`, 3` and 4` position surfaces respectively (α > β).

Figure 9 provides RMS errors in measuring lateral size of the object using the distance measuring method to define the highlights coordinates as a function to the lateral size a of an object and x coordinate value. For this we assumed that: 1) x = 30 m, h = 1 m (figure 9 (a)), 2) a = 0.5 m, h = 1.6 m (figure 9 (b)).

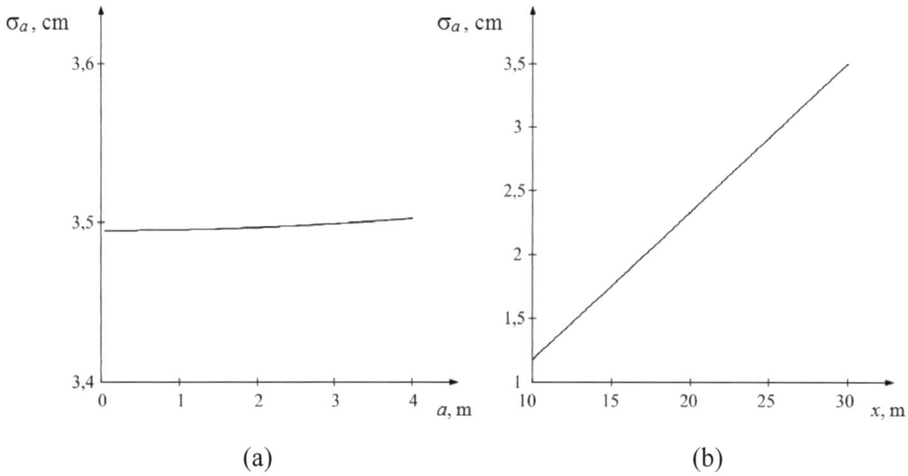

Figure 9. RMS errors in measuring lateral size of the object using the distance measuring method as a function to the lateral size and x coordinate value.

Figure 9 (a) shows that the higher lateral size value with x coordinate being constant has few influence on the RMS errors in estimating the lateral size. This can be explained by the fact that the distance between the transceiver positions along the y axe is comparable to the lateral size of an object, and angles between the osculating planes to the position surfaces nearly do not vary with the changes of the lateral size of an object. Figure 9 (b) shows that the higher is the x coordinate value with the lateral size being constant, the higher is the value of RMS errors in measuring lateral size, as the angle between the osculating planes to the position surfaces, drawn through the intersection point, is reduced.

5. Conclusion

Our model based on dimension measuring and surround view monitoring allows real-time monitoring of the environment. This model of vehicle radar system ensures detailed information about the traffic conditions and increases the traffic safety.

References

[1] Chernyak V S 1993 *Mnogopozitsionnaya radiolokacziya* Moscow p 416 (in Russian)
[2] *Implementing Digital Processing for Automotive Radar Using SoC FPGAs* 2013 (Altera Corporation) p 16
[3] Kondrat'yev V S, Kotov A F and Markov L N 1986 *Mnogopozitsionnyye radiotexnicheskiye sistemy* (Moscow) (in Russian)
[4] Myakinkov A V and Smirnova D M 2010 *Measurement of coordinates of the targets placed behind of radio-transparent barrier with multi-static ultra-wide band radar* Ultrawideband and Ultrashort Impulse Signals (Sevastopol, Ukraine) pp 1-3
[5] Ostrovitjanov R V and Basalov F A 1982 *Statisticheskaja teorija radiolokacii protjazhennyh celyei* (Moscow) p 55-60 (in Russian)

[6] 2012 Recommendation M.1452. *Millimetre Wave Vehicular Collision Avoidance Radars and Radiocommunication Systems for Intelligent Transport System Applications* (Geneva) p 9
[7] Schneider M 2005 Automotive Radar – Status and Trends *Proceeding of German Microwave Conference* (Ulm, Germany) pp 144–7
[8] Shishanov S V and Myakinkov A V 2016 *Bespilotnye transportnye sredstva: problemy i perspektivy* (Nizhnij Novgorod) pp 72–8 (in Russian)
[9] Shishanov S V and Myakinkov A V 2015 Sistema krugovogo obzora transportnyh sredstv na osnove sverhshirokopolosnyh datchikov *Izvestija vuzov Rossii. Radiojelektronika* No 2 pp 55-60 (in Russian)

IOP Conf. Series: Materials Science and Engineering **315** (2018) 012023 doi:10.1088/1757-899X/315/1/012023

The mathematical model of dynamic stabilization system for autonomous car

A M Saikin, S E Buznikov, N S Shabanov and D S Elkin

FSUE "NAMI", Moscow, Russia

E-mail: sergey.buznikov@nami.ru

Abstract. Leading foreign companies and domestic enterprises carry out extensive researches and developments in the field of control systems for autonomous cars and in the field of improving driver assistance systems. The search for technical solutions, as a rule, is based on heuristic methods and does not always lead to satisfactory results. The purpose of this research is to formalize the road safety problem in the terms of modern control theory, to construct the adequate mathematical model for solving it, including the choice of software and hardware environment. For automatic control of the object, it is necessary to solve the problem of dynamic stabilization in the most complete formulation. The solution quality of the problem on a finite time interval is estimated by the value of the quadratic functional. Car speed, turn angle and additional yaw rate (during car drift or skidding) measurements are performed programmatically by the original virtual sensors. The limit speeds at which drift, skidding or rollover begins are calculated programmatically taking into account the friction coefficient identified in motion. The analysis of the results confirms both the adequacy of the mathematical models and the algorithms and the possibility of implementing the system in the minimal technical configuration.

1. Introduction

At present, practically all leading foreign companies and domestic enterprises carry out extensive researches and developments in the field of control systems for autonomous cars and in the field of improving driver assistance systems [1]. The search for technical solutions, as a rule, is based on heuristic methods and does not always lead to satisfactory results. Thus, the known electronic stability control (ESC) and velocity control system (VCS) have non-removable deficiencies, which does not allow to effectively use their capabilities in the road and climate conditions of Russian Federation.

The electronic stability programs use cyclic operation (like ABS) with separate brake control. The cyclic operation initiates sharp decrease in braking efficiency and stability stabilization on uneven surfaces such as "washboard", on paving stones, as well as on ice and snow-covered surfaces. On sandy surfaces, when driving in a track, the rollover speed is lower than the drift speed even for cars with low center of gravity.

For vehicles with high center of gravity such as SUVs, minivans the situation occurs on smooth asphalt surfaces. For these vehicles, dynamic stabilization systems were developed based on predicting the rate of rollover, skidding and drift [2,6]. However, the lack of technology to identify the top values of the friction coefficients in changing driving conditions required an underestimation of their estimates and, accordingly, the predicted speeds of skidding and drift.

The purpose of the paper is to formalize the problem of improving vehicle's stability in terms of modern control theory, working-out adequate mathematical model for its solution, including the choice of software and hardware.

2. The formulation of vehicle's stability problem

At the conceptual level, vehicle's stability problem is reduced to predicting the excess of the limit speeds of skidding, drifting and rollover and to formation the control actions on traction and brakes, preventing the occurrence of these events.

At the content level, vehicle's stability problem follows from the solution of heading angle Ψ_m differential equation:

$$\dot{\Psi}_m = \omega_m + \Delta\omega_m, \tag{1}$$

where ω_m is the yaw rate of a car at a turn;

$\Delta\omega_m$ is the additional component of yaw rate at wheel drifts.

The solution of (1) with allowance $\omega_m = b^{-1}V_m \Psi_c$ in the discrete time can be represented in the form:

$$\Psi_m(k) = \Psi_m(k-1) + b^{-1}\int_{t_{k-1}}^{t_k} V_m(\tau)\Psi_c(\tau)d\tau + \int_{t_{k-1}}^{t_k} \Delta\omega_m d\tau. \tag{2}$$

The increment of heading angle $\Delta\Psi_m(k)$ at the kth step is:

$$\Delta\Psi_m(k) = \Psi_m(k) - \Psi_m(k-1) = b^{-1}\int_{t_{k-1}}^{t_k} V_m(\tau)\Psi_c(\tau)d\tau + \int_{t_{k-1}}^{t_k} \Delta\omega_m d\tau. \tag{3}$$

In the case of rear wheels drift (oversteering), $\Delta\omega_m$ coincides in sign with Ψ_c and increases the increment of the heading angle. In the case of front wheels drift (understeering), $\Delta\omega_m$ has the opposite sign Ψ_c and reduces the module of heading angle increment. Additional rotation with yaw rate $\Delta\omega_m$ occurs around the center of front wheels axle at rear wheels drift and around center of rear wheels axle at front wheels drift. The reason for these phenomena is the excess of the centrifugal force acting on the front and rear wheels, the frictional forces values in the lateral direction.

In the case of rollover, the torque created by the centrifugal force is greater than the returning torque, caused by the force of gravity acting on the center of mass. The centrifugal force is proportional to the square of the speed, which makes it possible to reduce the conditions for preventing drifts and overturning to the system of inequalities for the speed:

$$\begin{cases} -V_{Lm0} \le V_m \le V_{Lm0}; \\ -V_{Lm1} \le V_m \le V_{Lm1}; \\ -V_{Lm2} \le V_m \le V_{Lm2}, \end{cases} \tag{4}$$

where V_{Lm0} is rollover limit speed ($\mathrm{m \cdot s^{-1}}$);

V_{Lm1} is the limit speed of front wheels drift ($\mathrm{m \cdot s^{-1}}$);

V_{Lm2} is the limit speed of rear wheels drift ($\mathrm{m \cdot s^{-1}}$).

Combining inequalities (4) into one allows one to reduce the problem of preventing these events to the problem of dynamic stabilization:

$$V_{Lm}^{L} \le V_m \le V_{Lm}^{U}, \tag{5}$$

where $V_{Lm}^{U} = \min[V_{Lm0}, V_{Lm1}, V_{Lm2}]$; $V_{Lm}^{L} = \max[-V_{gr0}, -V_{gr1}, -V_{gr2}]$.

Obviously the fulfillment of inequality (5) with the corresponding V_{Lm}^{U} and V_{Lm}^{L} means that inequalities (4) will also be fulfilled, in this case, drift and rollover events will not occur.

From the standpoint of modern control theory, the solution of the dynamic stabilization problem on a finite time interval $(t_1 \div t_2)$ is estimated from the value of control quality quadratic functional, taking into account functional limitations on the technical and economic indicators of the system that implements control:

$$Q(t_2) = \int_{t_1}^{t_2} C_1 \, [V_m(\tau) - V_{Lm}^U(\tau)]^2 \, d\tau + \int_{t_1}^{t_2} C_2 \, [V_m(\tau) - V_{Lm}^L(\tau)]^2 \, d\tau \Rightarrow \min \quad , (6)$$

under $U \in U_{allow}$, $R = (R_H, R_S)^T \in R_{allow}$; $q_i(R) \le q_{i\,allow}$, $1 \le i \le l$, where

$$C_1 = \begin{cases} 0, if\ V_m < V_{Lm}^U; \\ C_1^* \gg 0, if\ V_m \ge V_{Lm}^U; \end{cases}$$

$$C_2 = \begin{cases} 0, \text{если}\ V_m > V_{Lm}^L; \\ C_{2i}^* \gg 0, \text{если}\ V_m \le V_{Lm}^L; \end{cases}$$

R is the vector of technical solutions;

R_H and R_S are vectors of hardware (H) and software (S) solutions;

R_{allow} is allowable set of technical solutions;

$q_i(R)$ is i^{th} component of technic and economic indicators vector;

$q_{i\,allow}$ is allowable value of the ith component.

The number of basic technical and economic indicators include the levels of power consumption, the influence of external factors, universality, fault tolerance, operating costs and the cost of hardware and software package.

The best solution to the problem in the proposed formulation is the algorithm for generating control actions that ensures dynamic stabilization of the object state vector, implemented in a software and hardware environment that satisfies the system of limitations on technical and economic indicators.

3. Results of constructing the mathematical model

The mathematical model of dynamic stabilization system includes equations of linear movement and limit speeds, equation system for wheel speeds on the turn, virtual sensor algorithms for car speed, turn angle, the additional component of yaw rate, for identification top values of friction coefficients between wheels and road surface, for control brakes by actuator.

3.1. The limit speeds of front, rear wheels drift and car rollover

The number of limit speeds in the general case includes next rates: rollover speed (V_{Lm0}), front drift speed (V_{Lm1}), rear drift speed (V_{Lm2}), tire cord breaking speed (V_{Lm3}), driving wheels slip speed (V_{Lm4}), speed of brake overheating (V_{Lm5}), traffic sign speed limit (V_{Lm6}), the speed of front obstacle (V_{Lm7}), driving speed with "limited use" spare tire (V_{Lm8}), driving speed at insufficient tire pressure (V_{Lm9}), etc.

The upper limit of safe speed is defined as $V_{Lm}^U = \min[V_{Lm0}, \dots, V_{Lmn}]$, and the lower limit is defined as $V_{Lm}^L = \max[-V_{Lm0}, \dots, -V_{Lmn}]$. For the positive direction of motion with $V_m \ge 0$ only the upper limit $V_{Lm}^U \ge 0$ has physical meaning.

The equation of rollover speed V_{Lm0} for horizontal surface is obtained from sufficient equilibrium condition (equality of tipping and returning torques):

$$V_{Lm0} = \sqrt{0.5abgh_m^{-1}|\Psi_C^{-1}|}, \quad (7)$$

where a and b are accordingly track and wheelbase of the car (m);

h_m is the height of car's mass center (m);

g is the acceleration of gravity (m · s^{-2});

Ψ_C is turn angle (rad).

The equation of front drift speed V_{Lm1} is obtained from III Newton Law (equality of frictional forces on front wheels in lateral direction and half of total centrifugal force):

$$V_{Lm1} = Re\sqrt{2[m_{12}gb - R_d\, a_{dT}]k_{sq}|\Psi_c^{-1}|}\,(\mathrm{m}\cdot\mathrm{s}^{-1}), \tag{8}$$

where $m_{12} = (m_1 + m_2) \cdot m_0^{-1}$ is relative mass distribution on front wheels (m_1 and m_2) to the total car's mass m_0 ;

R_d is the dynamic radius of driving wheels (m);

a_{dT} is traction-braking acceleration ($\mathrm{m}\cdot\mathrm{s}^{-2}$);

k_{sq} is top value of wheel frictional coefficient in lateral direction.

The equation of rear drift speed V_{Lm2} is also obtained from III Newton Law (equality of frictional forces on rear wheels in lateral direction and half of total centrifugal force):

$$V_{Lm2} = Re\sqrt{2[m_{34}gb + R_d\, a_{dT}]k_{sq}|\Psi_c^{-1}|}\,(\mathrm{m}\cdot\mathrm{s}^{-1}), \tag{9}$$

where $m_{34} = (m_3 + m_4) \cdot m_0^{-1}$ is relative mass distribution on rear wheels (m_3 and m_4) to the total car's mass m_0 .

Traction-braking acceleration a_{dT} for (8) and (9) is defined from equation of linear movement:

$$a_m = a_{dT} - k_x\, m_0^{-1}V_m^2 - k_{rf}g - \alpha_T\, g, \tag{10}$$

where $a_m = \mathrm{d}V_m/\mathrm{d}t$ is longitudinal acceleration of car's mass center ($\mathrm{m}\cdot\mathrm{s}^{-2}$);

k_x is the coefficient of frontal aerodynamic drag ($\mathrm{N}\cdot\mathrm{m}^{-2}\cdot\mathrm{s}^2$);

k_{rf} is the coefficient of rolling resistance between tires and road;

α_T is pitch angle (rad).

The values of tire lateral friction coefficient k_{sq} are related to the values of tire longitudinal friction coefficient k_{sd} by Kamm circle [3,7]:

$$(k_{sq}^2 + k_{sd}^2) \le (k_s^*)^2\,, \tag{11}$$

where k_s^* is the top (maximum) value of resulstant friction coefficient.

In this case, the value k_{sq} for one wheel is equal to:

$$k_{sq} = k_s^*\sqrt{1 - \left(\frac{k_{sd}}{k_s^*}\right)^2}\,, \tag{12}$$

where $k_{sd} = A \cdot m_0 \cdot a_{dT} \cdot F_N^{-1}$;

A is the distribution coefficient of tractive-braking forces per wheel;

F_N is the normal component of the force per wheel.

For the small values of module a_{dT} and A: $k_{sd} \ll k_s^*$ and $k_{sq} \approx k_s^*$.

The concept of optimal weighting follows from the equations of drift speed limits. The maximum speed at a turn for $V_{Lm1} < V_{Lm0}$ and $V_{Lm2} < V_{Lm0}$ is limited by $V_{min} = \min[V_{Lm1}, V_{Lm2}]$ and it's maximum value is achieved when $V_{Lm1} = V_{Lm2}$.

This condition is met when:

$$m_{12}gb - R_d\, a_{dT} = m_{34}gb + R_d\, a_{dT}.$$

Taking into account that $m_{34} = 1 - m_{12}$ we obtain:

$$m_{12} = a_{dT}R_d\, g^{-1}b^{-1} + 0.5 \text{ and}$$

$$m_{34} = 0.5 - a_{dT}R_d\, g^{-1}b^{-1}.$$

For Mercedes-Benz E-Class $b = 2.833$ m, $R_d = 0.3$ m, $a_{dT} = 3\ \mathrm{m \cdot s^{-2}}$ at second gear we obtain $m_{12} = 0.53$; $m_{34} = 0.47$ which coincides with the technical data of these models.

3.2. Virtual sensors for turn angle and the additional component of yaw rate

The indirect measurements of turn angle Ψ_c and additional component $\Delta\omega_m$ of yaw rate allow to exclude from the system specialized sensors of steering wheel angle and angular velocity sensor. The indirect measurements are based on the incorrect problem solution of calculation variables in the equation system (accepted indexation of variables: 1, 3 correspond to the front and rear wheels of the port side; 2,4 correspond to the front and rear wheels of starboard):

$$\begin{cases} V_1 = V_m + 0.5ab^{-1}V_m\,\Psi_c + \Delta V_{s1} + 0.5a\cdot\Delta\omega_m; \\ V_2 = V_m - 0.5ab^{-1}V_m\,\Psi_c + \Delta V_{s2} - 0.5a\cdot\Delta\omega_m; \\ V_3 = V_m + 0.5ab^{-1}V_m\,\Psi_c + \Delta V_{s3} + 0.5a\cdot\Delta\omega_m; \\ V_4 = V_m - 0.5ab^{-1}V_m\,\Psi_c + \Delta V_{s4} - 0.5a\cdot\Delta\omega_m, \end{cases} \tag{13}$$

where $\Delta V_{s1}, \Delta V_{s2}, \Delta V_{s3}, \Delta V_{s4}$ are longitudinal wheel slip speeds;

V_1, V_2, V_3, V_4 are linear wheel speeds.

The transformation of this problem to the correct one is possible when introducing additional determining conditions corresponding to the properties of the object [4,8,10]. So for the pair of wheels, the difference $V_i - V_j = \Delta V_{ij}$ $(i,j = 1,2; 3,4; 1,4; 3,2)$ is equal:

$$\Delta V_{ij} = ab^{-1}V_m\,\Psi_c + (\Delta V_{si} - \Delta V_{sj}) + a\cdot\Delta\omega_m. \tag{14}$$

Speed estimation for car's center of gravity is $\hat{V}_m = 0.5(V_i + V_j) = V_m + 0.5(\Delta V_{si} + \Delta V_{sj})$ and in the case $\Delta V_{si} = \Delta V_{sj} = 0$: $\hat{V}_m = V_m$.

The solution (14) with respect to Ψ_c taking into account the accepted assumptions is:

$$\Psi_c = \Delta V_{ij}\, a^{-1}bV_m^{-1} - bV_m^{-1}\Delta\omega_m. \tag{15}$$

Taking $\hat{\Psi}_c = \Delta V_{ij}\, a^{-1}bV_m^{-1}$, we obtain that:

$$\begin{cases} \hat{\Psi}_c = \Psi_c + b\cdot V_m^{-1}\cdot\Delta\omega_m; \\ \Delta\omega_m = b^{-1}\cdot V_m\cdot(\hat{\Psi}_c - \Psi_c). \end{cases} \tag{16}$$

Taking into account the additional defining properties of the object in the part of the value and sign $\Delta\omega_m$ in the form of conditions:

$$\begin{cases} \text{If } V_m < \min[V_{Lm1}, V_{Lm2}], \text{ then } \Delta\omega_m = 0; \\ \text{If } V_{Lm1} > V_{Lm2} \text{ and } V_m > V_{Lm2}, \text{then } sgn(\Delta\omega_m) = sgn(\hat{\Psi}_c); \\ \text{If } V_{Lm2} > V_{Lm1} \text{ and } V_m > V_{\mathrm{rp1}}, \text{then } sgn(\Delta\omega_m) = -sgn(\hat{\Psi}_c), \end{cases} \tag{17}$$

under $\Psi_c = \Psi_{Lm}sgn(\hat{\Psi}_c)$ we obtain:

$$\Delta\hat{\omega}_m = \begin{cases} 0, \text{if } V_m < V_{Lm1} \text{ and } V_m < V_{Lm2}; \\ b^{-1}V_m\,[\hat{\Psi}_c - \Psi_{Lm2}sgn(\hat{\Psi}_c)], \text{if } V_{Lm1} > V_{Lm2} \text{ and } V_m > V_{Lm2}; \\ -b^{-1}V_m\,[\hat{\Psi}_c - \Psi_{Lm1}sgn(\hat{\Psi}_c)], \text{if } V_{Lm2} > V_{Lm1} \text{ and } V_m > V_{Lm1}. \end{cases} \tag{18}$$

$\Psi_{Lm} = \min[\Psi_{Lm1}, \Psi_{Lm2}]$; $\Psi_{Lm1} = 2[m_{12}gb - R_d\, a_{dT}]k_{sq}V_m^{-2}$;
$\Psi_{Lm2} = 2[m_{34}gb + R_d\, a_{dT}]k_{sq}V_m^{-2}$.

3.3. Dynamic stabilization system
The dynamic stabilization system (DSS) includes ABS speed sensors, brake pedal actuator and control board with the program of processing and control.

The braking deceleration a_T at the step is formed as $[V_{m\,e} - V_{Lm\,e}] \cdot \Delta T^{-1}$, where $V_{m\,e}$ is extrapolated speed value of the car; $V_{Lm\,e}$ is the extrapolated value of the limit speed $V_{Lm\,e} = \min[V_{Lm0\,e}, V_{Lm1\,e}, V_{Lm2\,e}]$; $V_{Lm0\,e}, V_{Lm1\,e}, V_{Lm2\,e}$ are extrapolated with respect to the angle Ψ_c values of rollover speed, front drift speed and rear drift speed. To compensate for the delay of about 1 s, introduced by the brake actuator, the extrapolation time is set at 2 s.

Control actions are formed by power keys with PWM signals and *PWM_T* duty ratio in accordance with the calibrated characteristic $PWM\_T = f(a_T)$ of the brake system with the actuator. Differentiation of the control action is performed programmatically to give the cascade connection of the actuator-braking system the proportional element properties [5,9,11]. The maximum deceleration value a_{Tmax} is calculated depending on the identifiable top value of the sliding friction coefficient in order to prevent wheel blocking during braking. In the motion mode with activated adaptive cruise control, the set speed V_{mz} is defined as $V_{mz} = \min[V_{mz0}, V_{Lm0}, V_{Lm1}, V_{Lm2}, \dots, V_{Lm9}]$ and possible excesses $V_{m\,e}$ over $V_{Lm\,e}$ can't be significant.

3.4. Experimental results
DSS research tests were carried out for LADA Kalina car with electric traction in FSUE "NAMI" including NAMI's Testing Centre.

The total mass of the car with driver and passenger is $m_0 = 1280$ kg, mass distribution on front wheels is $m_{12} = 0.5$, the track is $a = 1.42$ m; the wheelbase is $b = 2.461$ m, the height of car's mass center is $h_m = 0.65$ m, $k_x = 0.51\ \mathrm{N \cdot s^{-2} \cdot m^2}$, winter studless tires are Michelin X-Ice 175/65 R14 82Q.

Test drives were conducted in a circle with the diameter of about 20 m on wet and dry asphalt, as well as during the passage of right and left turns with the radius of about 10 m with a change in the course angle by 90°. Figure 1 shows the time diagrams of car motion with disabled DSS at $k_s^* = 0.58$.

The analysis of the test results shows that the resulting turn angle Ψ_c , determined from the speeds of rear wheels with a fixed rudder position, decreases modulo as car speed increases, which corresponds to the side-drift effect due to non-zero lateral sliding of the wheels. The virtual sensor fixes positive $\Delta\omega_m$ corresponding to the drift of front wheels at speed close to $8\ \mathrm{m \cdot s^{-1}}$, positive a_{dT} and lateral acceleration (a_q) greater than $6\ \mathrm{m \cdot s^{-2}}$, which does not contradict the theoretical data.

Figure 2 shows the time diagrams of change car parameters at motion with enabled DSS, $k_s^* = 0.58$.

The analysis of the results shows that the actuation of the braking system occurs during predicted excess of the limit speed, which ensures the absence of non-zero $\Delta\omega_m$, speed stabilization at the level of $7\ \mathrm{m \cdot s^{-1}}$and lateral acceleration stabilization at the level of $6\ \mathrm{m \cdot s^{-2}}$.

IOP Conf. Series: Materials Science and Engineering **315** (2018) 012023 doi:10.1088/1757-899X/315/1/012023

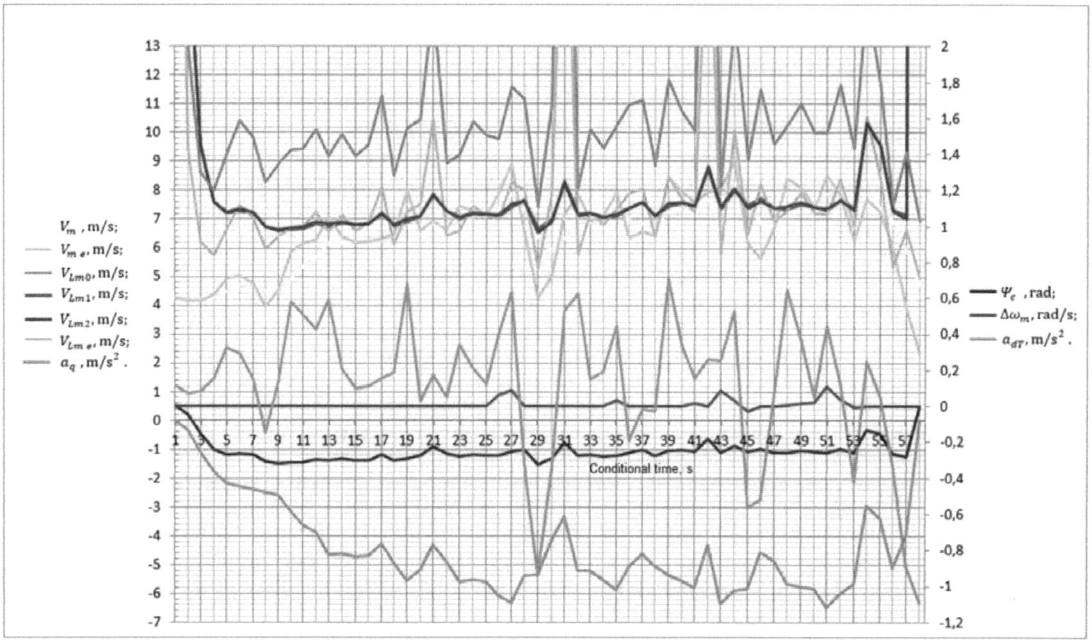

Figure 1. The time diagrams of car motion with disabled DSS on a circle.

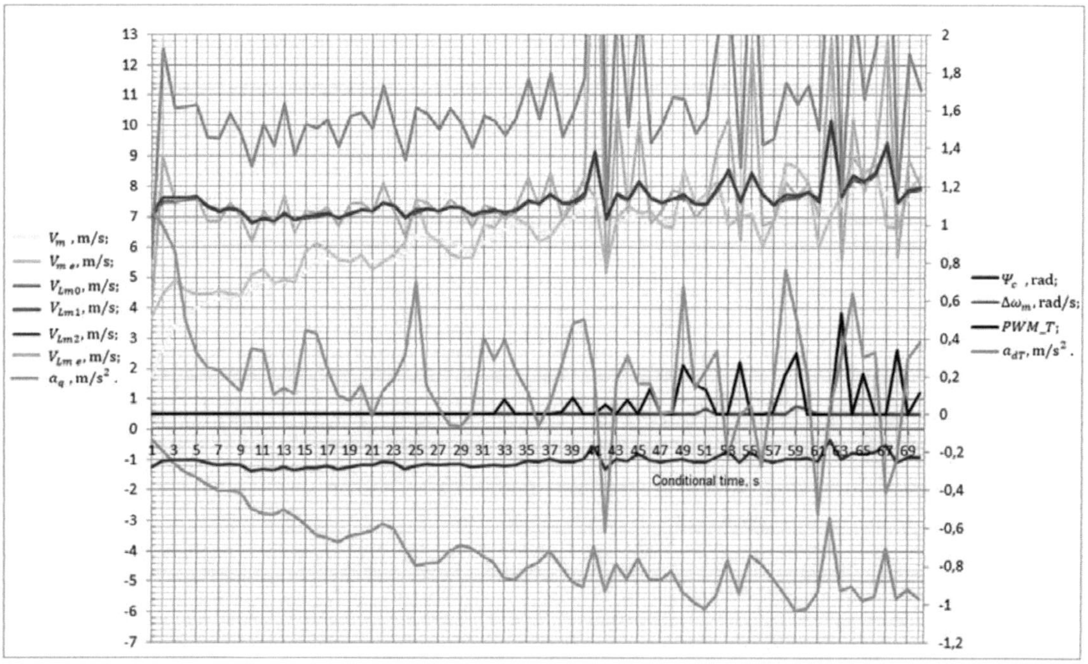

Figure 2. The time diagrams of car motion with enabled DSS on a circle.

4. Conclusion

The dynamic stabilization problem for autonomous car has been formulated in terms of minimizing the quadratic functional for control quality, taking into account functional limitations on technical and economic indicators.

The analysis of conducted research confirms both the adequacy of the mathematical models, the algorithms and the possibility of implementing the system in the minimal configuration of technical means.

Acknowledgments
This research article was prepared on the results of applied scientific research, which performed with financial support of Ministry of education and science of Russia. Agreement No. 14.625.21.0042 Unique project identifier RFMEFI62517X0042.

References

[1] Jeong E and Oh C 2017 Evaluating the effectiveness of active vehicle safety systems *Accident Analysis & Prevention* 85-96

[2] Ni L, Gupta A, Falcone P and Johannesson L 2016 Vehicle lateral motion control with perfomance and safety guarantees *IFAC Proceedings Volumes* 285-90

[3] Bakfish K and Hajnc D 2003 *New book about tires* (Moscow: Izdatel'stvo Astrel') – in Russian

[4] Buznikov S E 2009 The method of constructing information virtual sensors for car's active safety systems *Proc. of XVII Int. Conf. "The management problems of safety in complex systems"* (Moscow: Russian State University for the Humanities Press) pp 420-4 – in Russian

[5] Buznikov S E, Elkin D S, Shabanov N S and Strukov V O 2016 Task of safe automatic braking of the vehicle *Trudy NAMI* 44-52 – in Russian

[6] Buznikov S E, Elkin D S 2007 Identification of maximal values of sliding friction coefficients of the vehicle wheels: the certificate of official registration program for computer # 2007610818 *Rospatent*

[7] Saykin A, Bakhmutov S, Terenchenko A, Endachev D, Karpukhin K and Zarubkin V 2014 Tendency of Creation of "Driverles" Vehicles *Abroad Biosciences Biotechnology Research Asia* **11** p 241-6

[8] Saikin A, Buznikov S and Karpukhin K 2016 The Analysis of Technical Vision Problems Typical for Driverless Vehicles *Research Journal of Pharmaceutical, Biological and Chemical Sciences* **7 #4** p 2053-9

[9] Shadrin S, Ivanov A and Karpukhin K 2016 Using Data From Multiplex Networks on Vehicles in Road Tests, in Intelligent Transportation Systems, and in Self-Driving Cars *Russian Engineering Research* **36 #10** p 811-4 doi:10.3103/S1068798X16100166

[10] Dakroub H, Shaout A and Awajan A 2016. Connected Car Architecture and Virtualization SAE *Int. J. Passeng. Cars – Electron. Electr. Syst.* **9(1)** p 153-9 doi: 10.4271/2016-01-0081

[11] Shadrin S and Ivanov A 2016 Algorithm of autonomous vehicle steering system control law estimation while the desired trajectory driving ARPN *Journal of Engineering and Applied Sciences* **11 #15** p 9312-6

IOP Conf. Series: Materials Science and Engineering **315** (2018) 012024 doi:10.1088/1757-899X/315/1/012024

Integration of passive driver-assistance systems with on-board vehicle systems

V V Savchenko and S N Poddubko

The Joint Institute of Mechanical Engineering of the National Academy of Sciences of Belarus, Minsk, Republic of Belarus

E-mail: bats@ncpmm.bas-net.by

Abstract. Implementation in OIAS such functions as driver's state monitoring and high-precision calculation of the current navigation coordinates of the vehicle, modularity of the OIAS construction and the possible increase in the functionality through integration with other on-board systems has a promising development future. The development of intelligent transport systems and their components allows setting and solving fundamentally new tasks for the safety of human-to-machine transport systems, and the automatic analysis of heterogeneous information flows provides a synergistic effect. The analysis of cross-modal information exchange in human-machine transport systems, from uniform methodological points of view, will allow us, with an accuracy acceptable for solving applied problems, to form in real time an integrated assessment of the state of the basic components of the human-to-machine system and the dynamics in changing situation-centered environment, including the external environment, in their interrelations.

1. Introduction

Domestic driver assistance systems (ADAS) are being implemented in the Republic of Belarus and in the Russian Federation within the framework of science-and-technology programs at different levels. These are supported by relevant funds, targeted enterprises and performed by companies proactively. A sufficient number of current developments is intended for production of passive ADAS oriented to its use on domestic trucks. Such systems alert a driver only on several potentially dangerous cases (giving a visual and sound alarm message) and they do not affect a vehicle's dynamics, working automatically. As a rule, such systems use information obtained from cameras (vehicle monitoring in a track or related tasks), ultrasound sensors (blind spots monitoring), radars (ACC – adaptive cruise control systems), navigation modules for location identification, etc. installed on a vehicle unit. There are many mass-produced systems, for instance, driver's vigilance monitoring systems such as DVMS or VIGITON that monitor driver's psychophysiological parameters to interpret automatically a driver's current functional state and his ability to perform current activity algorithms to drive a car [1,2]. For instance, Mobileye systems of 5th and 6th series are advanced driver assistance systems having a built-in camera and EyeQ chips family to process obtained images; the systems alert a driver on several potentially dangerous cases and give the following warnings and notifications (namely, visual and sound alarm message): a warning of a dangerous frontal approach (FCW and UFCW), of potential front collision with a vehicle driving (standing) in front of; a pedestrian collision warning (PCW); lane departure warning (LDW); a warning of keeping a distance between the cars and of a dangerous approach with a forward vehicle

(HMW), etc. Moreover, the following articles may be referred to this group: speed limit information systems (SLI) that detect and classify different signs of a speed mode notifying a driver of exceeding of permissible speed limit indicated on the sign; traffic sign recognition systems (TSR) that detect and classify different road signs and notify of them the driver. In case of using the Mobileye systems, the driver has to perform all management tasks on his own and keep control over the vehicle, track car's conditions, fulfill all recommendations in terms of safe driving. In the User's Manual, the Manufacturer specifies, "it cannot guarantee and does not guarantee 100% detection of vehicles, pedestrians, road signs or traffic lanes and does not guarantee giving any sound and visual warning notifications related to it." Aside from that, recognition and responsiveness of the Mobileye systems may be adversely affected by road, weather and other conditions" [3]. However, 27 OEM Suppliers of the ADAS implement in their systems Mobileye components, EyeQ chips (6th generation is used nowadays) and the software.

As it is known, works in such direction including production of active ADA are carried out by large corporations developing components and/or safety shields: BOSCH (Germany), Knor-Bremse (Germany), WABCO (USA), etc. according to the market analysis performed by Semicast Research (figure 1), Continental company took a top position in an area of electronic components for automotive ADAS systems in 2015 [4].

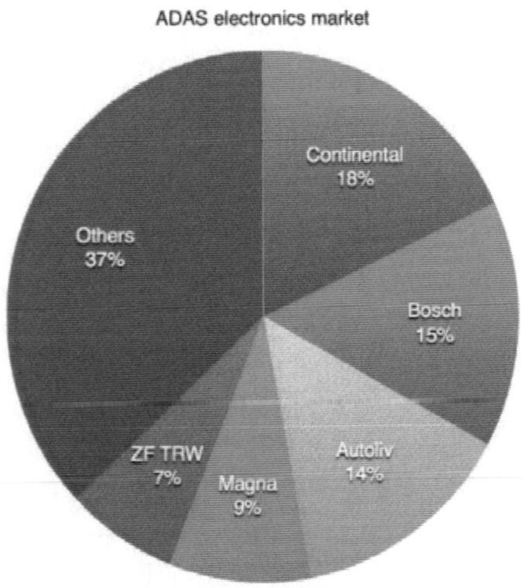

Figure 1. Analysis of the electronics suppliers sector in ADAS segment by Semicast.

2. Problem Statement

To develop a basic structure of an onboard information analysis system (OIAS) and a basic software architecture designed to create a vehicle float body for commercial vehicles. The software shall embody the following main functions: vehicle's location identification accurate to the nearest centimeters, first, to support ADAS in terms of vehicle monitoring in a track, if a traffic lane and a road border are not recognized by optical methods; communication maintenance with dispatching offices and emergency services; transfer of information, including obtained in an automatic mode, on system «human-to-machine» dynamics (current state of the major systems, units and aggregates, driver's state, warning on emergency cases, emergency call in case of an accident or if a driver has fallen unconscious or is dead), communication between passive ADAS and active safety systems or its integration to coordinate their joint operation in real-time mode in terms of accident prevention or mitigation.

3. Results

An information and analytical system with the indicated functions is being developed at the Institute jointly with the partners having the expertise relevant to the area under consideration, within the framework of the Scientific and Technical Undertaking of the Union State “Avtoelektronika”: the Limited Liability Company “NTLab-sistemy” included into the group of companies “NTLab” (Republic of Belarus, http://www.ntlab.com) develops and produces very-large-scale integration circuits (VLSI circuits), plug-in modules, as well as portable hardware and software packages, including navigation sets for usage in automotive vehicles. The company develops the products based on specialized VLSI circuits of in-house design; the Open Joint-Stock Company “Ekran” (Republic of Belarus, http://www.ekranbel.by), a developer and supplier of production active safety systems, on-board monitoring systems and their spare parts for CIS machine-building enterprises; the Closed Joint-Stock Company «Neurocom» (Russian Federation, http://www.neurocom.ru), creation and development of a monitoring methodology for tracking the functional state of operators in the «human-to-machine» systems and further development based on it, mastering the production and supply of a production range of production systems for railway and automotive transport.

Figure 2 depicts the VLSI circuits structure and its plugging-in by joint usage with other intelligent on-board systems on board of a motor vehicle (MV).

The on-board module for detection of driver’s dangerous states (developed by the Closed Joint-Stock Company «Neurocom») is connected to the system via CAN data bus. Module’s application enhances OIAS functional abilities significantly and specifies it as a prevention safety system that enables to prevent accident occurrence because of driver’s relaxation, unconsciousness or death. The module is unique due to the fact that the following features are used in its construction: a specialized chip set, a robust algorithm to recover an useful signal from noise, original algorithms for automatic interpretation of the driver’s functional state, short-range telemetering measurement to transfer information to the stationary part of the system. In total, it enables a high interpretation reliability of dynamics in a driver’s vigilance level complied with requirements specified for active safety systems. As prototype for the module under development, we may consider DVMS or VIGITON, which use monitoring and automatic analysis of electrodermal arousal parameters (the main criterium for automatic interpretation of the current dynamics in the driver’s relaxation level, the used method has the lowest, among the known ones, probability of dangerous failure (p – 0,0001), principles of building a system of safety monitoring of the operator’s (as a human being) state have been applied [1].

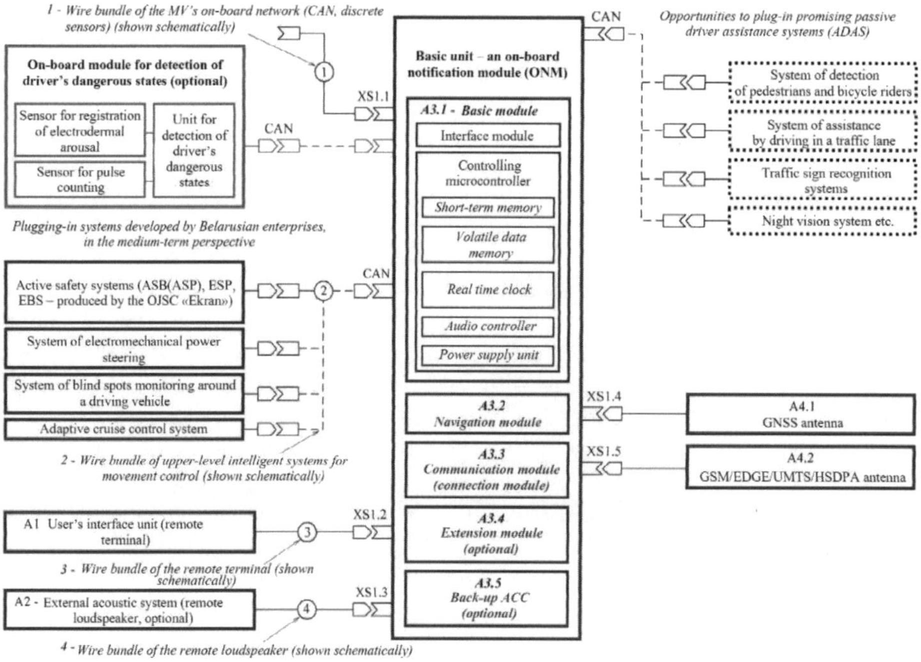

Figure 2. Structure of OIAS and its connection on MV’s board.

The technical specification to the on-board module, namely the definitions of the driver's dangerous states (deep relaxation, unconsciousness, death), include a probability of a dangerous error of the used registration method (an error of the 2nd class) in rate of no more than $4*10^{-4}$ and a probability flow of dangerous failures causing car accidents in rate of no lower than 10^{-9}.

Navigation module. To improve the accuracy and reliability of the navigation module, the use of jointly processed signals of the operating global navigation satellite systems (GNSS) such as GLONASS (in the frequency ranges of L1, L2), GPS (in the frequency ranges of L1, L2), GALILEO (in the frequency ranges of E1B, E1C, E5a, E5b) and BEIDOU (in the frequency range of B1-C) is provided. To further improving the accuracy of the coordinates determination, it is possible to use RTK (Real Time Kinematic) technology, which allows, due to simultaneous signals processing in real time obtained from GNSS and signals of correcting stations, gaining centimeter accuracy of determining the coordinates. The ability to integrate GNSS signal combining with information from sensors of the microelectromechanical systems and the odometer sensor of the vehicle's path allows the use of the "Dead reckoning" technology, which will reduce the error in calculating navigation parameters, and will also allow estimating the coordinates when GNSS signals are inaccessible, in specified time spans. The module supports automatic monitoring of the integrity (reliability) of navigation calculations and the elimination of inauthentic measurements, the implementation of robust algorithms of differential correction (RTK mode) and algorithms for combining GNSS data and inertial sensors (accelerometers) of in-house design, determination of navigation parameters in WGS-84 or PZ- 90 reference systems (by confidence coefficient of 0.95).

Figure 2 depicts the systems already being developed or planned in the near future in the Republic of Belarus (the electromechanical power steering system, the system of blind spots monitoring around a driving vehicle, ECP brake system and the adaptive cruise control system in the future). Mass production of such systems on board of vehicles is to expect in the short-term perspective. Plugging-in these systems is assumed to be performed via CAN data bus that may be implemented as a part of hardware taking into consideration the developed system.

There is also a significant potential associated with the use of promising passive ADAS. This includes systems of detection of pedestrians and bicycle riders, a system of assistance by driving in a traffic lane, a traffic sign recognition system and night vision system, etc. The operation of these systems is based on the rapidly developing systems of technical vision or (in a number of developments in terms blind spots monitoring, assistance while parking) scanning of space by ultrasonic sensors.

The basic module contains the necessary hardware features and a set of interfaces that provides the implementation of basic functions and technological modes.

The extension module supports additional functions related to interaction with other intelligent systems on board of a vehicle.

The software has a multi-level modular hierarchical structure. Individual modules operate asynchronously with respect to each other, which allows implementing the multi-tasking mode of operation in real time. The file structure used allows maximum isolation of the levels code and modules. The architecture of the basic software is shown in Figure 3.

Description of the levels of the basic software project. BASE level: at the BASE level, there is a module for software implementation of the operating modes of the system and a module of debug interface. The module for software implementation of the operating modes consists of subprograms ensuring the system's operability at the logical level. The module of the debug interface provides access to the variables of the software implementation module at the hardware-independent level. BASE level modules can directly access the hardware-dependent functions of the controller that have atomic access via the HAL level.

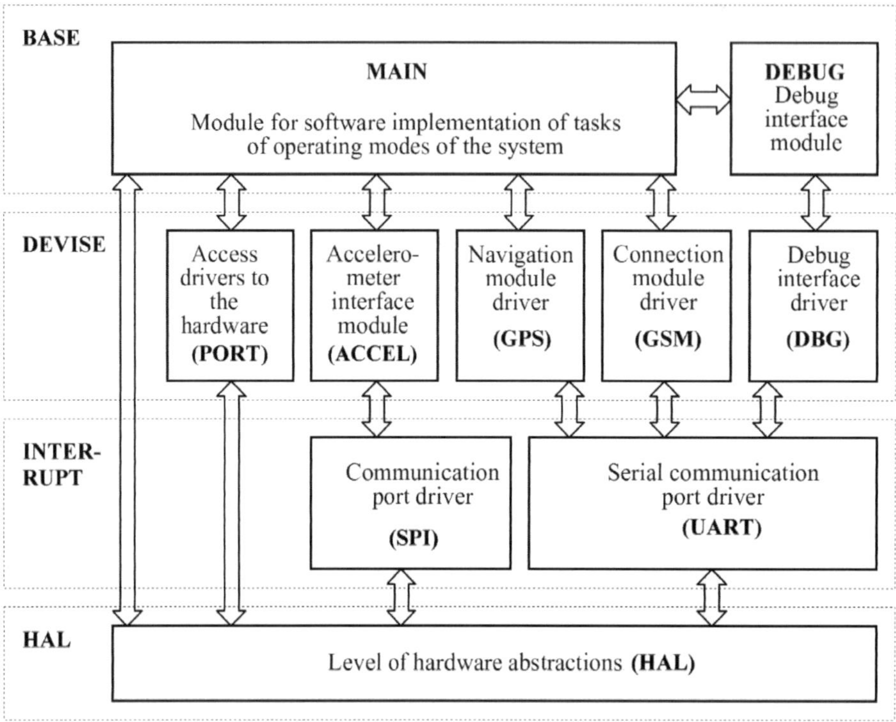

Figure 3. Architecture of the basic software.

DEVICE level: at DEVICE level, there are modules of external device drivers that provide information exchange at the level of protocols specific for each device: active safety systems, driver's functional state monitoring, ADAS, navigation, communications, etc.

The debug interface driver DEBUG provides access to the variables of the software implementation module at the protocol level. Access drivers to the PORT hardware allow calling hardware-dependent controller functions that require processing at the protocol level (generation of time delays, input of the state of controls with debounce filtering, etc.). The INTERRUPT level provides secure joint access to subprogram's data operating in the user mode and to the subprograms running in the corn mode (hardware interrupt handlers). The HAL level ensures the independence of the BASE, DEVICE and INTERRUPT software codes from the implementation of the hardware platform used.

Conclusion

Implementation in OIAS such functions as driver's state monitoring high-precision calculation of the current navigation coordinates of the vehicle where components, algorithms and software developed in the Union State are used, modularity of the OIAS construction and the possible increase in the functionality through integration with other on-board systems has a promising development future.

The development of intelligent transport systems (ITS) and their components allows us to set and solve fundamentally new tasks for the safety of human-to-machine transport systems, and the automatic analysis of heterogeneous information flows provides a synergistic effect.

Based on the results obtained, the task is to develop a concept of cross-modal (intersensory) interaction in the human-to-machine transport systems. The cross-modal interaction in human-to-machine transport systems is the interaction of hierarchically arranged multi-functional on-board monitoring and control systems (vehicle's and driver's technical components) between themselves and external systems (primarily ITS, dispatching and emergency services) with a targeted function to improve the efficiency of human-to-machine systems. The analysis of cross-modal information exchange in human-machine transport systems, from uniform methodological points of view, will allow us, with an accuracy acceptable for solving applied problems, to form in real time an integrated assessment of the state of the basic components of the human-to-machine system and the dynamics in changing situation-centered environment, including the external environment, in their interrelations.

Definition

Cross-modal interaction in human-to-machine systems is the transformation, synchronization and analysis of heterogeneous information flows in CAN, MOST, FlexRay data buses, in the future, vehicle Ethernet or its analog and V2V, V2I and V2P (P2V) protocols, in real time scale.

The purpose of the cross-modal interaction analysis is to obtain an integrated assessment (synthesis of the integrated assessment criteria) and automatic generation of control commands for the efficient functioning of vehicles with the support of the intelligent transport systems in real time, and maintenance of human-machine systems in accordance with the actual state.

Reference List

[1] Dementienko V V Driver vigilance remote monitoring system *Science Journal of Transportation* 2015 Especial Issue **6** pp 110–114

[2] Savchenko V V Development of methodology for operator's functional state monitoring in "human-to-machine" transport systems *Mechatronics, automatization, control* 2013 **6** pp 27–32

[3] ADAS [Digital resource] Access mode: https://www.mobileye.com/our-technology/adas/ (Reference date: 12.07.2017)

[4] Continental, Bosch takes a top position among ADAS suppliers [Digital resource]. Access mode: http://www.eenewsautomotive.com/news/continental-bosch-lead-adas-vendor-share-ranking (Reference date: 21.07.2017)

The development of ground unmanned vehicles, driver assistance systems and components according to patent publications

A M Saykin[1], G S Tuktakiev[2], A V Zhuravlev[3] and E P Zaitseva[4]

[1] Special Programs Department, FSUE “NAMI”, Moscow, Russia
[2] Department for intellectual property rights registration, FSUE "NAMI", Moscow, Russia
[3] Head of the tractors Department, FSUE "NAMI Moscow, Russia
[4] Engineer of Special Programs Administration, FSUE "NAMI", Moscow, Russia

E-mail: a.saykin@nami.ru

Abstract. The paper contains the analysis of the main trends in the patenting of ground unmanned vehicles, driver assistance systems (ADAS) and unmanned vehicle components abroad during the period from 2010 to 2016. The conclusion was made that the intensity of their patenting abroad increased.

1. Introduction

The development of industry and the economy in developed countries significantly depends on the further development of vehicles and systems that must ensure greater mobility of the population, the efficiency of cargo and passenger transport, the improvement of road safety, the reduction of the burden on the environment and improve the comfort of drivers and transport users. Effective way to solve these problems is the creation of ground unmanned vehicles and driver assistance systems.

This report presents the results of a patent search in ground unmanned vehicles, driver assistance systems (ADAS) and components for them, performed on Russian and foreign patents, scientific and technical materials for the period 2010 - 2016.

Last decade, the development of unmanned vehicles has been experiencing a technological boom in the automotive industry of all leading countries [1]. The most active work on the creation of unmanned vehicles is carried out in the United States, Germany, Japan, China, Great Britain, Sweden, France, and Korea [2, 3]. A significant number of projects on the creation of the unmanned vehicle are defense-related and for this reason, the results of the research are seldom published in the open press. Complex high technology solutions, mathematical apparatus, motion control algorithms, software, control systems sensors of unmanned vehicles in many countries are classified as dual-purpose products.

The main advantages of unmanned vehicles are:

- improvement of transport and environmental safety, minimization of road accidents and the number of human victims in them;
- reduction of time and costs for transportation of goods and passengers;
- reduction of fuel consumption, emission of harmful substances into the atmosphere;

- more efficient use of road capacity;
- expansion of car use opportunities for disabled people;
- possibility to transport goods in hazardous areas, during natural and technological disasters or military operations;
- improvement of passengers' comfort.

The most complex and knowledge-intensive in the ground unmanned vehicle is an automatic traffic control system (ATCS). Figure 1 shows a typical functional diagram of the main subsystems in the ATCS.

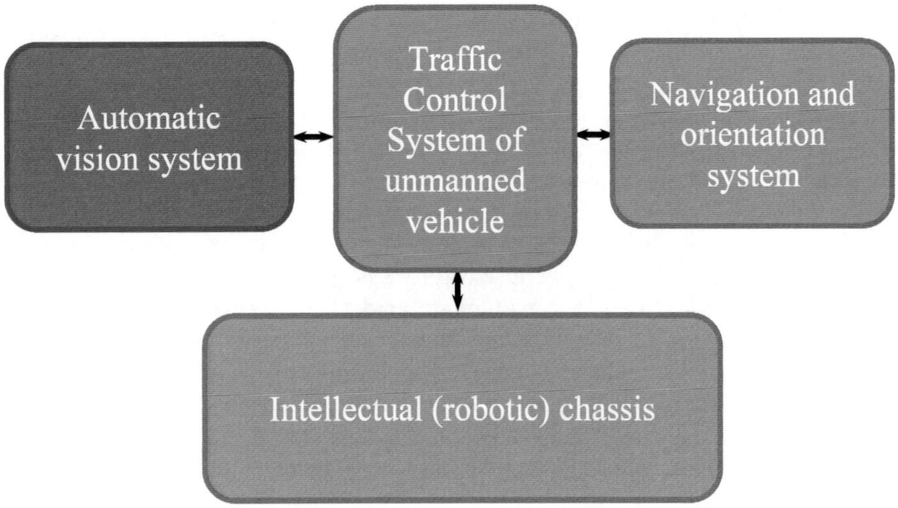

Figure 1. A typical functional diagram of the main subsystems in unmanned vehicle's automatic control system.

The navigation and orientation systems ensure the location of unmanned vehicle on the ground. The system of technical vision forms a picture of the environment, provides recognition of other vehicles, obstacles on the route, road markings, road signs, traffic lights [4]. The information management system is the core of the unmanned vehicle traffic control system. It is responsible for building the traffic route, emergency decision-making depending on the surrounding traffic situation and the formation of control signals for actuators. Actuators in accordance with the control signals provide a control effect on the units and vehicle systems, simulating the actions of the driver. Depending on the chassis design of the unmanned vehicle, the system of executive drives provides the operation of thrust control drives (fuel supply when using internal combustion engine), clutch, gear shift, steering gear, brake, starter, lighting devices and etc.

The variety of vehicle and aggregate design predetermines a large number of technical proposals for automating traffic control. Technical proposals are reflected in patent documents belonging to organizations and individuals from leading industrialized countries. Patent research in the field of "Electronic systems for traffic control of land unmanned vehicles" included analysis of patent documentation on methods and devices for automatic control of cars, agricultural and special equipment, mobile robots, etc.

Patent research was conducted in the following key areas:

1. Ground unmanned vehicles;
2. ADAS systems;
3. Components of the unmanned vehicle.

The purpose of patent studies was to assess the number and dynamics of patented developments in these areas by country and key developers. The patent search was conducted on the basis of the following electronic databases located on the Internet:

- database of the European Patent Organization (EPO) (http://worldwide.espacenet.com);
- database of the World Intellectual Property Organization (WIPO);
- database - register of inventions and published filed applications of the Industrial Property Federal Institute of the Russian Federation (http://www.fips.ru);
- database of Thomson Reuters Derwent World.
- based on the subject of the search and on the patent classifier analysis, the relevant classification clauses of the International Patent Classification (IPC) were selected.
- the analysis shows that the development of the unmanned vehicle is carried out in two main directions:
- on the way of introducing and expanding the functionality of various driver assistance systems (cars of all classes are being currently equipped with these systems);
- on the way of creating methods and control systems for fully unmanned vehicles (which are currently in the stages of prototypes, including operational ones).

A modern car contains many electronic assistance systems for the driver with varying degrees of the driving process automation. They include cruise control systems, traffic stability, collision avoidance, parking distance control, preventive preparation for an accident, etc. These systems take on a part of the vehicle management functions, for example, the functions of automatic speed control, acceleration, braking, turning, engine and transmission modes. Thereby in some conditions (for example, when driving on a highway) the car is able to move in an autonomous mode, and the driver can not interfere with the control process. Further improvement and expansion of the driver assistance systems functionality ensures an increase in the autonomy of the car, bringing it closer to a completely unmanned vehicle.

Figure 2 shows the dynamics of patenting in respect of unmanned vehicles, ADAS systems, and unmanned vehicle components abroad.

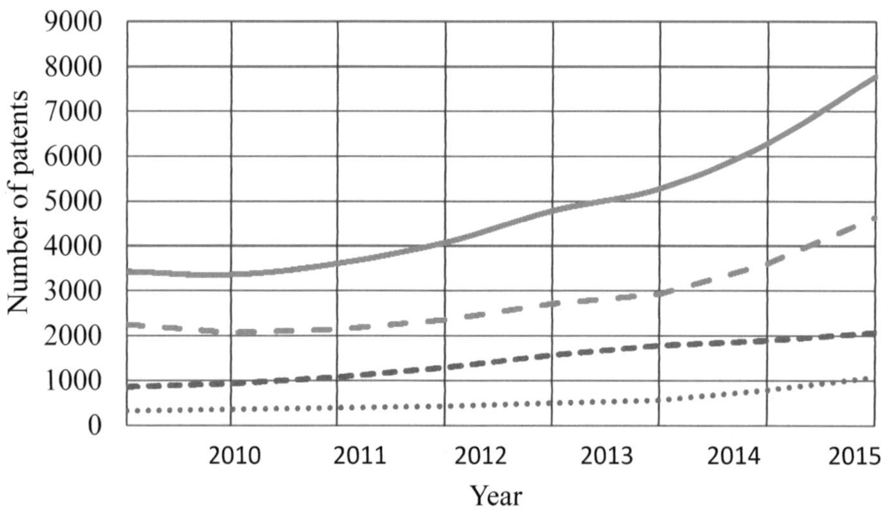

Figure 2. Dynamics of patenting in respect of ground unmanned vehicles abroad, ADAS systems and unmanned vehicle components in general and separately for the period from 2010 to 2016.

ADAS systems
Unmanned vehicle components
Ground unmanned vehicle
Total number of patents

IOP Conf. Series: Materials Science and Engineering **315** (2018) 012025 doi:10.1088/1757-899X/315/1/012025

Patenting of autonomous ground unmanned vehicles abroad is most active. At the same time, the number of patents for working methods and assistance systems for the driver, as well as for the components of the unmanned vehicles, is also constantly growing. Patenting by leading foreign developers is conducted, as a rule, by patents packages in the directions under consideration.

The number of issued patents and published patent applications for the period from 2010 to 2016 indicates the priority of the subjects among foreign automakers in comparison with other developments in the automotive industry.

There are many manufacturing companies among the developers. Toyota Corporation (Japan) is the universally recognized world leader in the patenting of ground unmanned vehicles. Then follow other corporations: Bosch (Germany), Denso (Japan), Hyundai (South Korea), FORD GLOBAL TECH LLC, GM (USA) - see figure 3.

In order to realize autonomous control over the movement of an unmanned vehicle, the development and patenting of traffic control systems are most actively carried out, which include: intelligent chassis control units, vision systems, information processing and transmission, navigation and orientation systems (figure 1). In general, the patented traffic control systems of the unmanned vehicle provide the following operations:

- receiving and processing the data from sensors;
- consolidation and harmonization of the received data;
- image processing;
- identification of obstacles, road conditions and vehicles, distance to them;
- positioning of the unmanned vehicle and determining the current state of the system;
- implementation of automatic control: speed, trajectory (course) of traffic, automatic response to objects surrounding;
- making decisions on management actions;
- control of executive devices;
- formation of a database for further analysis.

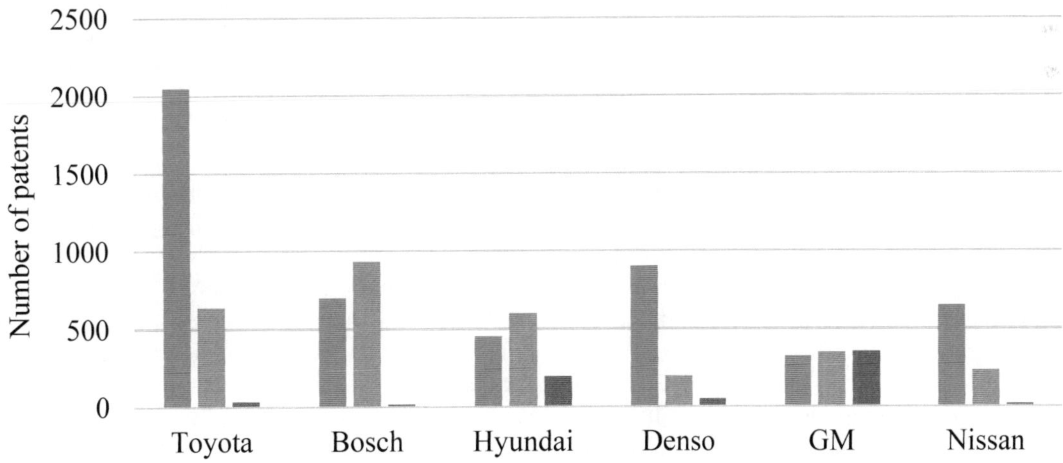

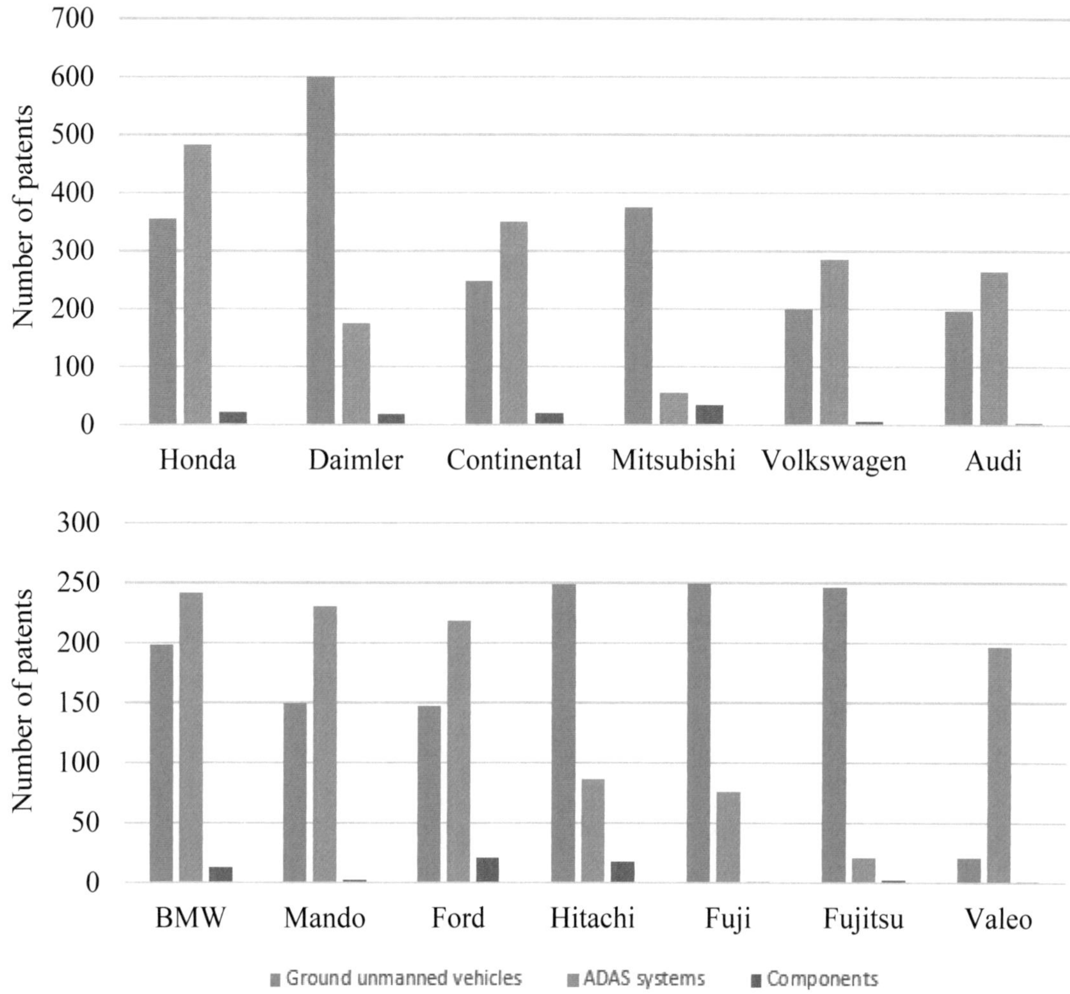

Figure 3. Quantitative data on the patenting in the field of unmanned vehicles by leading foreign automakers for the period from 2010 to 2016.

In the patented complex traffic control systems of the unmanned vehicles, the software usually includes two levels - the lower and upper level. Lower level software is responsible for interaction with sensors and actuators, and the upper one directly corresponds to the implementation of the unmanned vehicle motion control algorithm. For testing in the vehicle traffic control system, a group of patents provides for the use of algorithm modeling processes.

In the patenting of unmanned vehicles, Japan occupies the leading positions in the world - see figure 4. Leaders in this direction are the corporations Toyota (Japan), Denso (Japan), Bosch (Germany), Nissan (Japan) and Honda (Japan). The American company Google takes 19th place in this area, and FORD is on the 20th place. Representatives of Asian countries took eleven positions in the list of the leading 20 corporations.

The development and patenting of driver assistance systems (ADAS) were conducted ahead of ground unmanned vehicles developments. Main systems of driver assistance are:

- adaptive cruise control;
- warning of leaving the lane;
- helping in movement in the lane;

- assistance with changing lanes;
- frontal collision warning;
- warning of frontal collision with pedestrians and obstacles;
- emergency braking system;
- intellectual speed limit;
- parking assistance (parallel, perpendicular);
- helping in traffic jams;
- highway autopilot;
- control of the driver's condition;
- recognition of traffic signs;
- recognition of traffic light signals;
- adaptive control of headlight;
- night vision;
- side video surveillance;
- circular view;
- detection of vehicles during the return movement;
- tire pressure monitoring;
- automatic control of the conditioning system operation;
- automatic windshield wiper.

The largest number of patents for ground-based unmanned vehicles and driver assistance systems belongs to Bosch Corporation - see figure 4. Toyota, Hyundai, Daimler, Continental, GM are also major developers in this area. Out of five first places, three belong to German companies: Bosch is the leader in patenting; Daimler and Continental occupy the fourth and fifth places respectively. Toyota (Japan) is on the second and Hyundai (South Korea) is on the third place.

In the field of unmanned vehicle components following developments are patented: methods of work, devices and telecommunications systems, unmanned transportation management technologies, additional systems and means for improving road safety, information transfer and exchange systems, especially between vehicles, vehicles and the infrastructure of intelligent transport systems. The scale of patenting in this area is much smaller than the patenting of ground unmanned vehicles and active safety systems. However, these issues are important and promising for solving the problem of improving the road safety [5, 6].

General Motors Corporation (USA) takes the first place in patenting of unmanned vehicle components (figure 4), followed by Hyundai (South Korea), Marvell (USA), LG (South Korea) and Denso (Japan). In this area, special attention is paid to systems and methods for the transfer and exchange the information between vehicles, vehicles and the infrastructure of intelligent transport systems.

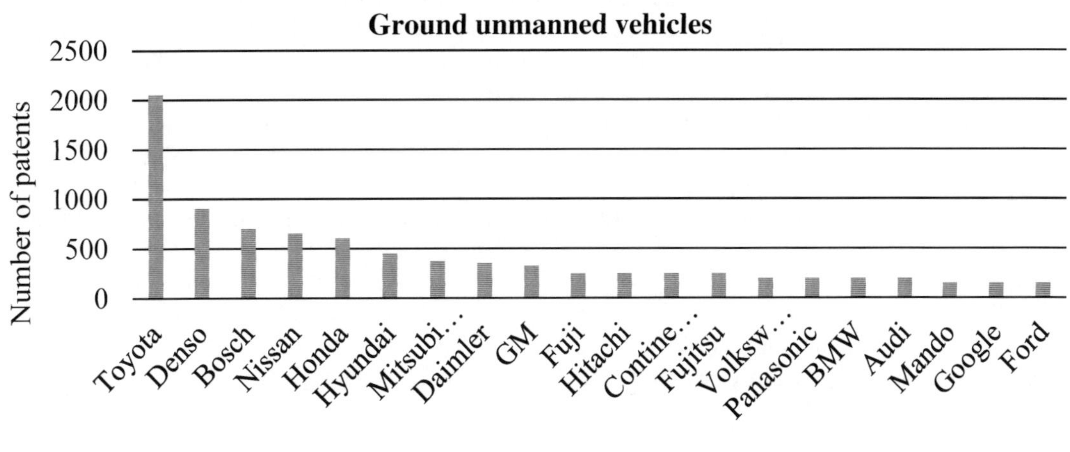

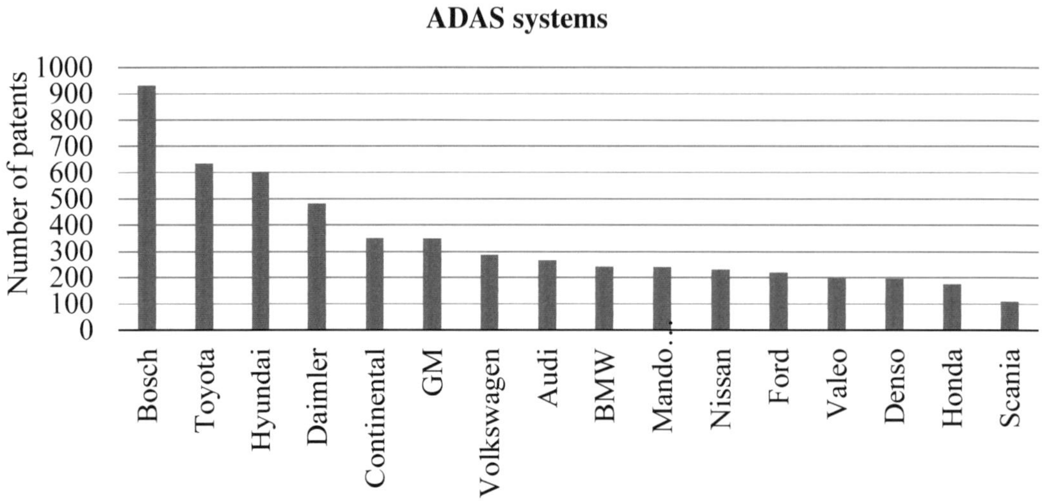

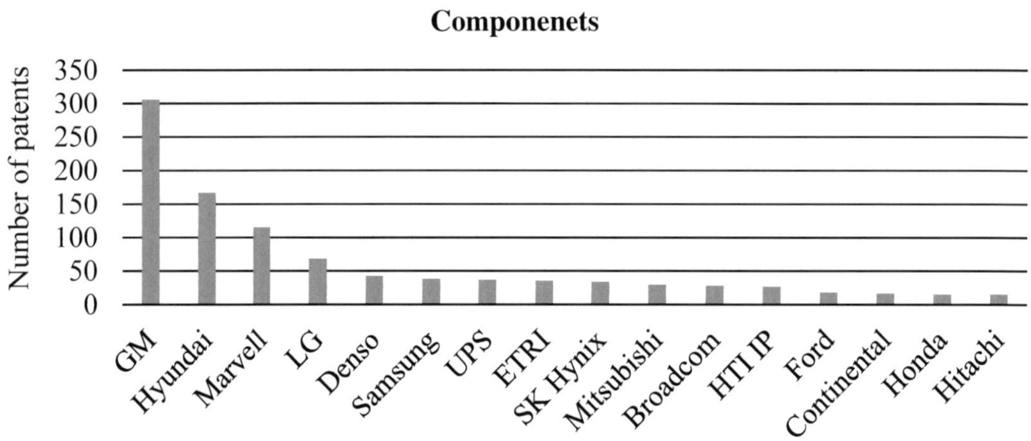

Figure 4. Ranking of foreign corporations on patenting of ground unmanned vehicles, ADAS systems and components during 2010 – 2016.

Summarizing the foregoing, it can be stated that for the period from 2010 to 2016, the number of foreign patents and published patent applications for ground unmanned vehicles, driver assistance systems and components for them exceeds 35,000. This branch of the automotive industry is rapidly developing and it is priority one. In the Russian Federation for the same period, the number of published patents and patent applications for the ground unmanned vehicles, driver assistance systems and components submitted by domestic developers does not exceed 80 patents. The owners of patents in the Russian Federation are both enterprises and organizations, as well as individuals, with no leader among them. In the Russian Federation FORD GLOBAL TECH LLC is actively patented their developments.

2. Conclusions

The total number of patents and patent applications abroad for ground unmanned vehicles, driver assistance systems and components for the period from 2010 to 2016 exceeds 35,000, which indicates the priority and prospects of this direction in the development of the automotive industry and transport systems. Patents in the territory of the Russian Federation are received and supported by the majority of foreign firms.

The number of patents and patent applications for ground unmanned vehicles, driver assistance systems and components in the Russian Federation of domestic developers for the same period does not exceed 80 patents. Owners of patents are enterprises, organizations, and individuals. The leader among them was not identified. It could be conclude, that the patenting in the Russian Federation of ground unmanned vehicles, ADAS systems and their components by domestic developers in comparison with foreign ones is extremely slow and does not have a systemic character. At the same time, every year the competition with patents of foreign firms is becoming more and more complicated.

Acknowledgement
This scientific paper was prepared based on the results of applied scientific research, the financial support was ensured by the Russian Federation, particularly by the Ministry of Education and Science, Agreement number is № 14.624.21.0049, unique identifier of the project is RFMEFI62417X0049.

References

[1] Saykin A, Bakhmutov S, Terenchenko A, Endachev D, Karpukhin K and Zarubkin V 2014 Tendency of driverles vehicles creation abroad *Biosciences Biotechnology Research Asia* **11** (Spl. Edn.) pp 241-246 (http://dx.doi.org/10.13005/bbra/1469)
[2] ISO 15765-4:2011 (E) Road vehicles - Diagnostic communication over Controller Area Network (Do CAN) - Part 4: Requirements for emissions-related systems Second edition 2011
[3] ISO 15031-5:2011 (E) Road vehicles - Communication between vehicle and external equipment for emissions-related diagnostics - Part 5: Emissions-related diagnostic services -Second edition 2011
[4] Saikin A, Buznikov S and Karpukhin K 2016 The analysis of technical vision problems typical for driverless vehicles *Research Journal of Pharmaceutical, Biological and Chemical Sciences* **7** (4) pp 2053-59
[5] Shadrin S, Ivanov A and Karpukhin K 2016 Using data from multiplex networks on vehicles in road tests, in intelligent transportation systems, and in self-driving cars *Russian Engineering Research* **36** (10) pp 811-14 (doi:10.3103/S1068798X16100166)
[6] Shadrin S and Ivanov A 2016 Algorithm of autonomous vehicle steering system control law estimation while the desired trajectory driving *ARPN Journal of Engineering and Applied Sciences* **11** (15) pp 9 312-16

Practical implementation of the concept of converted electric vehicle with advanced traction and dynamic performance and environmental safety indicators

K M Sidorov, V E Yutt, A G Grishchenko and T V Golubchik

Moscow Automobile and Road Construction State Technical University (MADI), Leningradsky prospect, 64, Moscow, 125319, Russia

E-mail: kmsidorov@gmail.com

Abstract. The objective of the work presented in this paper is to describe the implementation of the technical solutions have been developed, with regard to structure, composition, and characteristics, for an experimental prototype of an electric vehicle which has been converted from a conventional vehicle. The methodology of the study results is based on the practical implementation of the developed concept of the conversion of conventional vehicles into electric vehicles. The main components of electric propulsion system of the experimental prototype of electric vehicle are developed and manufactured on the basis of computational researches, taking into account the criteria and principles of conversion within the framework of presented work. The article describes a schematic and a design of power conversion and commutation electrical equipment, traction battery, electromechanical transmission. These results can serve as guidance material in the design and implementation of electric propulsion system (EPS) components of electric vehicles, facilitate the development of optimal technical solutions in the development and manufacture of vehicles, including those aimed at autonomy of operation and the use of perspective driver assistance systems. As part of this work, was suggested a rational structure for an electric vehicle experimental prototype, including technical performance characteristics of the components of EPS.

1. Introduction

Advanced technologies in the field of machine vision and object detection, positioning and geolocation, actuators and software - this is an incomplete list of necessary components of an autonomous vehicle. However, the EPS is no less important component of unmanned transport of the present and future. Due to advanced technical solutions in the field of electric drive, voltage converters and traction batteries, the vehicle is fully capable of meeting modern standards and requirements with respect to operational safety, environmental and energy efficiency, as well as the criteria for the implementation of automated control. In other words, an autonomous (unmanned) vehicle of the future is predominantly an electric vehicle (EV).

In light of existing environmental problems, interest in EVs continues to grow as their use is devoid of polluting emissions and the clean energy they provide as an alternative compensates for the energy and resources that are necessary in their construction. Upgrading the existing fleet of vehicles currently in use, inclusive of passenger vehicles and taxis as well as vehicles for municipal and utility use, is feasible with the technology available today. This would include converting the current fleet of

IOP Conf. Series: Materials Science and Engineering **315** (2018) 012026 doi:10.1088/1757-899X/315/1/012026

internal combustion vehicles into ones with an electric drive. This article presents the full process and results of this implementation process of converting [1] conventional internal combustion vehicles into EVs based on experimental prototype of EV.

2. Aspects of practical implementation

2.1. Aspects of conversion in electric vehicles and trends to solve them

The design of electric propulsion systems for modern electric-accumulating vehicles is inseparably associated with the required operating modes and operational features of the vehicle. The parameters and characteristics of the latter depend strongly on the structure of the electric or combined power system as well as the layout constraints of the base motor vehicle.

The main problem in creating an electric vehicle based on the mass produced vehicles determines the complexity of EPS configuration on board in the vehicle. The limited space available for placement of electrical components and allowed weight of the base motor vehicle, along with the basic principles of conversion [1], necessitates the use of optimized technical solutions and perspective element base as well as modern technologies and materials.

The significant reduction in unladen weight of the vehicle can be achieved mainly through the use of modern element base. This mainly concerns the chemical power sources, traction electrical machines and converter equipment, which shall meet the high specific performance, including specific power, energy and torque.

The presence of the speed-change gearbox in the EV will lead to a decrease in the overall efficiency of the electromechanical transmission and will also increase the weight of the overall propulsion system. The optimal solution would be a direct electric drive scheme that excludes the use of the speed-change gearbox, which also opens up more opportunities to improve traction and dynamic characteristics of EV compared to the base motor vehicle as well as reduces the weight of the propulsion system, eliminates the step change in torque and speed on the drive axle. The scheme of stepless transmission is the basic technical solution adopted for further consideration and implementation in the experimental prototype of the convertible EV.

2.2. Purpose and key characteristics of experimental prototype of electric vehicle

As part of the discussed project, the experimental prototype of electric vehicle converted on the basis of the car, should be considered as an object for implementation of the full technological cycle for conversion of conventional vehicles into electric vehicles.

The main design and technical and operational parameters implemented in the experimental prototype of EV include:

a) gross weight of no more than 1.75 tons;
b) wheel circuit of 4x2;
c) type of drive as rear wheel drive;
d) transmission is designed as stepless transmission (without use of gearbox);
e) type of electric power source is battery pack, based on lithium–ion battery cells;
f) type of traction motor-generator is the permanent magnet synchronous motor;
g) braking system is a standard system with use of electric braking and possibility of power regeneration;
h) type of traction motor cooling and a power voltage converter is liquid type;
i) maximum power of the traction motor-generator is not less than 140 kW;
j) peak allowable torque on the shaft of traction motor-generator is 700 N·m;
k) nominal voltage of electric power source is 360 V;
l) rated energy reserve of traction power source is 28 kW·h;
m) stand-alone driving range with zero toxicity is over 120 km.

The engine room of experimental prototype of electric vehicle is shown on figure 1.

Figure 1. The engine room of experimental prototype of electric vehicle.

2.3. Substantiation of choice of base motor vehicle for experimental prototype of electric vehicle

The base motor vehicle referred to here is the conventionally designed vehicle based on the internal combustion engine and converted into an EV on its chassis. Though a propulsion system (propulsion, transmission) is the only thing subject to change in the vehicle. The basic systems of the base vehicle remain unchanged relative to the initial performance and should run normally under the conditions arising from the dismantling of the internal combustion engine.

The car of a popular brand was selected as an alternative option of the base vehicle for the orientation in the applied research and developments. This choice is based on:

- difficulties in terms of technical and functional equipment, on-board electrical equipment, which is capable fullest provide the basis to solve the problems of applied researches, including the adaptation of the standard vehicle staff car systems;
- the possibility of rational distribution of electric drive on-board;
- parameters of standard internal combustion engine, that determine significant opportunity to improve the characteristics of the base vehicle through the use of traction electric drive system;
- presentable exterior, able to garner attention, including of young people, to the results of the development, problems of the environment and the consumption of hydrocarbon fuels.

2.4. Structure and composition of EPS of experimental prototype of EV

In accordance with UN Vehicle Regulation No. 100, the basic functions of electric propulsion system necessary for the implementation in the electric vehicle shall include the production and transformation of:

- mechanical energy for motion of the vehicle;
- electric energy to feed systems ensuring this motion;
- On the basis of the functional characteristics it can generate a generalized structural diagram of the power circuit of electric power system of EV shown in figure 2;
- This EPS structure is implemented in the experimental prototype of EV and it is composed of:
- electric power source (battery, BAT);
- traction motor-generator (TMG);
- traction inverter and TMG control system (TI);
- shutoff protection unit for traction motor (SPU);
- HV power commutation and distribution unit for electric power source (PCDU);
- on-board charging device (OBCD);
- HV air conditioning and heating system (HVAC);

- DC-DC converter for high-voltage battery (DC-DC) and low-voltage power supply for on-board electric equipment;
- auxiliary low-voltage battery (AB);
- vehicle low-voltage electric equipment (LVEE);
- devices for functioning the standard systems of base motor vehicle (DFSS);
- EPS component cooling system (CS);
- control, coordination and monitoring devices for EPS components – high level(main) control system (MCS);
- operational speed controllers of EV (OSC);
- battery management system (BMS);
- Out of scope of the electric propulsion system (EPS):
- mechanical transmission (MT);
- driving wheels (DW).

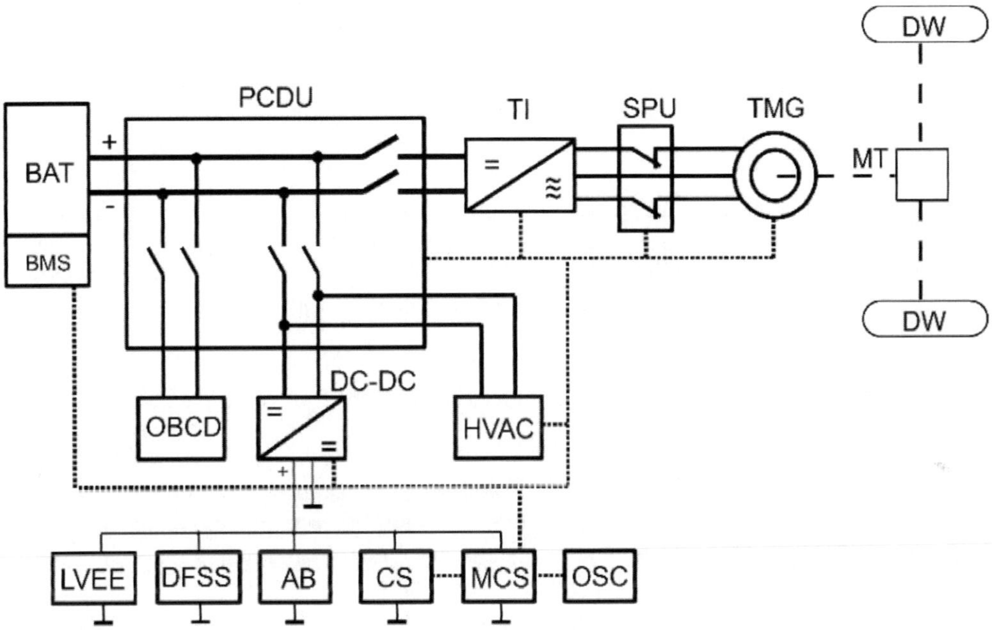

Figure 2. The generalized structural diagram of the power circuit of EPS of EV.

2.5. EPS component layout on experimental prototype of convertible EV

It was highlighted earlier that the particular importance at converting to an electric vehicle shall be given to the rational distribution of the most weight significant components such as traction battery and the electric motor. This task is complicated by the fact that, on the one hand, the weight of traction battery and traction motor shall be distributed on the vehicle, taking into account the allowable mechanical loads on the chassis and body elements, on the other hand, the need to minimize the length of the EPS power and control of electrical circuits. In this paper, the solution of this task was carried out at the expense of battery modular design, consisting of stand-alone functional units – unified battery modules (UBM) and the special design of traction motor.

The layout of EPS components on the base motor vehicle is provided in figure 3 (where MSD - manual service battery disconnect).

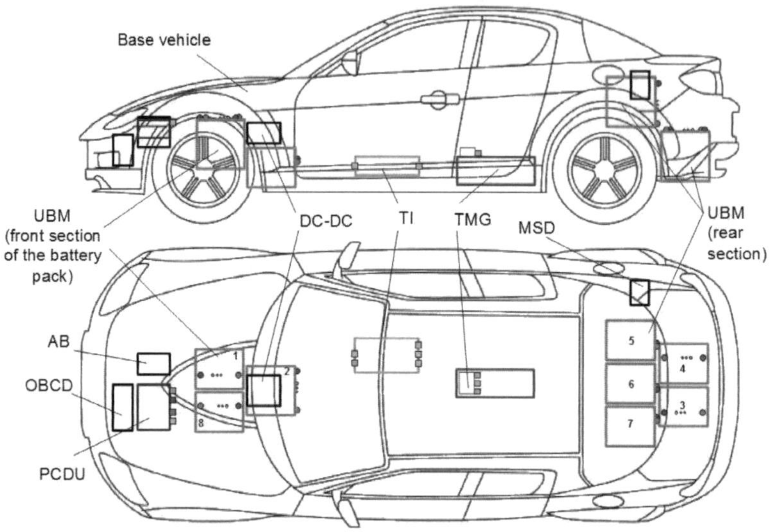

Figure 3. The layout of EPS components on the base motor vehicle.

The design features of base motor vehicle, requirements for placement of equipment required to implement characteristics of electric equipment of EV and the developed basic principles of conversion were crucial while elaborating the EPS key components. The technical description of the key components of developed and manufactured EPS components of experimental prototype of EV is given below.

3. Description of EPS components for experimental prototype of convertible EV

3.1. Traction inverter

The purpose of traction inverter is conversion of voltage as constant for the effective value of the voltage in the adjustable frequency and magnitude of the AC three-phase voltage to control the rotational speed and torque of the traction motor-generator in traction and braking modes.

The main functional features of traction inverter are as follows:

1. The conversion of input DC voltage to three-phase voltage, controlled in magnitude and frequency.
2. The smooth lead-out (in terms of driving comfort of EV) of traction motor to preset current (shaft torque).
3. Automatic maintenance of the preset current (torque) in voltage and power regulation zones of traction motor.
4. Smooth control of torque and rotation speed of traction motor in low power loss mode.
5. Change of the direction of rotation of traction motor.
6. The electric (regenerative) braking the vehicle by translating the traction motor in generator mode with a smooth control of braking torque.
7. Shutoff protection of traction motor in excess of maximum values of force multipliers and input voltage.

3.2. HV power commutation and distribution unit (PCDU)

The purpose of HV power commutation and distribution unit is power commutation in the traction power system and distribution of high voltage between the high voltage devices.

Key functions of PCDU:

1. The remote connection (disabling) of electric propulsion system to electric power source on two terminals (positive and negative).

2. The preliminary battery charge of filter capacitors with limitation of charging current.
3. Automatic disabling of the electric propulsion system from power voltage of electric power source at emergency or abnormal conditions of EPS components.
4. High-voltage distribution of the electric power source between traction inverter, DC-DC converter, HV package switchgear.
5. Integration of devices for monitoring the electrical characteristics of the operating modes of the traction battery and monitoring insulation in EPS.

The PCDU design is shown in figure 4.

3.3. Traction motor-generator

The prospective contactless traction motor-generator is used as a part of the electric propulsion system of the experimental prototype of EV.

Key features of traction motor:

- Type: permanent magnet synchronous motor (PMSM);
- Peak torque: 700 Nm;
- Maximum power on shaft, not less than 140 kW;
- Maximum phase current: 695 A;
- Weight: 97 kg;
- Length: 582 mm;
- Stator diameter, max.: 237 mm;
- Type of cooling: liquid;
- Material of permanent magnets: SmCo;
- Specific power: 1.52 kW/kg;
- Specific torque: 7.22 Nm/kg.

Mechanical characteristic of traction motor is provided in figure 5.

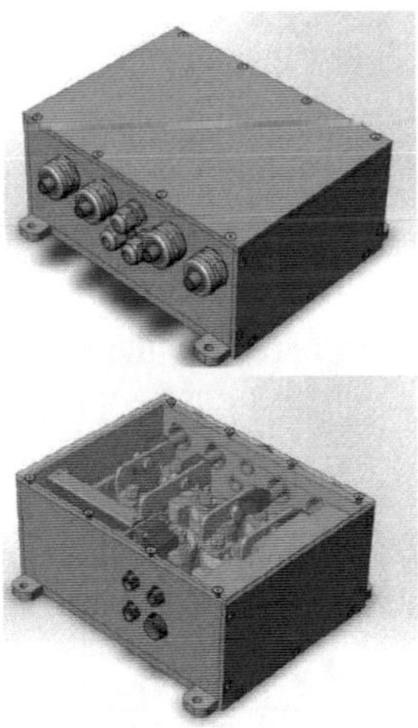

Figure 4. The PCDU design.

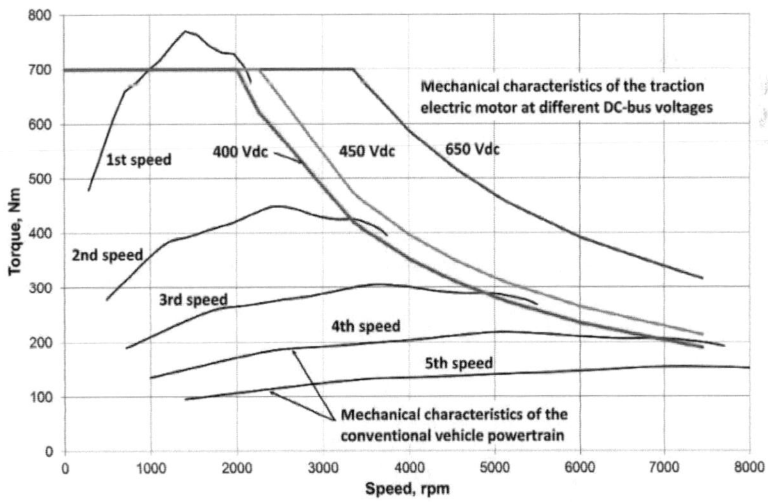

Figure 5. Combined mechanical characteristics of the traction electric motor of EV and convention vehicle powertrain.

The selected technical design of the traction motor is determined by specific feature of conventional design vehicle, which is a central tunnel. The general view of traction motor is shown in figure 6. The used traction motor has a small diameter with an increased length of the active part of the stator. This feature allows a favorable placement of traction motor in the narrow central tunnel of vehicle in close proximity to the differential and drive axle.

Figure 6. The general view of traction electric motor with mounting.

3.4. Unified battery module

The electric power source is the main on-board power source for motion of an electric vehicle. Key features of unified module of traction battery (UBM):

- Configuration: 12s2p;
- Cells type: Li-ion (NMC+LMO);
- Rated voltage: 45 Vdc;
- Maximum discharge current (10 s): 480 A;
- Maximum charge current: 160 A;
- Rated capacity: 80 A·h;
- Rated energy reserve: 2,6 kW·h;
- Dimensions: 255 x 338 x 316 mm;
- Weight of UBM: 33 kg.

On the basis of unified module of traction battery, it can be built an electric power source of any configuration and preset characteristics.

The characteristics of traction battery pack (TBP) for electric equipment of the electric vehicle:

- Number of connected in-series UBMs: 8 pcs;
- Rated voltage: 360 Vdc;
- Rated energy reserve: 28,8 kW·h;
- Maximum energy reserve: 32,3 kW·h;
- Weight of battery pack: 264 kg.

Technically, UBM is based on the principle of modularity and unification to generate on the basis of a functional and complete stand-alone unit of the electric power source of the preset configuration and characteristics.

Structurally, the UBM consists of composite plastic elements-frames that generate an UBM housing (see figure 7).

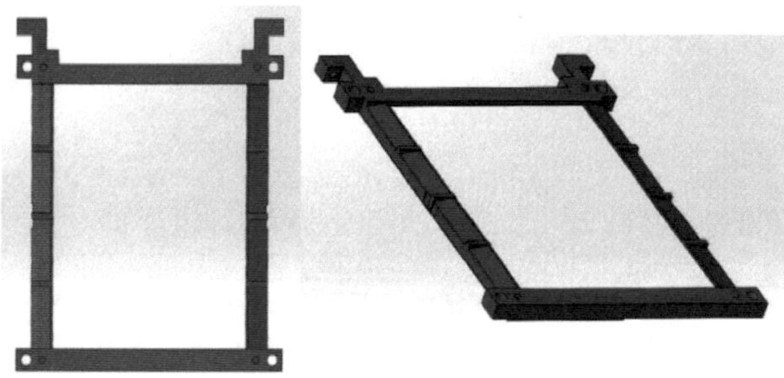

Figure 7. The composite plastic frame of UBM.

Figure 8. The UBM assembly section.

One UBM section consisting of two frames encloses two prismatic battery cells (see figure 8). The frame design determines the following functions:

- building-up the composite protective housing;
- fixation of battery cells in module;
- access for heat sink from battery cell.

The UBM sections are combined together using the tightening pins into a single structure, protected by aluminum plates on the edges.

The electrical connection of the battery electrodes is carried out in a special compartment formed by the configuration of the plastic frame. The same compartment encloses battery status control devices (BMS board) and is provided to connect the power cables to the power terminals of the UBM.

The heat sink of each battery in module is done through special air ducts, formed by a plastic frame. If necessary, forced air circulation can be arranged through air ducts using cooling fans fitted on the UBM or nozzles laid down from the external cooling system. The design of the UBM is shown in figures 9 and 10.

Figure 9. The concept of UBM.

Figure 10. Traction battery modules of Convertible EV.

3.5. Mechanical transmission components of EV

The design of the mechanical transmission assemblies provides:

- minimization of weight and dimensions of traction electric drive;
- minimization of mechanical losses in the transmission of torque;
- lack of intermediate devices of torque conversion and rotation speed (gears, gearbox);
- design reliability and relative ease of manufacture at a maximum preservation of standard units and aggregates of the base motor vehicle.

The general view of traction electric drive for electric equipment of the electric vehicle is provided in figure 11.

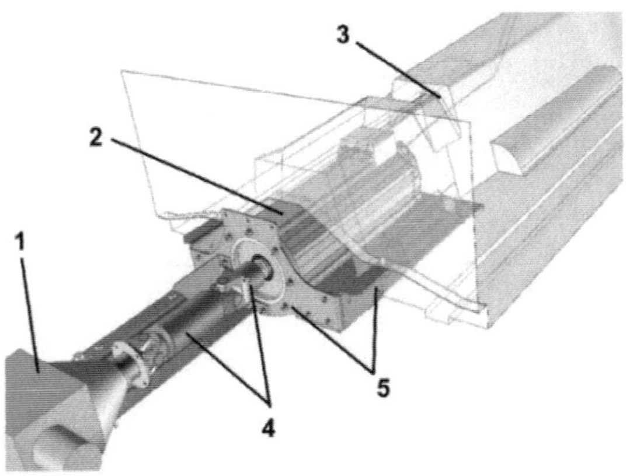

Figure 11. The general view of traction electric drive for electric equipment of EV.

There are shown: 1 is main gear (differential) of the base motor vehicle; 2 is traction motor-generator; 3 is central tunnel of base motor vehicle; 4 is mechanical transmission elements; 5 is structural elements to install electric equipment.

4. Conclusions

The practical implementation of the developed full technological cycle concept for conversion of conventional vehicles to EVs is based on the solution of a set of particular tasks concerning optimization of EPS structure, type, and characteristics of its constituent components, as well as the characteristics of their design and placement in the vehicle.

An idea of conversion is not new, however, today's existing concepts and technical solutions eliminate the creation of truly effective solutions in relation to consumer and operational properties of an electric vehicle. The latter is possible only by following the specific criteria and principles, among which it is especially worthwhile to highlight the principle of conservation of traction and dynamic performance and functions of standard on-board systems of the base motor vehicle. The modern electric vehicle of today shall be distinguished by:

- use of AC traction motors-generators (squirrel-cage motors or permanent magnet synchronous motors);
- use as a component of the traction battery of battery cells based on lithium-ion technology with values of specific energy of at least 140 Wh/kg;
- modular design of the electric power source;
- use of advanced methods of algorithms [2] to control electric machines as well as liquid cooling in converted electrical equipment;
- lack of speed-change gearbox;
- lack of cabin heaters operating on hydrocarbon fuels or based on traditional thermoelectric power resistive heaters.

The above features distinguish this developed concept of a full technological cycle for conversion of conventional internal combustion vehicles to electric vehicles and the presented results of the practical implementation of this concept.

The results in the presented work can be applied to both passenger vehicles and special municipal vehicles and buses and present a reality in which it is feasible to upgrade the existing motor transport systems currently in use in public transportation.

Acknowledgments

The article is part of the applied scientific research (RFMEFI57714X0156) conducted at Moscow Automobile and Road Construction State Technical University (MADI) and sponsored by the Ministry of Education and Science of the Russian Federation.

References

[1] Sidorov K, Golubchik T, Yutt V et al. 2016 December *Indian Journal of Science and Technology* The concept of converted electric vehicle with advanced traction and dynamic performance and environmental safety indicators: theoretical basis **9** (48) 1-12 (DOI: 10.17485/ijst/2016/v9i48/101220)

[2] Ospanbekov B, Sidorov K, Golubchik T et al. 2016 December *Indian Journal of Science and Technology* Study of energy indicators and features of propulsion system main components of electric vehicle using mathematical simulation **9** (48) 1-12 (DOI: 0.17485/ijst/2016/v9i48/102923)

Increasing the stability of the articulated lorry at braking by locking the fifth wheel coupling

G I Skotnikov, M M Jileykin and A I Komissarov

Department of Wheeled Vehicles, Bauman Moscow State Technical University, Moscow, 105005 Russian Federation

E-mail: skotnikovg@bmstu.ru

Abstract. The jackknifing of the articulated lorry is determined by the loss of stability with respect to the vertical axis of the fifth wheel coupling, which can be caused by the failure of the brake system, the displacement of the center of mass of the semitrailer or tractor from the longitudinal axis of the vehicle, the road parameters (longitudinal and transverse slopes), the difference in the friction coefficients under the sides of the articulated lorry. In this regard, the issue of creating devices that prevent the jackknifing, and their control systems is important. A method is proposed for maintaining the stability of the movement of articulated lorry when braking both on a straight line and in a turn by blocking the relative rotation of the tractor and the trailer. Blocking occurs due to the creation of a stabilizing moment in the direction opposite to the angular rate of folding. To test the developed algorithm for locking the fifth wheel coupling, a mathematical model of the spatial motion of the articulated lorry was developed, including the models of interaction of an elastic tire with a rigid terrain, suspension systems, transmission, steering, fifth-wheel coupling. The efficiency and effectiveness of the coupling locking control system is proved by comparing the results of the simulation of a straight-line braking and braking in turn. It is shown that the application of the control system significantly increases the stability of the road train.

1. Introduction

At emergency braking of articulated lorry even on a road with a high coefficient of friction, stability loss in the form of jackknifing is possible. Jackknifing is the loss of the stability of the towing vehicle and (or) the semi-trailer when braking, when the angle between the longitudinal axes of the tractor and the trailer (the folding angle) exceeds the value necessary for driving along the trajectory specified by the driver. Jackknifing can be caused by the failure of the brake system, the displacement of the center of mass of the trailer or the tractor from the longitudinal axis of the vehicle, road parameters (longitudinal and transverse slopes), difference in the friction coefficients under the sides of the articulated lorry. This type of loss of stability may lead to the lane departure, which can in turn lead to collision with the other vehicles, including those moving in the opposite direction. Modern active safety systems, such as the anti-lock braking system "ABS", the electronic stabilization system "ESP" and others significantly improve the stability of the articulated lorry, but do not exclude the danger of jackknifing, in this connection, the issue of developing systems and devices which would eliminate this effect [1] is of great importance.

IOP Conf. Series: Materials Science and Engineering **315** (2018) 012027 doi:10.1088/1757-899X/315/1/012027

As the analysis of patents has shown, there are several ways of fixing the relative rotation of the tractor and the semi-trailer around the pin axis: by equipping the fifth wheel coupling with elements the position of which significantly limits the folding angle; by equipping the fifth wheel coupling with friction mechanisms that limit the folding speed to a full stop in a wide range of folding angles. The drawbacks of the first embodiment are the impacts caused by the selection of the gaps between the fixing elements, as well as the impossibility of locking the coupling device during cornering with a predetermined folding angle, so the second method is more preferable.

The aim of the work is the development of a system that increases the stability of the movement of an articulated lorry when it brakes both on a straight line and in a turn by locking the fifth wheel coupling.

2. Description of the developed algorithm

The control system of the locking mechanism of the fifth wheel coupling is a closed loop system. External disturbance f (figure 1) causes the jackknifing, while the folding speed increases. The angular speed $\dot{\theta}_2 - \dot{\theta}_1$ of rotation of the semitrailer relative to the tractor around the axis of the kingpin is measured by the sensor and compared with zero.

Based on the received signal h_brake from the comparison unit and the signal of the driver pressing the brake pedal, the proportional regulator generates a control action, which by the locking mechanism is converted into a stabilizing moment M_{st}, which prevents the jackknifing of the articulated lorry. The magnitude of the moment is proportional to the angular velocity of the folding. One of the possible ways to meet the requirements set by the developed control system is to install a mechanism similar to the wheel brake on the fifth wheel coupling of the tractor. In this case, the direction of the stabilizing moment will be provided by the direction of the frictional forces, and the magnitude - by the degree of pressing of the brake pads.

The folding rate control as contrasted to the angle control used by many researchers makes it possible to stabilize the movement of the articulated lorry not only when braking on a straight line, but also when braking on a curved trajectory, when the probability of loss of stability is much higher.

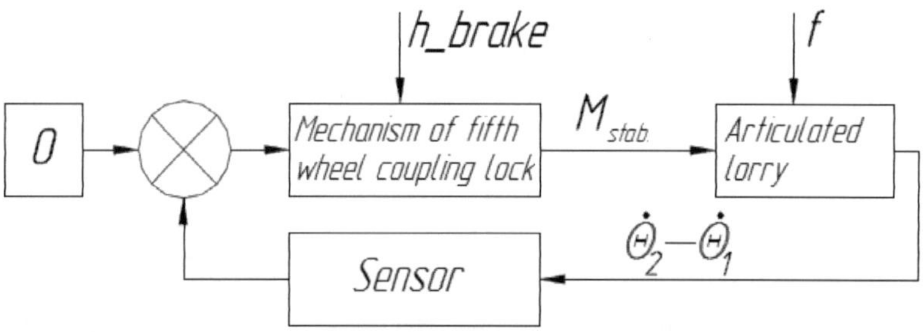

Figure 1. Block diagram of the control system of the locking mechanism of the fifth wheel coupling.

The law of operation of the control system is determined by the following equation:

$$\begin{cases} h\_\text{brake} = 0 & M_{stab} = 0; \\ h\_\text{brake} = 1 & M_{stab} = P(\dot{\theta}_2 - \dot{\theta}_1); \end{cases} \quad (1)$$

0 value of h_brake means that the brake pedal is not pressed, 1 – that it is pressed; P – coefficient of proportional regulator.

Verification of the efficiency and efficiency of the algorithm was carried out by simulation, for this, a mathematical model of the spatial motion of the road train was used.

3. Description of the mathematical model.

The spatial motion of an articulated lorry, consisting of a three-axle vehicle – a tractor, and a two-axle semi-trailer (figure 2), is regarded as the motion of two bodies connected by a hinge. The simulation model includes the model of interaction of an elastic tire with a rigid terrain, models of dependent and balanced tractor and semitrailer suspensions, model of the tractor transmission, steering, braking system and fifth wheel coupling.

In accordance with paragraphs 5.2.1.22 and 5.2.2.13 of GOST R 41.13-99 (UNECE Regulation No. 13), motor vehicles of categories M_2, M_3, N_2 and N_3 with not more than four axles, trailers of categories O_3 and O_4 should be equipped with anti-lock systems, so the simulation of braking a road train without an anti-lock system (ABS) does not make sense [2]. In the brake system model, the ideal ABS algorithm is realized, according to which the braking torque T_i on the i-th wheel, taking into account the work of the ABS, will be calculated in accordance with the following equation [3]:

$$\begin{cases} T_i = \frac{\omega_{wi} r_k}{V_X} M_{Ti}, & V_X \neq 0; \\ T_i = 0, & V_X \neq 0; \end{cases} \tag{2}$$

The kinematic parameters and external disturbances are coupled by a set of differential equations.

Using the theorems of variation of translational and angular momentum of a rigid body in the projections on the movable coordinate system axis, we obtain the equations of motion for the tractor and semi-trailer [4,5]:

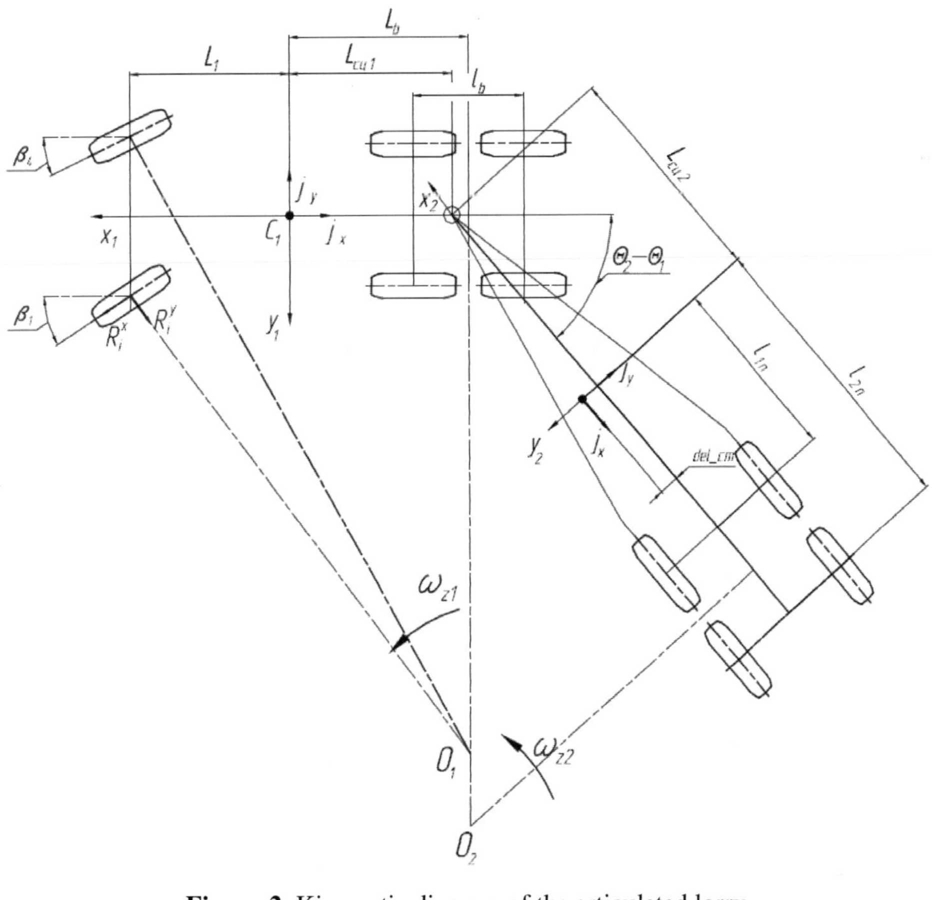

Figure 2. Kinematic diagram of the articulated lorry.

IOP Conf. Series: Materials Science and Engineering **315** (2018) 012027 doi:10.1088/1757-899X/315/1/012027

$$\begin{cases}(m_t+2N_tm_{kt})\dot{V}_{cxt}+(m_t+2N_tm_{kt})(\omega_{yt}V_{czt}-\omega_{zt}V_{cyt})=G_{xt}+F_{xt}+\sum_{i=1}^{2N_t}R_{it}^{x}+R_{tX_2fwc}\\ (m_t+2N_tm_{kt})\dot{V}_{cyt}+(m_t+2N_tm_{kt})(\omega_{zt}V_{cxt}-\omega_{xkt}V_{czt})=G_{yt}+F_{yt}+\sum_{i=1}^{2N_t}R_{it}^{y}+R_{tY_2fwc}\\ m_t\dot{V}_{czt}+m_t(\omega_{xkt}V_{cyt}-\omega_{yt}V_{cxt})=G_{zt}+F_{zt}+\sum_{i=1}^{2N_t}P_{it}^{z}+R_{tZ_2fwc}\\ I_{xt}\dot{\omega}_{xt}+\omega_{yt}\omega_{zt}(I_{zt}-I_{yt})=M_{xt}(F)+\sum_{i=1}^{2N_t}M_{xt}[P_{it}]+R_{tY_2fwc}(h_{ct}-h_{fwc})\\ I_{yt}\dot{\omega}_{yt}+\omega_{zt}\omega_{xt}(I_{xt}-I_{zt})=M_{yt}(F)+\sum_{i=1}^{2N_t}M_{yt}[P_{it}]+R_{tZ_2fwc}L_{cfwc}+R_{tX_2fwc}(h_{ct}-h_{fwc})\\ I_{zt}\dot{\omega}_{zt}+\omega_{xt}\omega_{yt}(I_{yt}-I_{xt})=M_{zt}(F)+\sum_{i=1}^{2N_t}M_{zt}[R_{it}]+\sum_{i=1}^{2N_t}M_{nkit}+R_{tY_2fwc}L_{cfwc}-M_{st}\end{cases} \quad (3)$$

$$\begin{cases}(m_{st}+2N_{st}m_{kst})\dot{V}_{cxst}+(m_{st}+2N_{st}m_{kst})(\omega_{yst}V_{czst}-\omega_{zst}V_{cyst})=G_{xst}+F_{xst}+\sum_{i=1}^{2N_{st}}R_{ist}^{x}+R_{stX_2fwc}\\ (m_{st}+2N_{st}m_{kst})\dot{V}_{cyst}+(m_{st}+2N_{st}m_{kst})(\omega_{zst}V_{cxst}-\omega_{xkst}V_{czst})=G_{yst}+F_{yst}+\sum_{i=1}^{2N_{st}}R_{ist}^{y}+R_{stY_2fwc}\\ m_{st}\dot{V}_{czst}+m_{st}(\omega_{xkst}V_{cyst}-\omega_{yst}V_{cxst})=G_{zst}+F_{zst}+\sum_{i=1}^{2N_{st}}P_{ist}^{z}+R_{stZ_2fwc}\\ I_{xst}\dot{\omega}_{xst}+\omega_{yst}\omega_{zst}(I_{zst}-I_{yst})=M_{xst}(F)+\sum_{i=1}^{2N_{st}}M_{xst}[P_{ist}]+R_{stY_2fwc}(h_{cst}-h_{fwc})+R_{stZ_2fwc}del_{cst}\\ I_{yst}\dot{\omega}_{yst}+\omega_{zst}\omega_{xst}(I_{xst}-I_{zst})=M_{st}(F)+\sum_{i=1}^{2N_{st}}M_{st}[P_{ist}]+R_{stZ_2fwc}l_{1c}+R_{stX_2fwc}(h_{cst}-h_{fwc})\\ I_{zst}\dot{\omega}_{zst}+\omega_{xst}\omega_{yst}(I_{yst}-I_{xst})=M_{zst}(F)+\sum_{i=1}^{2N_{st}}M_{zst}[R_{ist}]+\sum_{i=1}^{2N_{st}}M_{nkist}+R_{stY_2fwc}l_{1c}+R_{stX_2fwc}del_{cst}+M_{st}\end{cases} \quad (4)$$

ω_x, ω_y, ω_z	- projections of the angular velocity vector on the axes of coordinate system *OXYZ;*
$\dot{\omega}_x, \dot{\omega}_y, \dot{\omega}_z$	- projections of the angular acceleration vector on the axes of coordinate system *OXYZ;*
V_{cx}, V_{cy}, V_{cz}	- projections of the velocity of center of mass vector on the axes of coordinate system *OXYZ*;
$\dot{V}_{cx}, \dot{V}_{cy}, \dot{V}_{cz}$	- projections of the acceleration of center of mass vector on the axes of coordinate system *OXYZ*;
G_x, G_y, G_z	- projections of the gravity vector on the axes of coordinate system *OXYZ*;
F_x, F_y, F_z	- projections of the external force vector on the axes of coordinate system *OXYZ;*
R_i^x, R_i^y, R_i^z	- projections of the interaction force between the ground and the wheel on the axes of coordinate system *OXYZ;*
$R_{X_2fwc}, R_{Y_2fwc}, R_{Z_2fwc}$	- projections of the reaction vector in the fifth wheel coupling;
P_i	- forces of the suspensions of the wheels;
$M_x(F)$, $M_y(F)$, $M_z(F)$	- projections of the moment on the force of external influence on the axes of coordinate system *OXYZ;*

$M_x(R_i)$, $M_y(R_i)$, $M_z(R_i)$	- projections of the moment of the interaction force between the ground and the wheel on the axes of coordinate system *OXYZ;*
M_{st}	- the stabilizing moment created by the locking mechanism of the fifth wheel coupling;
I_x, I_y, I_z	- moments of inertia relative to the axes the axes of coordinate system *OXYZ;*
M	- mass of sprung parts;
m_k	- mass of unsprung parts;
h_c	- the height of the center of mass of the sprung parts in the fixed coordinate system;
h_{fwc}	- the height of the center of the fifth wheel coupling in the fixed coordinate system;
L_{cfwc}	- longitudinal coordinate of the fifth wheel coupling in the fixed coordinate system;
l_{1c}	- longitudinal coordinate of the fifth wheel coupling in the semi-trailer coordinate system;
del_{cst}	- lateral displacement of the center of mass of the semi-trailer relative to the longitudinal axis;
N	- number of axles of the semi-trailer or truck.

* index *t* refers to the tractor, index *st* refers to the semi-trailer.

4. Verification of the efficiency and effectiveness of the developed algorithm

To check the efficiency of the control system of the locking mechanism of the fifth wheel coupling, the following computational experiments were carried out: "Emergency braking on a straight line", "Emergency braking in a turn" for the articulated lorry equipped with a locking mechanism and for the same lorry without the locking mechanism. All experiments were carried out with the characteristics of the road, corresponding to the asphalt-concrete coating, namely: the maximum longitudinal and transverse friction coefficients $\mu_{X\max} = 0.7$; $\mu_{Y\max} = 0.7$, the coefficients of $\mu(S)$ equation $S_1 = 0.1$, $S_0 = 0.05$, the coefficient of rolling resistance $f = 0.01$ [6].

The jackknifing of the articulated lorry occurred in the case of the lateral displacement of the center of mass of the semi-trailer from the longitudinal axis by 30 mm.

The parameters of the tractor and the trailer are shown in Tables 1 and 2.

Table 1. Characteristics and parameters of the tractor.

Sprung mass	7850 kg
Axle mass	500 kg
Wheel mass	120 kg
Moment of inertia relative to the transverse axis	56892 kg·m^2
Moment of inertia relative to the longitudinal axis	19238 kg·m^2
Moment of inertia relative to the vertical axis	40000 kg·m^2
Moment of inertia of the axle	190 kg·m^2
Moment of inertia of the wheel	16.5 kg·m^2
Wheelbase	4450 mm

Distance from the center of mass to the 1st axle	1905 mm
Distance from the center of mass to the equalizer axle	1803 mm
Distance between the 2nd axle and the 3rd axle	1430 mm
Distance from the center of mass to the fifth wheel coupling	1830 mm
Wheel track	2020 mm
Distance between the suspension springs	1616 mm
Center of mass height	1430 mm
Loaded tire radius	530 mm

Table 2. Characteristics and parameters of the semi-trailer.

Sprung mass	29400 kg
Axle mass	300 kg
Wheel mass	120 kg
Moment of inertia relative to the transverse axis	75000 kg·m^2
Moment of inertia relative to the longitudinal axis	30000 kg·m^2
Moment of inertia relative to the vertical axis	10000 kg·m^2
Moment of inertia of the axel	210 kg·m^2
Moment of inertia of the wheel	16.5 kg·m^2
Wheelbase	8900 mm
Distance from the center of mass to the 1st axle	3500 mm
Distance from the center of mass to the equalizer axle	4900 mm
Distance from the center of mass to the fifth wheel coupling	4000 mm
Wheel track	2100 mm
Distance between the suspension springs	1680 mm
Center of mass height	2300 mm
Loaded tire radius	530 mm

The simulation of emergency braking was carried out at the initial speed of 90 km/h. The diagram of pressing the brake pedal is shown in figure 3.

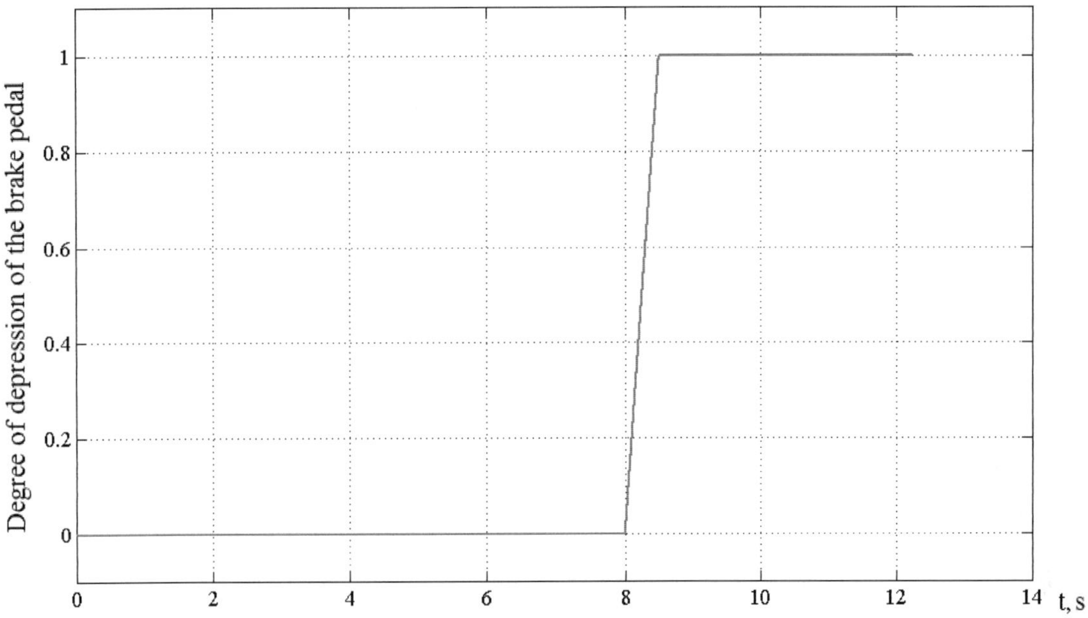

Figure 3. Brake pedal actuation.

Figure 4 shows the time histories of the tractor speed and the speed of the centers of mass of the 1st and 6th wheels when rolling without sliding. Figure 5 presents similar graphs for the center of mass of the trailer and the centers of mass of the 1st and 3rd wheels. Figures 6 and 7 show the time histories of the jackknifing angle and the jackknifing rate. Figure 8 shows the position of the articulated lorry after a complete stop (black solid lines depict the dimensions of the lane in accordance with GOST R 52399-2005, red line is the trajectory of the centers of the tractor and the semi-trailer).

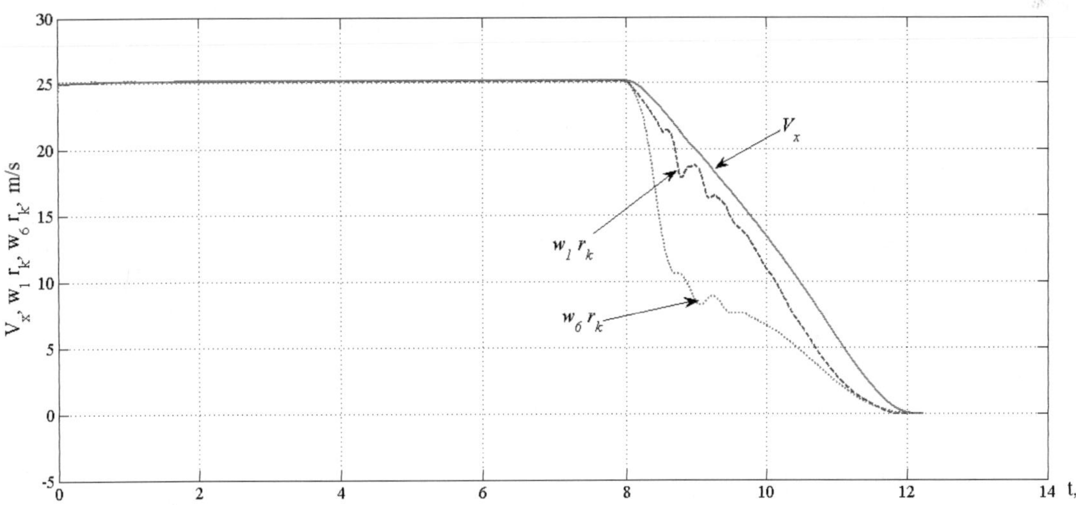

Figure 4. Longitudinal velocity of the tractor center of gravity and the speeds of the centers of mass of the wheels without sliding.

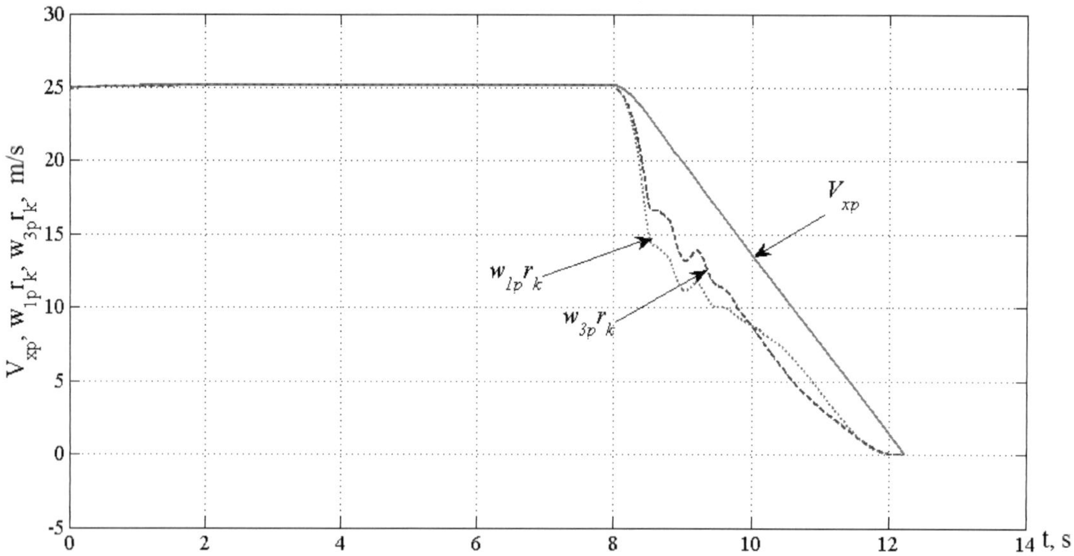

Figure 5. Longitudinal velocity of the semi-trailer center of gravity and the speeds of the centers of mass of the wheels without sliding.

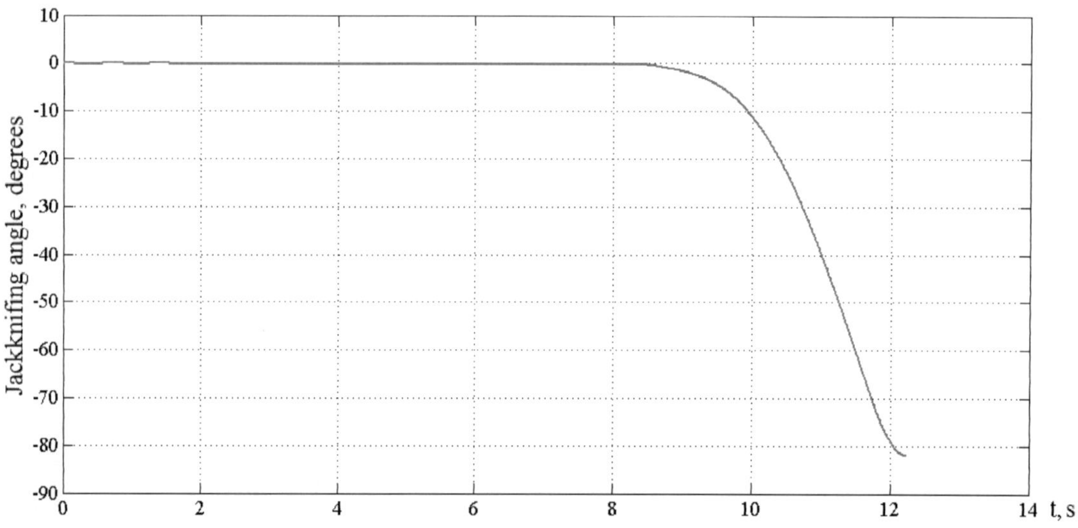

Figure 6. Jackknifing angle.

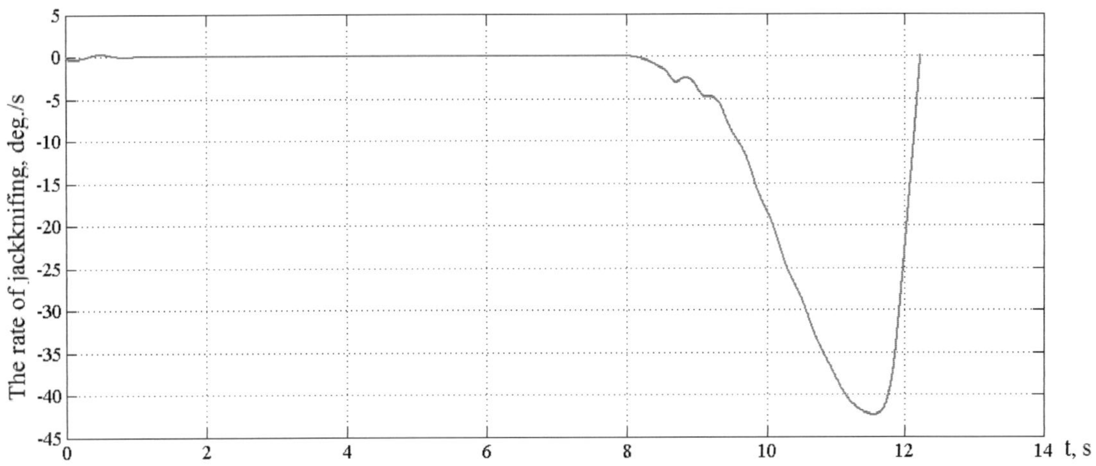

Figure 7. Jackknifing rate.

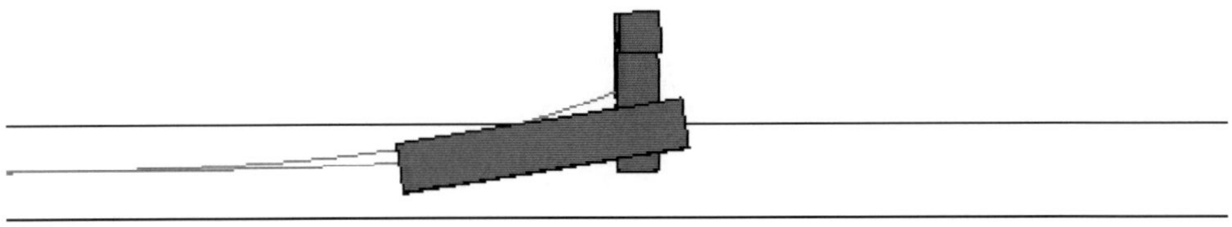

Figure 8. Position of the articulated lorry after a complete stop.

The simulation of the braking of the articulated lorry with the coupling locking device was carried out under identical conditions.

Figures 9 and 10 are similar to figures 4, 5. Figures 11 and 12 show time histories of the jackknifing angle and the jackknifing rate. Figure 13 shows the position of the articulated lorry after a complete stop (the elements of the figure are similar to ones of figure 8).

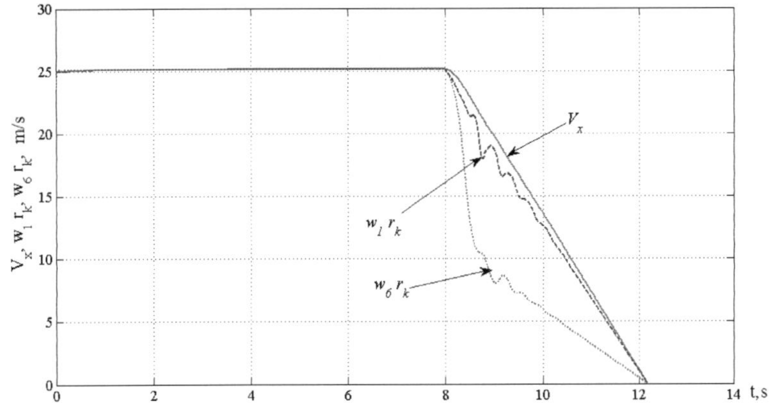

Figure 9. Longitudinal velocity of the tractor center of gravity and the speeds of the centers of mass of the wheels without sliding.

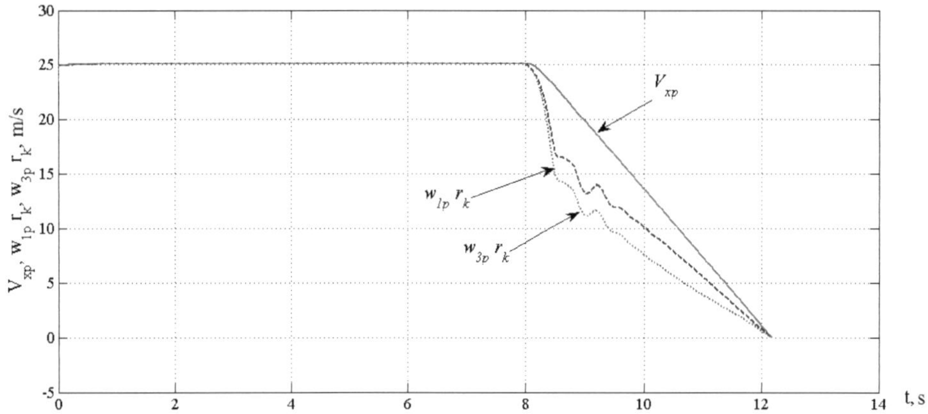

Figure 10. Longitudinal velocity of the semi-trailer center of gravity and the speeds of the centers of mass of the wheels without sliding.

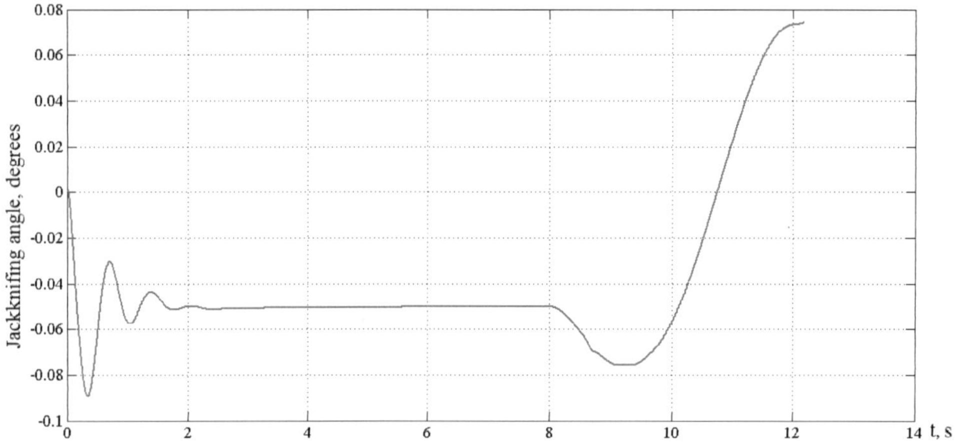

Figure 11. Jackknifing angle.

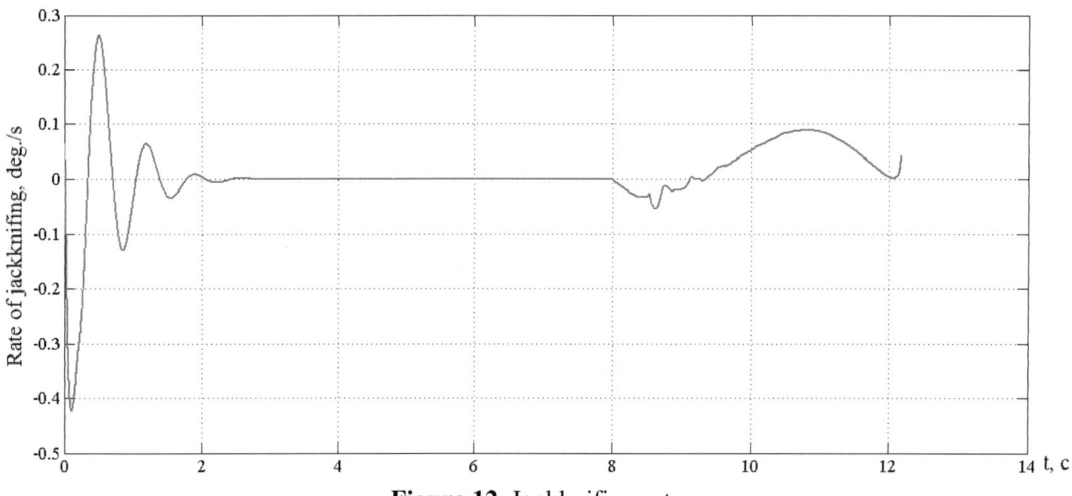

Figure 12. Jackknifing rate.

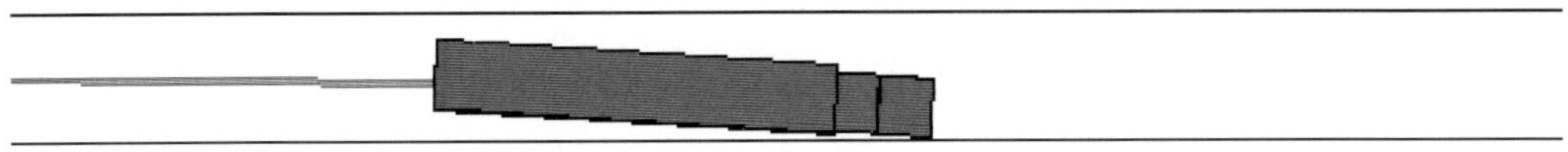

Figure 13. Position of the articulated lorry after a complete stop.

From the presented results, it can be concluded that in case of emergency straight-line braking of an articulated lorry without a locking device, there is a risk of jackknifing. Usually, the heavier semitrailer "pushes" the tractor to the adjacent lane (with the jackknifing angle reaching 80°) despite the ABS system installed on the vehicles of this type. These results are confirmed by the situations occurring in real-life operation, which once again proves the adequacy of the developed mathematical model.

The use of this locking system allows keeping the articulated lorry within the lane without the need of the driver control inputs. The jackknifing angle decreases by 3 orders of magnitude to 0.08°.

Simulation of emergency braking of the articulated lorry in a turn was carried out with the following initial conditions: initial speed 50 km/h, maximum average angle of the steered wheels 8°, turn radius 50 m. Time histories of the brake pedal actuation and the average angle of the steered wheels are shown in figure 14, 15.

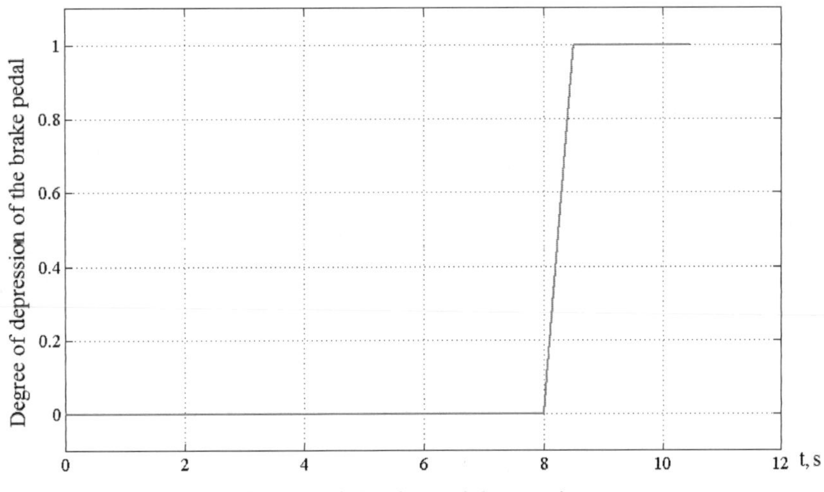

Figure 14. Brake pedal actuation.

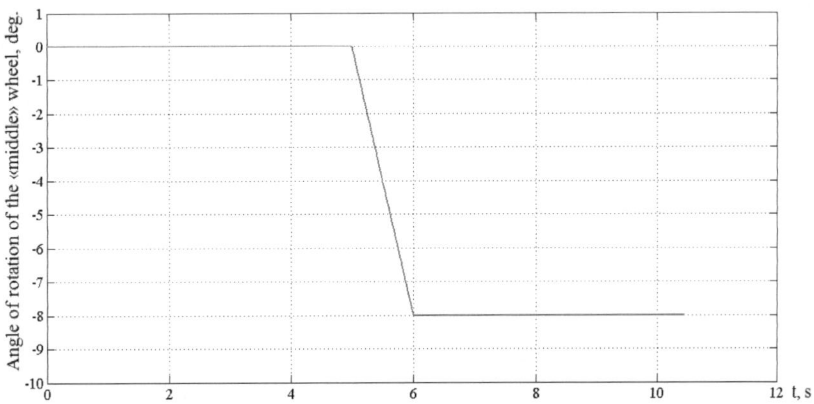

Figure 15. Average angle of the steered wheels.

IOP Conf. Series: Materials Science and Engineering **315** (2018) 012027 doi:10.1088/1757-899X/315/1/012027

In order to make sure that the articulated lorry steadily moves along the trajectory with the given initial conditions, the motion was simulated without braking. Figure 16 shows the trajectory and the marking of the corridor of motion.

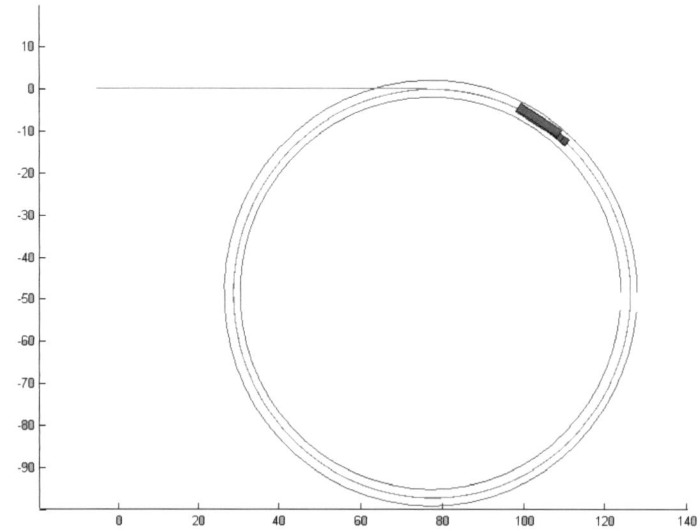

Figure 16. Trajectory of the articulated lorry in a turn.

Figures 17 and 18 show the time histories of the rate of the tractor and semi-trailer centers of mass and the speed of rolling without sliding the centers of mass of the wheels during emergency braking of the articulated lorry in a turn without activating the locking of the fifth wheel coupling. In figures 19 and 20 show the time histories of the jackknifing angle and the rat. Figure 21 shows the position of the articulated lorry after a complete stop. Figures 22 and 23 show graphs similar to those shown at figure 17, 19, but with the operating device locking the fifth wheel coupling. Figures 24 and 25 show the angle of jackknifing and the rate of jackknifing. Figure 26 shows the position of the articulated lorry after a complete stop.

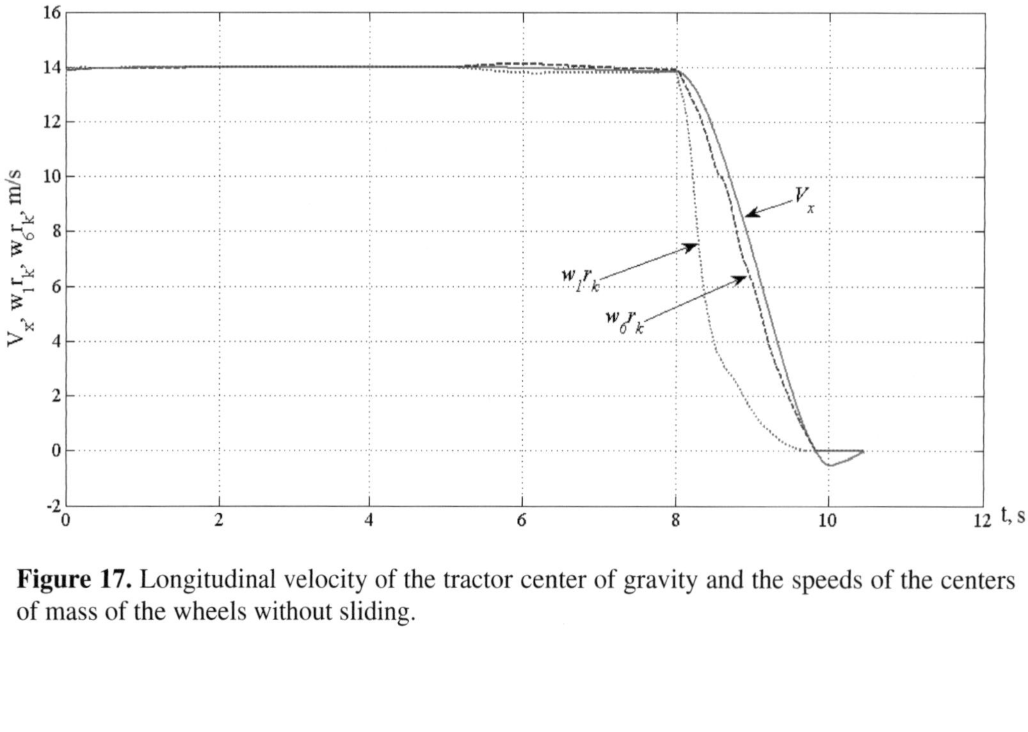

Figure 17. Longitudinal velocity of the tractor center of gravity and the speeds of the centers of mass of the wheels without sliding.

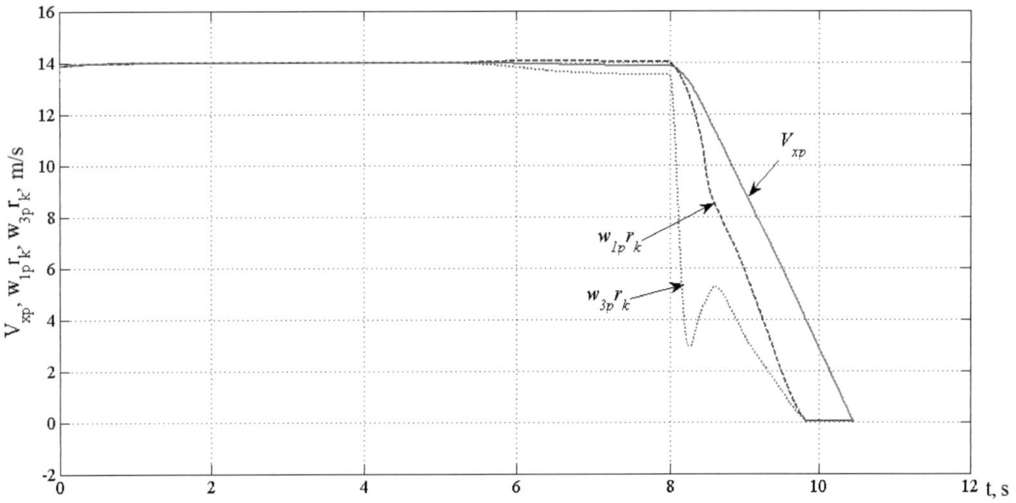

Figure 18. Longitudinal velocity of the semi-trailer center of gravity and the speeds of the centers of mass of the wheels without sliding.

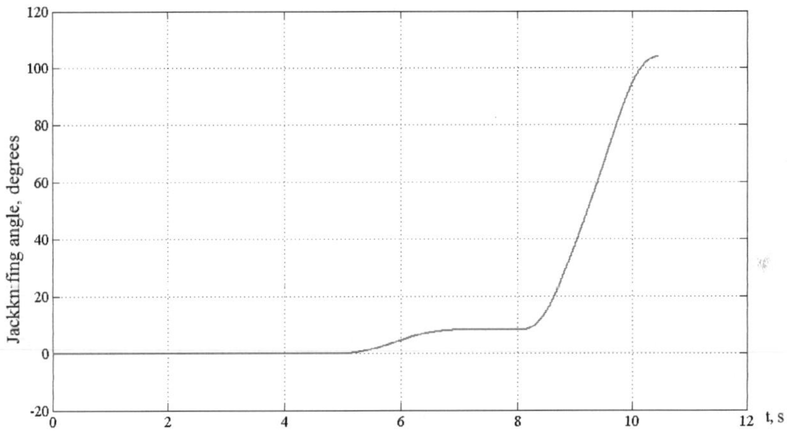

Figure 19. Jackknifing angle.

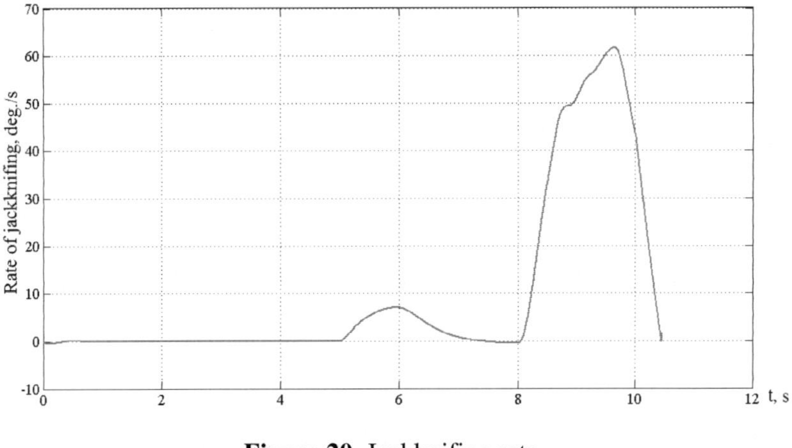

Figure 20. Jackknifing rate.

IOP Conf. Series: Materials Science and Engineering **315** (2018) 012027 doi:10.1088/1757-899X/315/1/012027

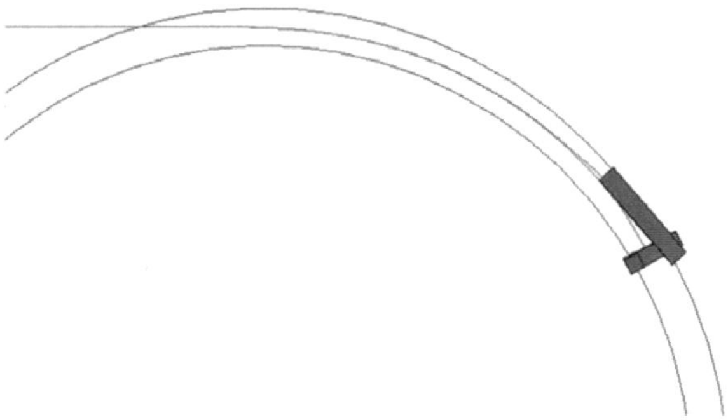

Figure 21. Position of the articulated lorry after a complete stop.

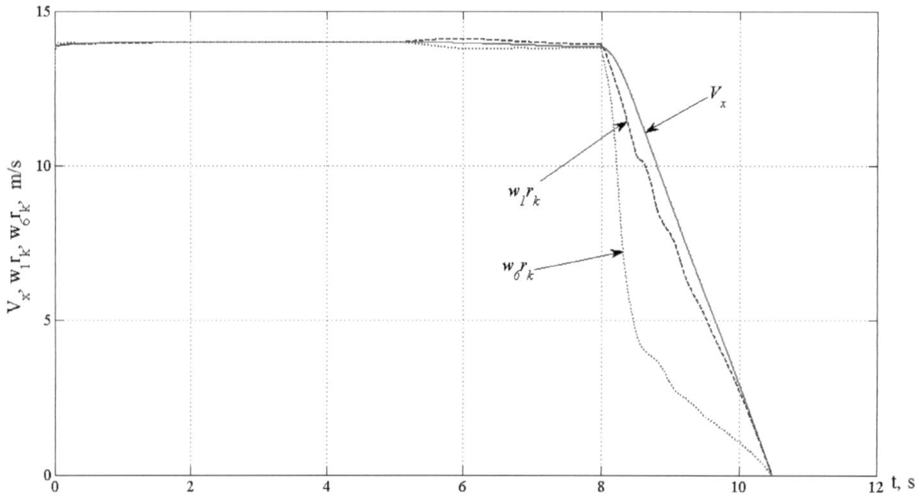

Figure 22. Longitudinal velocity of the tractor center of gravity and the speeds of the centers of mass of the wheels without sliding.

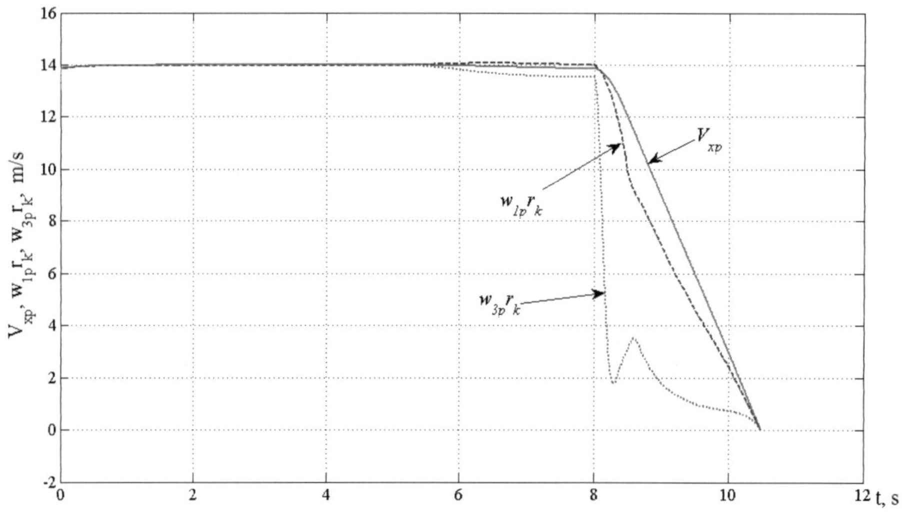

Figure 23. Longitudinal velocity of the semi-trailer center of gravity and the speeds of the centers of mass of the wheels without sliding.

IOP Conf. Series: Materials Science and Engineering **315** (2018) 012027 doi:10.1088/1757-899X/315/1/012027

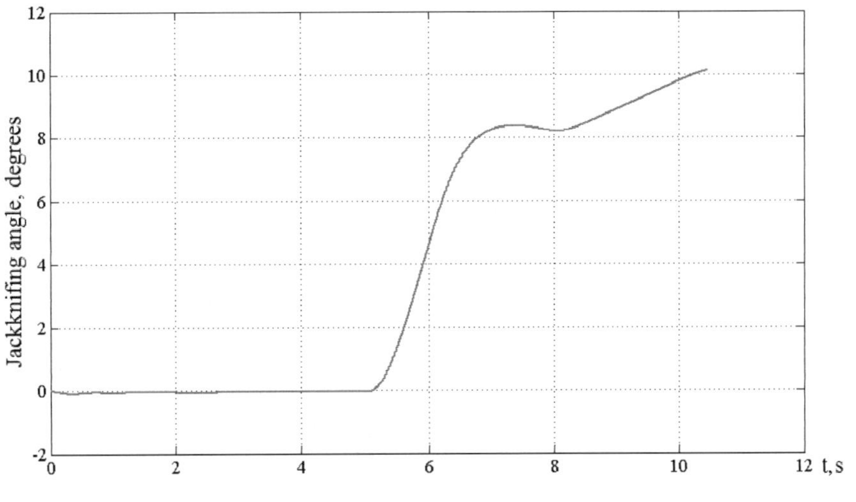

Figure 24. Jackknifing angle.

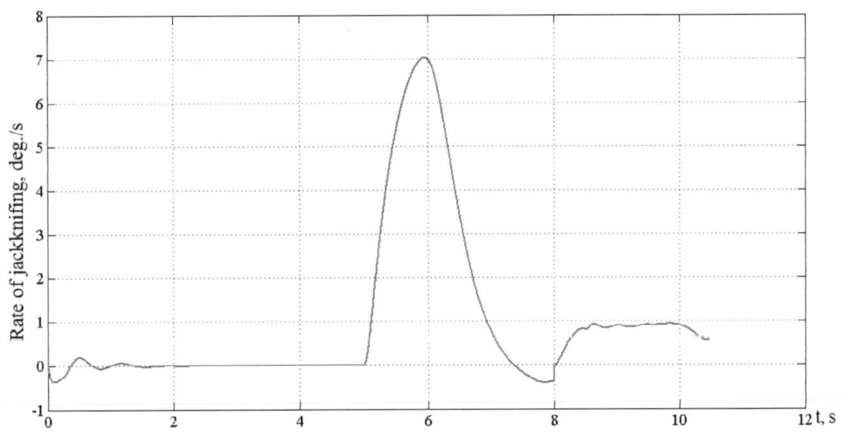

Figure 25. Jackknifing rate.

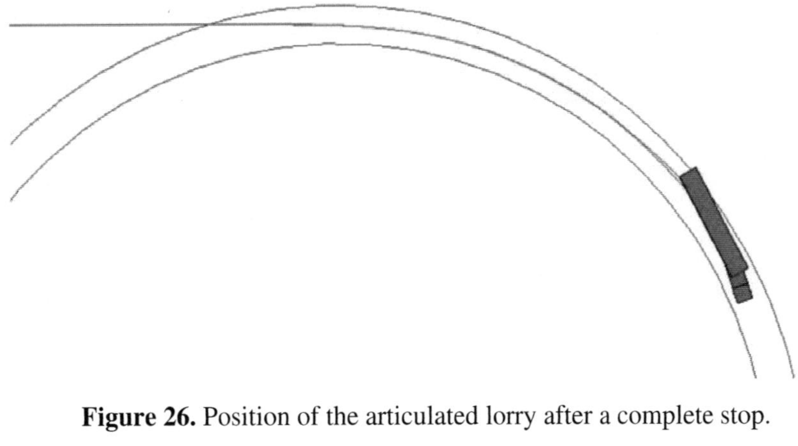

Figure 26. Position of the articulated lorry after a complete stop.

5. Conclusion

The simulation results showed efficiency of use of the fifth wheel coupler locking device for prevention of jackknifing at braking not only on a straight line, but also on a curvilinear trajectory. With the locked fifth wheel coupler, the articulated lorry slows down in a turn practically preserving the trajectory of movement due to the fixation of the jackknifing angle that has arisen at the entrance to the turn. In the absence of locking, the vehicle folds (the jackknifing angle reaches 100°) and leaves the traffic lane, thereby creating conditions for a traffic accident. It is worth noting that when braking in turn with a critical initial speed on the verge of tipping with the locked fifth wheel coupler, the trailer may skid even on the road with a high coefficient of friction. To exclude this adverse effect, it is necessary to introduce a system limiting the vehicle speed in turn thus preventing not only the drift of the articulated lorry but also the overturning.

References

[1] Bouteldja M Cerezo V 2011 *Jackknifing warning for articulated vehicles based on a detection and prediction system* (Indianapolis: Submitted to the 3rd International Conference on Road Safety and Simulation, September 14-16) http://onlinepubs.trb.org/onlinepubs/conferences/2011/RSS/3/Bouteldja,M.pdf
[2] GOST R 41.13-99 (UNECE Regulation No. 13) *Uniform provisions concerning the vehicles of* categories *M, N and O with regard to braking*
[3] Jileikin M M 2016 *Theoretical bases of increase of efficiency and controllability of the wheeled vehicle based on fuzzy logic* (Moscow: MGTU im. N.E. Baumana Press) p 238
[4] Jileikin M M Skotnikov G I 2015 *Development of principles of increasing of stability of motion of multilink tractor trains* (Moscow: Traktoryi i selhozmashinyi vol 10) pp 19 – 23
[5] Ankinovich G G, Verzhbitskiy A N, Jileykin M M, Skotnikov G I 2016 *The development of principles to improve the stability of articulated lorries against roll-over in the turn* (Moscow: Izvestiya VUZov. Mashinostroenie vol 2) pp 28 – 35
[6] Larin V V 2010 *Theory of motion of all-wheel drive vehicles* (Moscow: MGTU im. N.E. Baumana Press) p 391

IOP Conf. Series: Materials Science and Engineering **315** (2018) 012028 doi:10.1088/1757-899X/315/1/012028

Mathematical model of rolling an elastic wheel over deformable support base

V N Volskaia[1], M M Zhileykin[2] and A Y Zakharov[3]

Department of Wheeled Vehicles, Bauman Moscow State Technical University, ul. 2-ya Baumanskaya 5, Moscow, 105005 Russian Federation

E-mail: [1] volskaja52@mail.ru; [2] jileykin_m@mail.ru; [3] zakharov-al@mail.ru

Abstract. One of the main direction of economic growth in Russia remains to be a speedy development of north and northeast regions that are the constituents of the 60 percent of the country territory. The further development of these territories requires new methods and technologies for solving transport and technological problems when off-road transportation of cargoes and people is conducting. One of the fundamental methods of patency prediction is imitation modeling of wheeled vehicles movement in different operating conditions. Both deformable properties of tires and physical and mechanical properties of the ground: normal tire deflection and gauge depth; variation of contact patch area depending on the load and pressure of air in the tire; existence of hysteresis losses in the tire material which are influencing on the rolling resistance due to friction processes between tire and ground in the contact patch; existence of the tangential reaction from the ground by entire contact area influence on the tractive patency. Nowadays there are two main trends in theoretical research of interaction wheeled propulsion device with ground: analytical method involving mathematical description of explored process and finite element method based on computational modeling. Mathematical models of interaction tire with the ground are used both in processes of interaction individual wheeled propulsion device with ground and researches of mobile vehicle dynamical models operated in specific road and climate conditions. One of the most significant imperfection of these models is the description of interaction wheel with flat deformable support base whereas profile of real support base surface has essential height of unevenness which is commensurate with radius of the wheel. The description of processes taking place in the ground under influence of the wheeled propulsion device using the finite element method is relatively new but most applicable lately. The application of this method allows to provide the most accurate description of the interaction process of a wheeled propulsion devices and the ground, also this method allows to define tension in the ground, deformation of the ground and the tire and ground's compression. However, the high laboriousness of computations is essential shortcoming of that method therefore it's hard to use these models as part of the general motion model of multi-axis wheeled vehicles. The purpose of this research is the elaboration of mathematical model of elastic wheel rolling over deformable rough support base taking into account the contact patch deformation. The mathematical model of rectilinear rolling an elastic wheel over rough deformable support base, taking into account variation of contact patch area and variation in the direction of the radial and tangential reactions also load bearing capacity of the ground, is developed. The efficiency of developed mathematical model of rectilinear rolling an elastic wheel over rough deformable support base is proved by the simulation methods.

IOP Conf. Series: Materials Science and Engineering **315** (2018) 012028 doi:10.1088/1757-899X/315/1/012028

1. Introduction

One of the main direction of economical growth in Russia remains to be a speedy development of north and northeast regions that are the constituents of the 60 percent of the country territory. The further development of these territories requires new methods and technologies for solving transport and technological problems when off-road transportation of cargoes and people is conducting. In this way the movement of vehicle is difficult and impossible in some cases. The existing models of wheeled, tracked and rotary screw vehicles don't correspond to present functional, efficiency, reliability and ecological requirements for wheeled propulsion devices when operating in north regions of the country where supporting surfaces are weak-bearing. In this way there are technical, economic and social need in creating and using vehicles with pneumatic wheeled propulsion devices including ultra-low pressure devices that satisfy the requirements.

It should be noted that our country doesn't have necessary collection of energy efficient off-road vehicles currently. The existing off-road machines performed according to old traditional schemes and produced serially by industry don't meet the requirements defining the efficiency and environmental friendliness of wheeled propulsion devices in difficult climatic operating conditions. In current situation the problem of wheeled vehicles patency prediction of movement over supporting surface with weak-bearing properties is relevant. [1,2]. One of the fundamental methods of patency prediction is simulation method of wheeled vehicle movement in different operational conditions that is based on mathematical models of interaction the elastic tire with deformable support base unevenness.

Established [3, 4] that the traction patency wheeled vehicle mutually affect both the deformation properties of tire and physical and mechanical characteristics of soil: normal deflection of the tire and the depth gauge; the change in the reference area of the contact patch depending on load and air pressure in the tire; the presence of hysteresis losses in the material of the tire that affect rolling resistance caused by friction in the contact patch on the ground; creating a tangent reactions of the soil over the entire area of contact.

Along with the definition of the dependencies to describe "load-deformation of the ground " and "load-deformation mover", not less important task is the selection of modelreset vertical deformations of soil in contact with the pneumatic tire. On the correct choice of the type and kind of approximating dependence is largely determined by both qualitative and quantitative aspects of General solution of the problem of contact interaction of elastic mover with deformable ground [5].

Currently in the theoretical study of the interaction of propellers with the ground formed two main directions: analytical method involving a mathematical description of the process under consideration [6 – 8] and finite element method (FEM), based on computer simulation [9 – 12]. In the study of the interaction of wheel elements with the soil, an analytical method has found wide application. Developed mathematical models of the interaction of the tyre with the ground allow to solve various tasks. These models are used as in the study of processes of interaction of a single wheel mover with the soil array and at research of dynamic models of mobile machines operating in specific road-soil (RSC) and climatic conditions. One of the most significant shortcomings of these models is the description of the interaction of the wheel with a smooth deformable base, while the real profile of the support surface has a significant height of irregularities is comparable with the radius of the wheel.

The study of the processes occurring in the soil mass under the influence of the wheel mover, the finite element method are relatively new, currently the largest application. This method is better than others provided numerical procedures for the study of mathematical models of objects. The most important advantage is the availability of implicit methods of integration of systems of differential equations. The application of this method is the most accurate to describe the process of interaction of wheel elements with the soil, to determine the stresses in the soil mass deformation of tire and soil compaction. In contrast to analytical methods form the contact patch of the tyre with an elastic support surface is the result of the simulation taking into account the independent soil characteristics and the constructive and operational parameters of the engine. However, a significant disadvantage of this method is its high computational complexity, therefore, with currently available computers to use these models as part of the overall model of the motion of multi-wheeled vehicles difficult.

The aim of this study is to develop a mathematical model of rolling elastic wheels on deformable rough reference base, taking into account the deformation of the contact patch.

2. A mathematical model of the rolling

Key assumptions

- Normal pressure in the contact patch evenly distributed.
- The force of interaction of the wheel with its support base directed in the opposite direction from the slip speed.

Coordinate systems and basic calculation dependencies

The proposed model uses two different coordinate systems figure 1 due to the structure and form of the equations of motion of the object.

First, the stationary coordinate system (SCS) $O_2X_2Y_2Z_2$, is used to simulate the specified ground traffic conditions. The origin of the system of point O_2 coincides with the beginning of the simulated track.

To determine the forces acting on the car from the ground we introduce a micro mobile coordinate system (MMCS) under which we understand the system $O_TX_TY_TZ_T$, the center of which O_T coincides with the geometric center of rotation, the axis of O_TX_T coincides with the projection of the longitudinal axis of symmetry of the wheel on the support surface. Consider the scheme of movement of the wheel in figure 1.

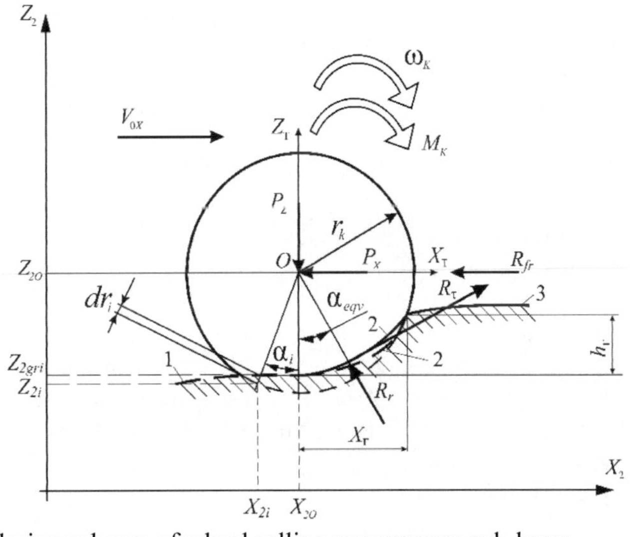

Figure 1. A design scheme of wheel rolling over uneven sub-base: 1 – strain profile of soil; 2 – deformable soil profile; 3 – the undeformed profile of the soil.

For model development we use the results of [13]. On the lower semicircle of the undeformed profile of the wheel will select a certain number of points, n, the position of which will determine the angle α_i between the vertical line dropped from the center of the wheel on the axis X_2, and a ray connecting a point of the undeformed profile of the wheel with its center figure 1. The number of points is chosen based on a compromise between accuracy of the model and its performance. Define the coordinates X_{2i} and Y_{2i} of selected points of the profile at the SCS

$$X_{2i} = X_{20} + r_k \sin \alpha_i;$$
$$Y_{2i} = Y_{20} + r_k \sin \alpha_i;$$
$$-\frac{\pi}{2} \le \alpha_i \le \frac{\pi}{2};$$

where X_{20}, Y_{20} are the coordinates of the center of the wheel O_B SCS. The vertical coordinate Z_{2i} *i*-th point of the undeformed profile of the wheel in the SCS, we define by the formula

$$Z_{2i} = Z_{20} - r_k \cos \alpha_i.$$

where Z_{20} is the vertical coordinate of the wheel center at the SCS. The formation of the longitudinal profile Z_{2gri} is carried out according to the following algorithm:

$$Z_{2gri} = \begin{cases} Z_{2gri}^{undef}, \text{if } X_i \ge X_{20} + X_r \\ Z_{2gri}^{undef} - (1 - \frac{h_r}{X_r^2} X_i^2), \text{if } X_{20} \le X_i < X_{20} + X_r \\ Z_{2gri}^{undef} - h_r, \text{if } X_i \le X_{20} \end{cases} \quad (1)$$

where X_i is the current coordinate of the X_i-th point of the profile gauge in the SCS; h – depth gauge; Z_{2gri}^{undef} - the vertical coordinate of the undeformed profile of the gauge (simulated in advance by a known method described, for example, in [14]). Xr the amount determined by the formula

$$X_r = \sqrt{r_k^2 - (Z_{20} - h_r)^2}$$

where r_k – free radius of the wheel. The d_{ri} deflection of the tire in the radial direction for the *i*-th point of the undeformed profile is determined from the following ratios

$$dr_i = \begin{Bmatrix} 0, Z_{2gri} \le Z_i \\ (Z_{2gri} - Z_{2i}) \cos \alpha_i, Z_{2gri} \ge Z_{2i} \end{Bmatrix}$$

where Z_{2gri} – vertical coordinate of the profile of the support base for the *i*-th point of the wheel. Thus, to determine the reactions of interaction of wheels with the supporting surface Rx and R_z in MMCS in the presence of several areas of "overlap" profile substructure undeformed contour of the wheel it is necessary to determine the equivalent angle α_{ecv} the point of application of the total radial reaction Rr in the radial direction and the tangential reaction $R\tau$ figure 1.
Define α_{ecv} as a weighted average value

$$\alpha_{eqv} = \frac{\sum_{i=1}^{n} \alpha_i dr_i}{\sum_{i=1}^{n} dr_i}$$

Radial reaction R_r is the sum of two components: elastic R_{ry} and damping $R_{rд}$: $Rr = R_{ry} + R_{rд}$. R_{ry} depends on the equivalent deflection of the tire

$$d_{eqv} = \frac{\sum_{i=1}^{n} dr_i}{n_k}$$

where n_k is the number of points in the undeformed profile in contact with the support surface. Rrd depends on the speed of the deflection of the tire in the radial direction. We define the projection of velocity of points of the contour of the wheel on the axis X_T and Z_T:

$$V_{iX_T} = \omega_k (r_k - dr_i) \cos \alpha_i + V_{0X_T};$$
$$V_{iZ_T} = \omega_k (r_k - dr_i) \sin \alpha_i + V_{0Z_T},$$

where ω_k – angular velocity of rotation of the wheel; V_{0XT} and V_{0ZT} – projection of the velocity vector of the wheel center (point O) on the axis X_T and Z_T:, respectively. The vector of linear speed of the i-th point of the undeformed profile of the wheel in the radial direction

$$V_{ri} = V_{iX_T} \sin \alpha_i + V_{iZ_T} \cos \alpha_i.$$

The speed of profile deformation of the i-th point in the radial direction

$$\frac{d}{dt}(dr_i) = \dot{Z}_{2gri} \cos \alpha_i - V_{ri}.$$

Equivalent speed of deflection

$$\frac{dr_{eqv}}{dt} = \sum_{i=1}^{n} \left(\frac{d}{dt}(dr_i) \right)$$

Further, knowing the elastic and damping characteristics of the tire in the radial direction, we find Rr. Tangential reaction $R_\tau = \mu_S R_r$, where - coefficient of friction-partial slip [15].

$$\mu_s = \mu_{s\alpha \max} \times \left(1 - e^{-\frac{s_K}{s_0}} \right)$$

μ_{samax}- coefficient of friction full slip for a given angle α rotation of the velocity vector of sliding, S_k the coefficient of slipping, S_0 is a constant. This expression is valid for non cohesive soils. The value μ_{samax} specifies the maximum value of the function $\mu_S(S_k)$, and, combined with the constant S_0 is the gradient of the $\mu_S(S_k)$ function at the origin, reflecting the properties of the soil at small slides.
The coefficient of slipping

$$S_k = \frac{V_{0X_T} - \omega_k (r_k - dr_{eqv})}{\omega_k (r_k - dr_{eqv})}$$ -for the traction mode of wheel rolling;

$$S_k = \frac{V_{0X_T} - \omega_k (r_k - dr_{eqv})}{V_{0X_T}}$$ - for brake or driven mode of wheel rolling.

Reaction of interaction of wheels with the supporting surface R_{XT} and R_{ZT} in MMCS

$$R_{X_T} = R_\tau cos\alpha_{eqv} - R_r sin\alpha_{eqv} - R_{fr}$$

$$R_{Z_T} = R_\tau sin\alpha_{eqv} + R_r cos\alpha_{eqv}$$

where R_{fr} – force frontal ground resistance of rolling wheels (bulldozer effect), which is determined by the method described [16].

$$R_{fr} = 0{,}5 b_t \left(r_w - h_r - dr_{eqv} \right)^2 q_p \gamma,$$
$$q_p = a_p \cdot (0{,}01 \cdot 90°)^{b_p}$$
$$a_p = 1 + a_{fa} \cdot (0{,}1 \cdot \varphi_S^*)^{b_{fa}}$$

$$b_p = a_{fb} \cdot (0{,}1 \cdot \varphi_S^*)^{b_{fb}}$$

$$a_{fa} = 0{,}3082 - 0{,}0709 \cdot \frac{\varphi_a^*}{\varphi_S^*}$$

$$b_{fa} = 2{,}0751 + 1{,}3354 \cdot \frac{\varphi_a^*}{\varphi_S^*}$$

$$a_{fb} = 0{,}5756 + 0{,}1024 \cdot \frac{\varphi_a^*}{\varphi_S^*}$$

$$b_{fb} = 1{,}0608 + 0{,}0619 \cdot \frac{\varphi_a^*}{\varphi_S^*}$$

where b_t – bus width, γ- is the specific weight of the soil, M N/m$^{3;}$ φ_S^*, φ_a^* –respectively angle of internal thorns of the soil and the friction angle of the soil on the wall of tyre deg. Graphs to determine φ_S^*, φ_a^* presented in [16]. Dynamics of rotation of the wheel shown in figure1 and described by equations

$$J_W \dot{\omega}_W = M_W - R_{X_T}(r_k - dr_{eqv}) - M_f - M_h - M_T;$$

$$M_f = R_r(r_k - dr_{eqv}) sin\alpha_{eqv} cos\alpha_{eqv};$$

$$M_h = R_{Z_T} h_r$$

where $\boldsymbol{M_w}$ – torque applied to the wheel; $\boldsymbol{M_T}$ – braking torque to the wheel; $\boldsymbol{M_f}$ – moment of rolling resistance of the wheel due to the shift reaction $\boldsymbol{Rr}$; $\boldsymbol{R_{XT}}$ is the projection of force interaction of wheel with grunting MMCS on the axis of the $\boldsymbol{OX_T}$; $\boldsymbol{M_h}$ - resistance moment caused by the vertical deformation of the soil [16].

3. Determination of the parameters of the contact patch and the track width h_r

The components of the length of the contact patch within the right (L_1) and left (L_2) lower quarters of the circular arc undeformed contour of the tire figure 1, respectively, can be defined as follows. Dependence $Z_{2gr}=f(X)$ within the bottom right quarter of the circular arc undeformed contour of the tyre is described by the formula (1). Then

$$L_1 = \int_0^{X_r} \sqrt{1 + (f'(X))^2} dX = \frac{1}{2A}\left[AX_r\sqrt{AX_r^2 + 1} + \sqrt{A} sinh^{-1}(X_r\sqrt{A})\right] \quad (2)$$

where $= \frac{h_r(t_{i-1})}{X_r^2}$; $\sinh(x) = \frac{e^x - e^{-x}}{2}$ is the hyperbolic sine. To calculate expression (2) value $h_r(t_{i-1})$ needs to take from the previous integration step to avoid the "algebraic loop". The initial value can be taken $h_r|_{t=0} = 0$. Due to the lack of analytical expressions $Z_{2gr}=f(X)$ within the lower left quadrant circular arc undeformed contour of the tire replace the integral (2) end for n amount of plots the undeformed profile of the tire for which $dr_i \neq 0$ when $-\frac{\pi}{2} \leq \alpha_i \leq 0$

$$L_2 = \sum_{i=1}^{n} \sqrt{(Z_{2gri} - Z_{2gri-1})^2 + (X_{2gri} - X_{2gri-1})^2}$$

The area of the contact patch F_t figure 2 define by the formula

$$F_t = 2\int_0^{\pi} b(\alpha)[r_k - dr(\alpha)]\,d\alpha,$$
$$b(\alpha) = b_t\left(1 + \frac{dr(\alpha)}{r_k}\right) \tag{3}$$

Where b_t the width of the tyre.
The discrete analogue of the formula (3) can be written as

$$F_t = b_t \sum_{i=1}^{n_k-1}\left(2 + \frac{dr_{i+1} + dr_i}{r_к}\right)(r_к - dr_i)(\alpha_{i+1} - \alpha_i)$$

Define p_{0z} normal pressure in the contact patch. By the formula of Bernstein-Letoshnev [16]

$$p_{0z} = c_{гр} h_r^{\mu}, MPa$$

where c_s MPa – coefficient of soil deformation; μ - is the density of the soil. On the other hand

$$p_{0z} = \frac{P_z}{F_t 10^6}, МПа$$

where P_z - vertical wheel load. Then

$$h_г = \left(\frac{P_z}{c_s F_t 10^6}\right)^{\frac{1}{\mu}}$$

4. The influence of lugs on the parameters of the motion of a wheel

In the presence of the cleats must also calculate the shear forces in the areas of the projections and depressions, as well as the removal of soil from the zone of contact in case of intense slipping of the wheel. Additional vertical penetration dh_{vp} the center of the wheel in the ground caused by the excavation of soil from the area of contact is calculated by the formula [16]

$$dh_{vp} = \frac{t_{lt} h_{gro} s_k}{t_{gro}(1 - s_k)},$$

$$t_{gro} = \frac{2\pi r_k}{n_{gro}},$$

where t_{lt}– length trench lug; h_{gro}– grouser height; t_{gro}– step lug; n_{gro} the total number of lugs on the tyre circumference. Then the total vertical depth h_Σ wheel center

$$h_\Sigma = h_r + dh_{vp}.$$

We assume that if $h_\Sigma \geq r_k$, you lose the mobility of the machine due to the hanging of the machine body on the ground. To determine the additional shear forces $R_{\tau j}^{gro}$ in the areas of the projections and depressions of the cleats figure 3 use the results obtained in [16]:

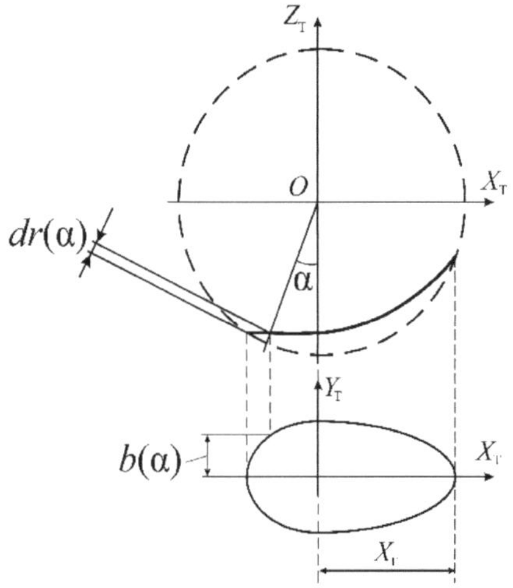

Figure 2. Calculation scheme for defining the area of the contact patch.

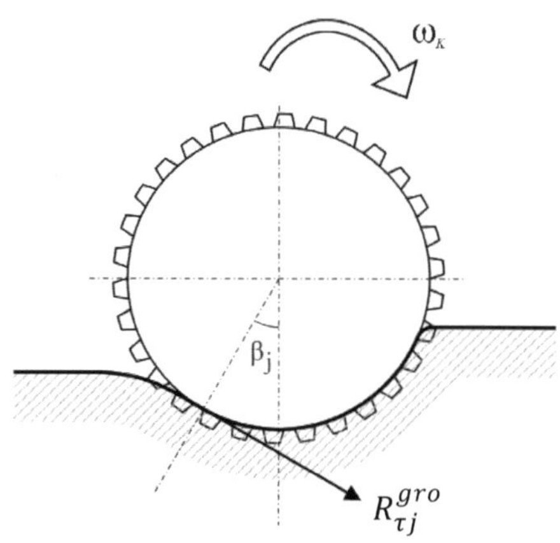

Figure 3. Calculation scheme for determination of the tangential forces $R_{\tau j}^{gro}$ in the areas of the projections and depressions of the cleats.

$$R_{\tau j}^{gro} = F_{gro} c_s 10^6 exp\left[-\frac{(|e_x| - e_{xm})^2}{0{,}05 e_{xm}} tg\varphi_s^*\right],$$

$$e_x = S(\Delta t) - \int_{t_1}^{t_2} \omega_k(t)(r_k - dr_{eqv})dt,$$

$$\Delta t = t_2 - t_1.$$

where F_{gro} – frontal area of the projection of the lug; e_{xm} – the maximum shift of the soil in which the cohesion of soil particles is not broken; e_x – current shift of the soil; $S(\Delta t)$ is the path traversed by the center of the wheel during the time Δt; $\omega_k(t)$ is the current angular speed of rotation of the wheel.
Total tangent force $R_{\tau X_T}^{gro}$ grsv projection on the axis X_TMMSC for all for all n grouser in the zone of contact of the wheel with its support base

$$R_{\tau X_T}^{gro} = \sum_{j=1}^{n} R_{\tau j}^{gro} cos\beta_j,$$

$$n = \frac{L_1 + L_2}{2\pi r_K} n_{gro}.$$

Then the expression for determining the longitudinal reaction R_{X_T} wheels with a support base in a projection on the axis X_T MMSC

$$R_{X_T} = R_\tau cos\alpha_{eqv} - R_r sin\alpha_{eqv} - R_{fr} + R_{\tau X_T}^{gro}.$$

5. Account the bearing capacity of the Foundation soil in horizontal direction under the action of vertical loads

The magnitude of the horizontal reaction R_{X_T} may be limited by two factors [16]: a wheel slip on the surface of the compacted soil after overcoming the forces of adhesion and loss of bearing capacity from the shear weight of the soil in the direction of action of the horizontal reaction. And if the coupling properties are accounted for by the factor μ_S, the loss of bearing capacity can be assessed by comparing the existing shear stresses τ with the maximum allowable stresses τ_{max}. The maximum shear stress in the contact patch is calculated using expression pendant

$$\tau_{max} = p_{0z} tg\varphi_s^* + c_s 10^6.$$

The current tangent stress τ define by the formula [16]:

$$\tau = p_{0z} tg\varphi_s^* \left[1 - exp\left(-\frac{|e_x|}{0{,}1e_{xm}}\right)\right] + c_s 10^6 exp\left[-\frac{(|e_x| - e_{xm})^2}{0{,}05e_{xm}} tg\varphi_s^*\right],$$

Finally, the expression for R_{X_T} takes the form

$$R_{X_T} = k_\tau (R_\tau cos\alpha_{eqv} - R_r sin\alpha_{eqv} - R_{fr} + R_{\tau X_T}^{gro}),$$

$$k_\tau = \begin{cases} 1, if\ \tau \le \tau_{max} \\ \frac{\tau_{max}}{\tau} if\ \tau > \tau_{max} \end{cases}.$$

6. Verifying mathematical models

To verify the developed methodology was developed by the program model the motion of a single wheel in the software package MATLAB/SIMULINK. Parameters of the Foundation soil according to [16] are given in table 1.

Table 1. Parameters of the Foundation soil according.

Parameters of the Foundation soil according	Marking	Value
The coefficient of deformation of the soil	c_s	0,18 MPa
The density of the soil	μ	0,77
The specific gravity of the soil	γ	0,0145 MN/m^3
The angle of internal friction of soil	φ_s^*	37^0
The maximum shift of the soil in which the cohesion of soil particles is not broken	e_{xm}	0,05 m
The coefficient of friction full slip	$\mu_{s\alpha\,\max}$	0,8
Constant	S_0	0,05

The main characteristics of the single wheel is presented in table 2.

Table 2. Characteristics of the single wheel.

Characteristics of the single wheel	Marking	Value
Weight	m	600 kg
Free radius	r_k	0,7 m
Width	b_t	0,6
The total number of tire grousers	n_{gro}	60
Tire grouser height	h_{gro}	0,015 m
The length of the bottoms of the tire grousers	t_{lt}	0,015 m
The angular velocity of rotation of the wheel	ω_k	4 c^{-1}

In figure 4 shows the dependence of the height profile Z_{2gr} subgrade from the longitudinal coordinate X_2 of the fixed coordinate system. In figures 5 – 9 show, respectively, depending on the time of

IOP Conf. Series: Materials Science and Engineering **315** (2018) 012028 doi:10.1088/1757-899X/315/1/012028

movement *t* of the following parameters: depth gauge, h_Γ , the height of the wheel center z_κ equivalent to the deflection of the tire dr_{ecv}, the length of the contact patch $L = L_1 + L_2$ equivalent angle α_{ecv}.

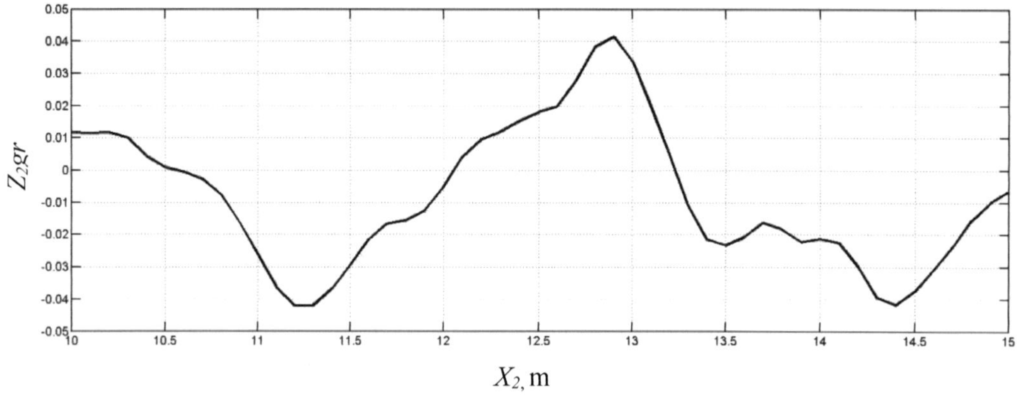

Figure 4. The dependence of the height profile Z_2gr subgrade from the longitudinal coordinate X_2.

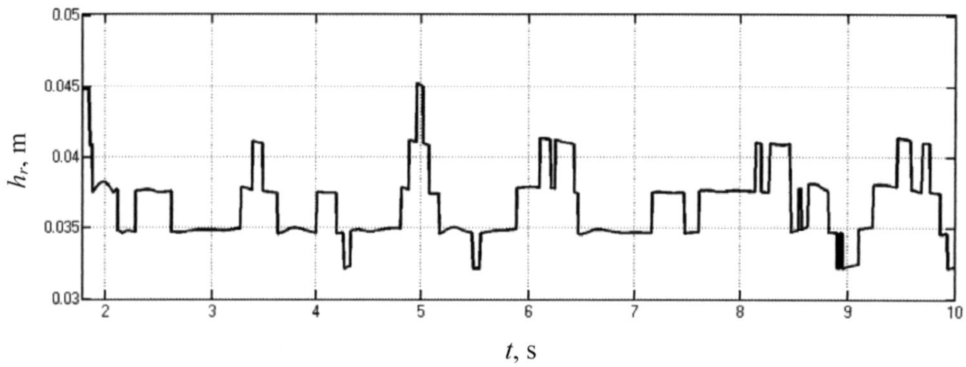

Figure 5. The dependence of the depth gauge $h_г$ time of motion *t*.

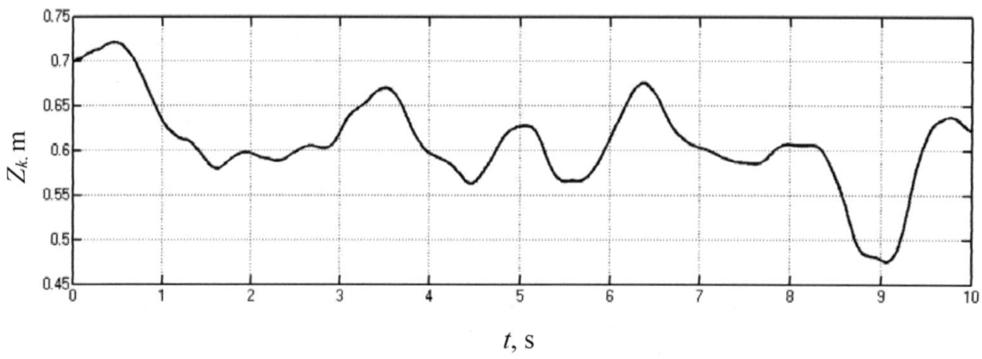

Figure 6. The dependence of the height of the wheel center z_κ time of motion t_t.

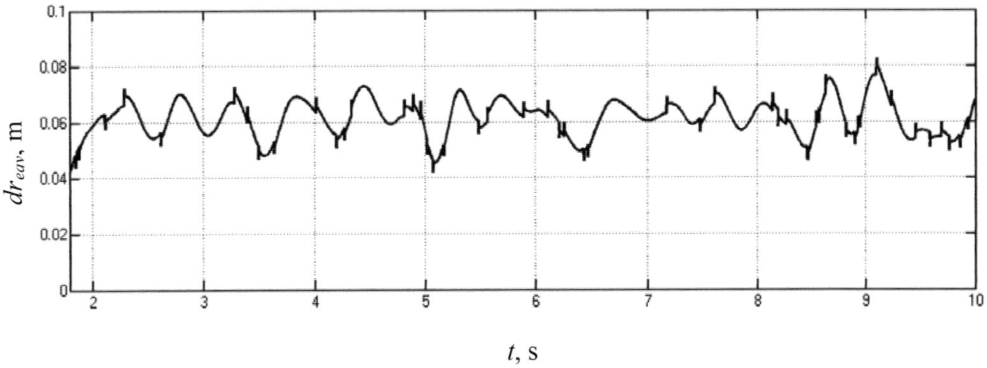

Figure 7. The dependence of the equivalent deflection of the tire dr_{eqv} the time of motion t.

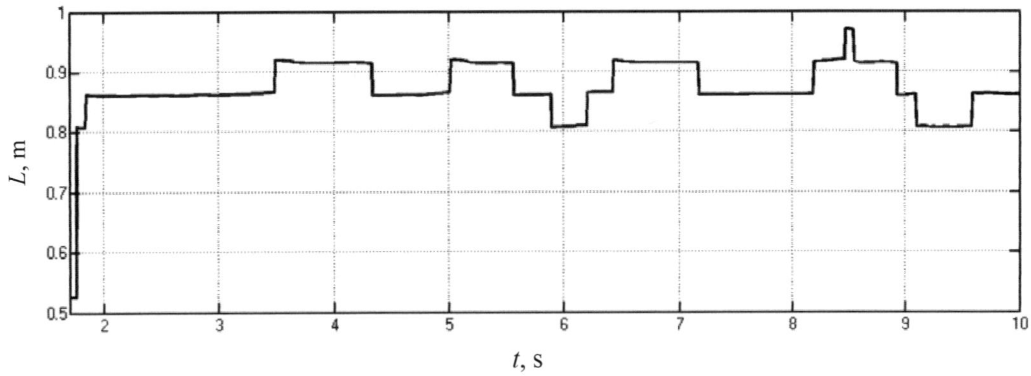

Figure 8. The dependence of the length of contact patch L the time of motion t.

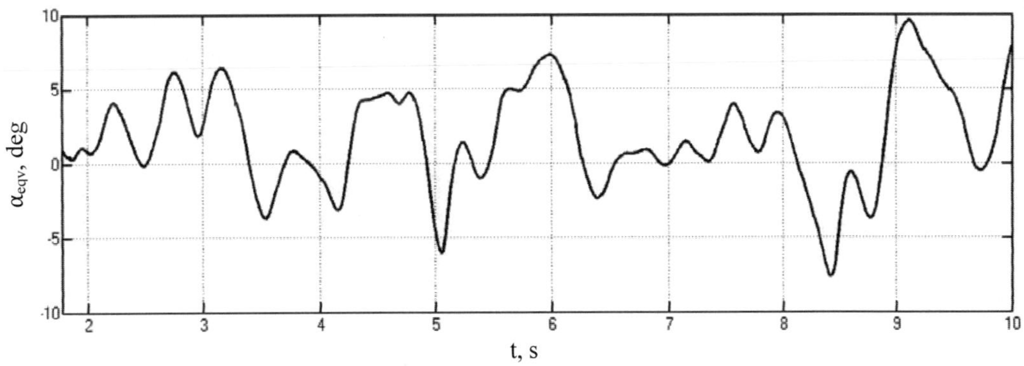

Figure 9. The dependence of the equivalent angle α_{ecv} time of motion t.

Shown in figures 5 – 9 dependence allow to judge about the efficiency of the proposed mathematical model of the elastic wheel movement on uneven deformable support base.

7. Conclusions

- The mathematical model of rectilinear elastic wheel rolling on a rough support deformable base, taking into account the change in the area of the contact patch, the direction of the radial and tangential reactions, and the bearing capacity of the soil.

- Methods of simulation prove the efficiency of the developed mathematical model of rectilinear rolling elastic wheels on uneven deformable support base.

References

[1] Kotliarenko V I 2008 *The main directions of improving the patency of wheeled vehicles* (Moscow: Publisher MGIY)

[2] Shukhman S B, Pliev I A and Malyarevich V E 2008 Ways to improve the environmental properties of multi-wheel drive vehicles operated in the far North *The Automotive Industry*, No 10 pp 15-7

[3] Wong J Y 2001 *Theory of Ground Vehicles* (Wiley IEEE)

[4] Pryadkin V I 2011 Assessment of the impact of the highly flexible tyre on the rolling surface *Forestry* No 5 pp 42-3

[5] Pirkowskiy J V 2001 *The theory of the motion of four-wheel drive vehicle (applied problems of optimization of chassis design)* proc. manual for schools (Moscow)

[6] Ageikin Y S, Volska N S and Chicheken I V 2010 *The flow of cars* (Moscow: Publisher MGIY) pp 34-56

[7] Babijchuk A E, Ageikin Y S and Volska N S 2013 The technique of determination of power losses in the rolling of the wheel mover with the type of drivetrain and the air pressure in the tires of the car *AAI* No 3 pp 24–7

[8] Lepeshkin A V and Petrov S E 2012 Mathematical model of interaction between elastic wheels with deformable bearing surface in the case of steady rectilinear rolling *Conf. AAI "Car and tractor const. in Russia: devel.t prior. and training"* (Moscow) *Publisher MSTU «MAMI»* pp 141 – 9

[9] Shoop S, Kestler K and Haehnel R 2006 Finite element modeling of tires on snow *Tire Science and Technology* V **34** No 1 p 2–37

[10] Botero J, Gobbi M and Mastinu G 2005 A new mathematical model of the traction force in pneumatic tire snow chain systems *AssociazioneItalianaPerL'AnalisiDelleSollecitazioni / Dipartimento di Meccanica, Politecnico di Milano* p 10 www.aiasonline.org/AIAS2005/Articoli/art084.pdf

[11] Belkin A E and Narsky N L 2004 Finite element analysis of the contact of the tires with the supporting surface on the basis of the shell model *Vestnik MSTU. Series Engineering* No 3 pp 14–28

[12] Odintsov O A 2008 Development of a method of solving nonlinear contact problems of stationary rolling tires *Diss. kand. tech. sciences. Publisher BMSTU*

[13] Zhileikin M M and Padalkin B V 2016 The mathematical model of rolling elastic wheels on rough rigid base *Izvestiya vuzov. Mechanical engineering* No 3 pp 24–9

[14] Polungian A A 2008 *Design of four-wheel drive wheeled vehicles* (Moscow: Publisher BMSTU) V. **1** p 496

[15] Ellis J R 1975 *Vehicle Handling* (Moscow: Mashinostroenie) p 216

[16] Larin V V 2010 *Theory of motion of four-wheel drive wheeled vehicles* (Moscow: Publisher BMSTU) p 391

IOP Conf. Series: Materials Science and Engineering **315** (2018) 012029 doi:10.1088/1757-899X/315/1/012029

Intelligent systems of the vehicles' suspension

D Yurlin

Chassis Department, FSUE «NAMI», Moscow, Russia

E-mail: dmitry.yurlin@nami.ru

Abstract. The article is devoted to the current condition of car's active suspension system. It presents the tendencies in development of the active systems of suspension system, adjustable elements incorporated in them and the companies succeeded in designing such systems. It also mirrors the problem of impact of active systems on car's safety and their importance for the driver. Advantages and disadvantages of the most common types of active elements are being described, analyzed and compared. The author concludes about the perspectives of these systems' development.

1. Current level of car's active systems. The essence of active systems of suspension system for the driver

The modern system of car design has improved strongly since the last 25 years. The most car parts are equipped with individual control systems, behavior algorithms and active actuators. The systems can decide themselves on changing widely the components' properties improving their efficiency, safety and comfort of a car.

From the technical point of view the most significant development occurred in chassis system, transmission, engine and systems of active security. But for the driver or passenger car's comfort, stability and handling remain the most important characteristics. Thus one of the key aspects during the implementation of a car active system is its' influence on the mentioned characteristics. From this point of view the work of suspension system is important for driver's comfort, his or her tiredness and control over the vehicle. This is the essence of active suspension system as a driver assistant system.

Strong vibrations that affect passengers in cars with badly adjusted suspension lead to quick driver's tiredness and shortens the working time. The same is true to the control over the vehicle. Such chassis' disadvantage as enlarged dynamic corridors makes driver stay highly concentrated and constantly correct the path during driving. Meanwhile even well adjusted suspension not always can solve the problem of driver's tiredness because any chassis' passive adjustment is a compromise between stability, control and driving comfort. Thus at some driving regimes this compromise can provoke some negative effects in car's behavior and driver's perception. Similar situations can occur in different states of a car. For example the suspension system is usually adjusted for a situation when a car carries two passengers (or even one) and a small amount of baggage. But if the car drives with full load (five passengers and baggage) all elements of suspension system work under high load beyond their well-balanced characteristics. There also exists a possibility of accumulating of usage of suspension system elements' (such as shock absorber, bushing and spring). In case when usage of these elements doesn't affect the vehicle's safety such situation is considered to be acceptable but characteristics of stability, control and driving comfort will be significantly worse than the same characteristics of a new car. In this situation the tiredness of the passengers will be higher, the driver's control over the vehicle will decrease regardless the basic chassis adjustment. All these problems can be solved by using modern active suspension systems that not only adopt the system's characteristics to the usage conditions but also can compensate for some elements' usage from the point of view of their effect on chassis' work. Thus modern intellectual systems of car suspension provide the driver with significant assistance in driving a vehicle.

2. The suspension system's configuration

The suspension system includes: springs, damping elements, control arms and stabilizing system (mostly – a torsion bar that connects the car's body with the wheels of the opposite side). It is called a stabilizer bar and acts as an extra spring element of suspension. In car industry automatic control systems are used in all basic suspension's components. In this case different technological methods are used to change component's characteristics. These methods have different types of control input and ways of effecting the suspension's characteristics

3. Adjustable spring elements of the suspension

Spring elements were the first suspension parts to become an active system. Springs were replaced with air springs that let change the road clearance and stiffness depending on the car's load. The initial systems were regulated via the switch situated in the cabin, later they received special self-operating algorithms and a control program. The example of pneumatic suspension on figure 1.

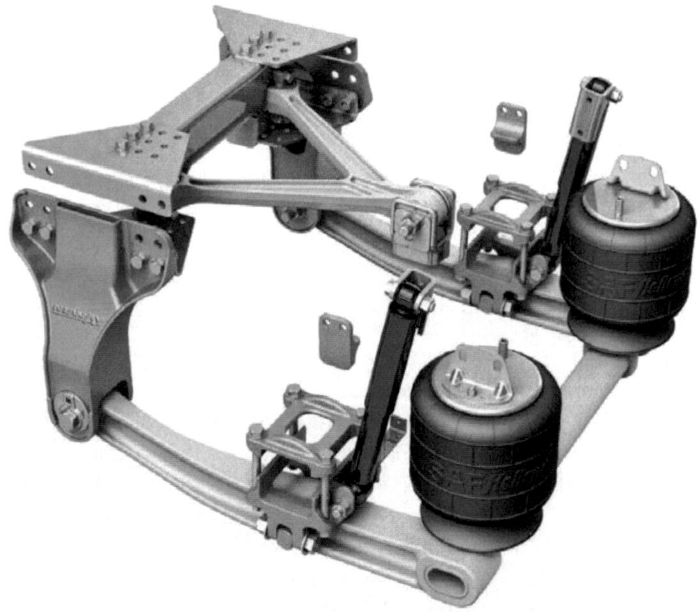

Figure 1. Example of pneumatic suspension.

Besides adjustable road clearance pneumatic suspension has such advantages as progressive performance rate and stiffness' automatic change in dependence on the load. Modern pneumatic suspension resists roll and shifts flexibility from soft to stiff regime. In case of changing road clearance it happens by adding or removing of gas in pneumatic element but in this case valves open additional gas volumes. Suspension stiffness declines when the gas cylinder enlarges under unchanged pressure and increases when its' volume is reduced (figure 2).

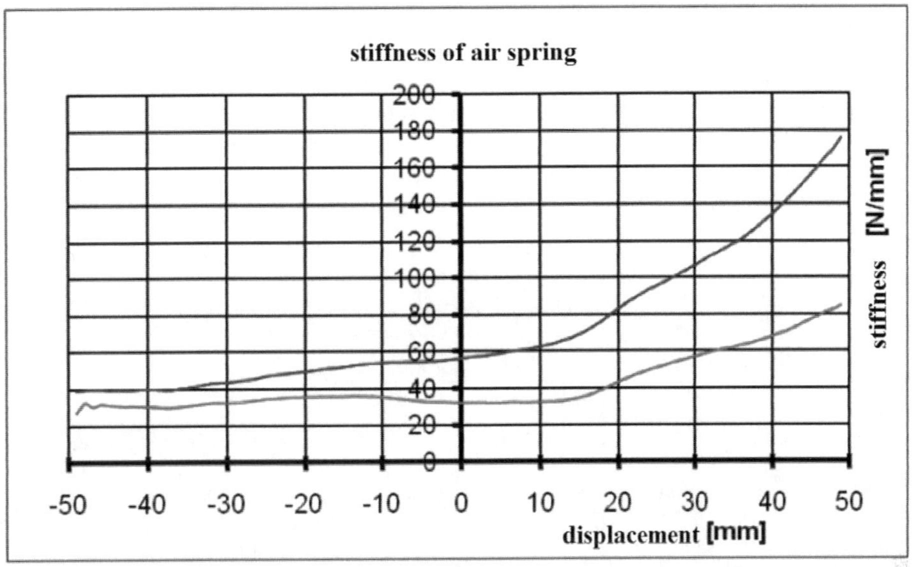

Figure 2. Changing the suspension stiffness' characteristics while changing pneumatic element's volume.

Gas cylinders separating valve actuates in few milliseconds and in case of determination of car's angular motion the system can reduce the pneumatic element's volume and increase the roll stiffness of the suspension. After the effect of heeling force is over the volume returns into initial position in accordance with lesser stiffness to decrease vibrations that affect passengers. This construction is used by Porsche and Mercedes (figure 3 [2]). Continental is acknowledged to be the world's leader of it's production.

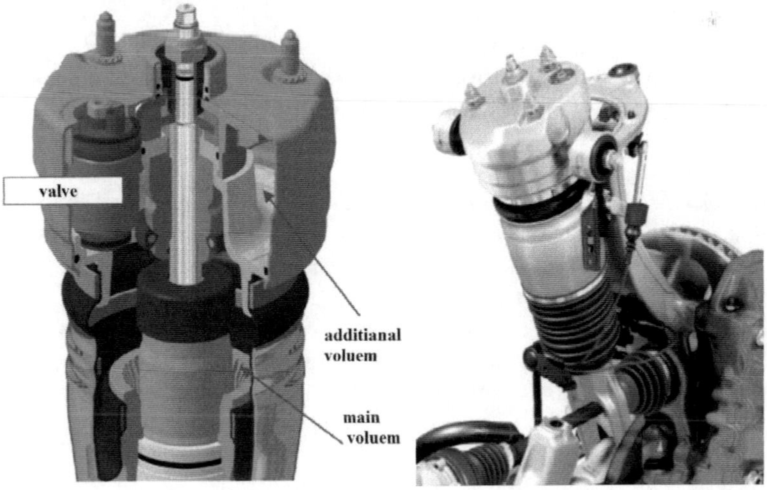

Figure 3. Construction of pneumatic suspension with disconnectable gas volume [2].

Except pneumatic elements car industry used hydraulic spring elements that compress liquid while working. It becomes possible by putting in a cylinder containing liquid a special needle that fills some volume and makes liquid compress. Such constructions have two crucial disadvantages: overheating and unreliability of seals. These problems don't let produce constructions satisfying modern requirements to cars' lifetime.

One should also mention electromagnetic elastic elements of a suspension. Rather popular is their concept worked out by BOSE (famous by its' sound systems). The company used its' experience in producing sound systems in suspension by replacing standard springs and damping elements with magnet coils – linearly electric engines that replace not only elastic elements but also damping ones (general view of the construction – on figure 4). The published video demonstrates exceptional properties of the system. It lets keep the car stable regardless of a maneuver and the road quality and implement a number of special functions. For example, the car can overcome such obstacles like a curb. A car jumps its over by throwing itself up, lifting the wheels before the obstacle and dropping them right after it. The car undergoes incommensurable lesser acceleration than during a run over an obstacle.

Figure 4. Construction of electromagnetic elastic elements of a suspension by BOSE.

The system was introduced 15 years ago but serial production didn't start. The reason was in expensiveness of the system, it's big weight and enormous power consumption. Nevertheless the technology seems to be rather promising and serial production can be started after solving the problem of car's power equipment.

Modern technologies used in spring elements advance car's consumer properties significantly. Pneumatic suspension reduce vibrations that affect passengers and improve ride comfort. Anti-roll system and electromagnetic suspension form systems of driver's active assistance and let pass problematic roads with low-quality surface without driver being tired.

4. Adjustable damping elements of suspension

The first after elastic elements damping elements were automatized. Their function is to damper vibrations of a car's body and unsprung mass but it often contradicts maintaining comfort or car's stability so it needs very careful setting. Characteristics of passive damper must correspond with certain frequency and load regimes that shift during driving. But controlled dampers have a range for regulation and an algorithm for its' automatic implementation what lets to fit the driving situation.

There are two main technologies to change damper's characteristics: adjustable valve with controlled orifice (figure 5) and a system with fixed orifice and controlled liquid viscosity.

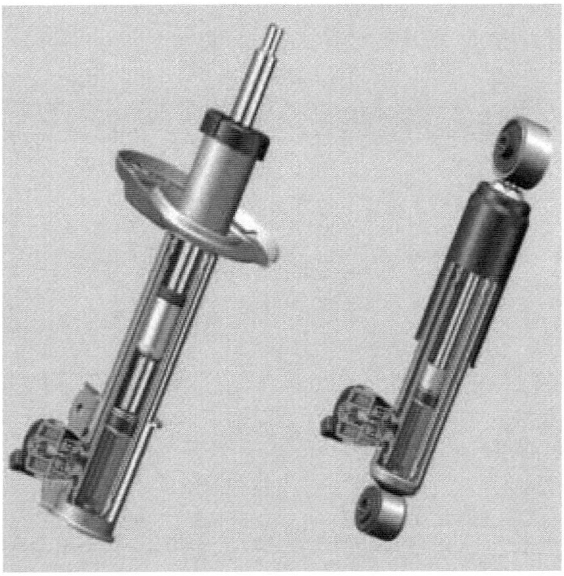

Figure 5. SACHS' dampers with controlled orifice.

The second type mostly uses magnetorheological liquid that changes its' viscosity under influence of electromagnetic field by orientation of diffused metal inclusions (figure 6) or electrorheological effect [3,4] when strong electric field is created near orifice valve and makes particles of special oils polarize and enhance viscosity. Electrorheological systems are rather promising but as electromagnetic suspension need more power than modern cars can provide.

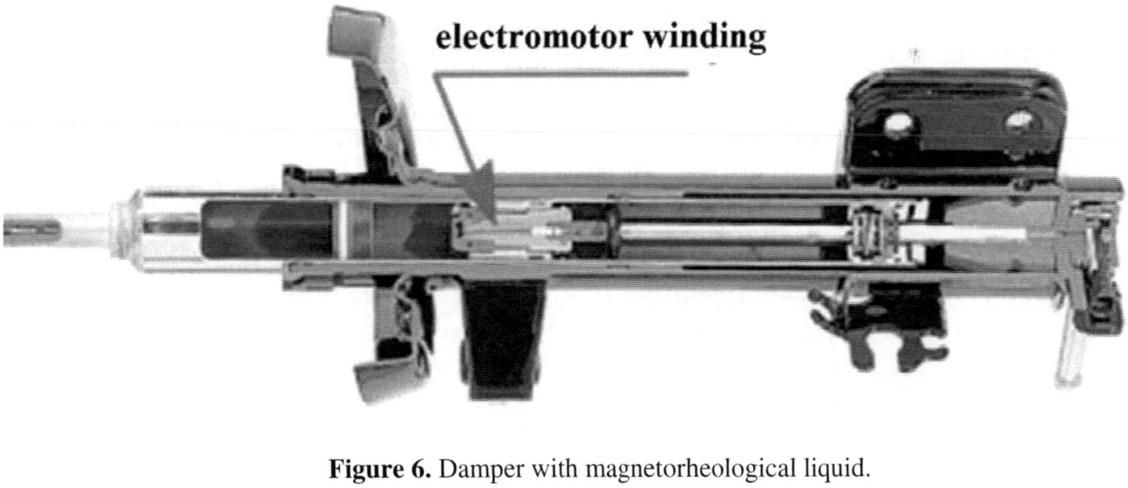

Figure 6. Damper with magnetorheological liquid.

In spite of existing difficulties in using damping technologies modern car industry succeeds in these systems' integration and has no problems with controlling their properties. Regular damper characteristic (force-piston's speed graph) usually look as it is shown on figure 7A and modern dampers with controlled orifice let change it in dynamics almost in any direction (as shown on figure 7B).

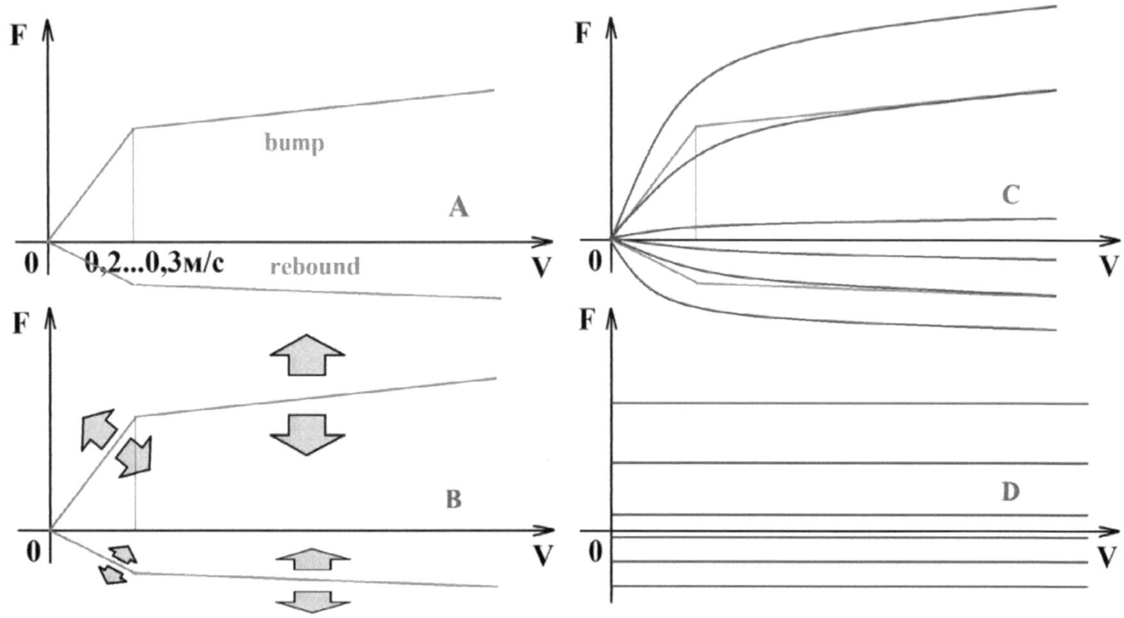

Figure 7. Damper characteristics.

The received characteristic is shown on figure 7C. The characteristic can take any position between boundary curves under different control current. As shown on figure 7C, dampers with controlled orifice have adjustment limits under low speed of a piston. In case of using magnetorheological liquid any characteristic can be received (as shown on figure 7D) because it doesn't have such problems. Disadvantages of these dampers are expensive liquid and problems while working at low temperatures.

The aim of damping elements as driver assistant system is eliminating vibrations that harm passengers. Damping systems with advanced programmable algorithms make car body almost immovable on a regular road. But in reality it is more comfortable for passengers to feel car body's moving than to drive in a motionless capsule. Thus modern systems keep frequency of comfortable 1-1,2 Hz and perspective damping systems (such as electrological dampers and electromagnetic suspension) make car body's behavior comfortable for any driver.

5. Adjustable systems of car's roll stabilization

The first after damping adjustment being integrated tendencies shifted towards automatization of car's roll stabilization systems. Passive roll stabilizer looked like a beam joining car body with wheels on one axle. The stabilizer tried to "untwist itself" in initial state by putting wheels on one axle and car body in zero roll. Stabilizer increases roll stiffness of suspension provided by elastic elements but excessive stiffness leads to losing road traction of one of the wheels what reduces car's stability and passing ability on smooth surface. Thus the first phase of damper automatization led to usage of clutch on twisted beam (figure 8). It let switch stabilizer on on asphalt road and switch off in tough road situations. Later clutch received an electric motor (figure 9) that could not only cut stabilizer's stiffness but also boost it by making an extra torque. Schaeffler possess most developed systems but still doesn't use them on cars with gross weight more than 3 tons.

Figure 8. Stabilizer bar with brake clutch.

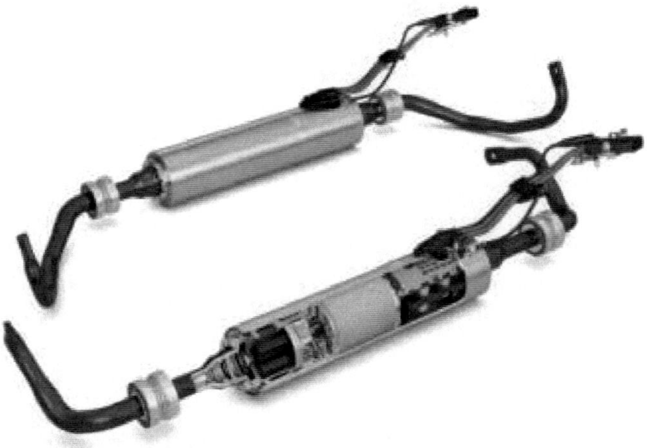

Figure 9. Stabilizer bar with electric motor by Schaeffler.

There is a special type of roll stabilization called "system of dampers' interconnection" that divides the force between suspensions to reduce car body's roll. Sometimes such systems get extra actuators that generate additional force that tries to compress and decompress the suspension. Tenneco uses such systems (figure 10).

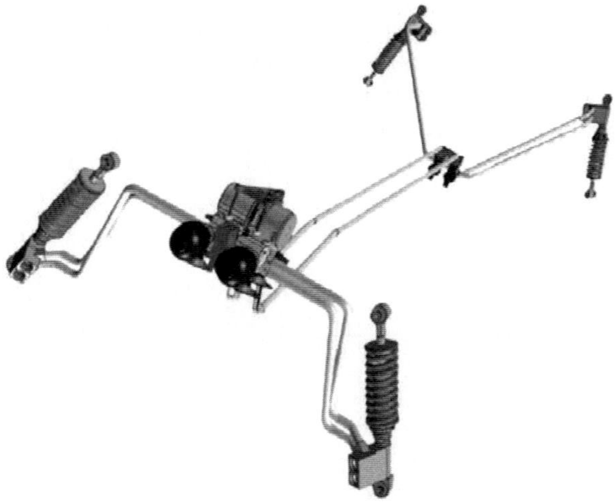

Figure 10. Roll stabilization system by Tenneco [2].

6. Adjustable systems of suspension arms

Suspension arms is very difficult to automatize because it has very strict safety and reliability requirements, it undergoes high load and has low characteristics elasticity in area of car's stability. Nevertheless stock cars' suspension arms already have active systems. As an example – steering back axle of the latest versions of Audi Q7 (figure 11).

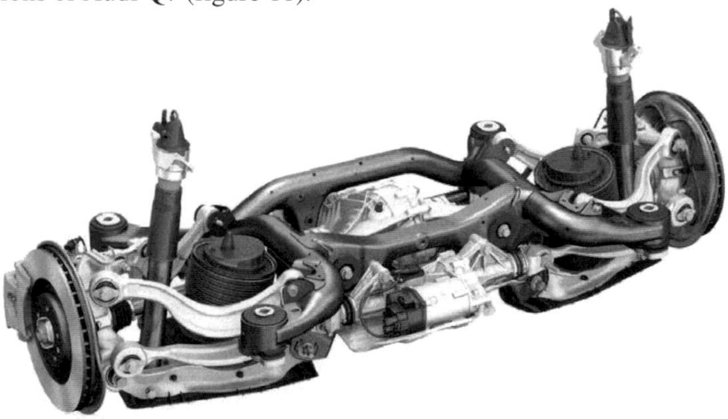

Figure 11. Audi Q7 rear suspension with steering system.

By using extra actuator on rear suspension the car could change automatically rear wheel slip angle during driving. Earlier rear steering axle was used in BMW 7 Series but in exploitation it turned out to be less reliable what with high price didn't let the car gain popularity. Moreover modern systems of suspension design let produce elastokinematic suspension schemes that provide high stability and handling properties without using steering axle.

Steering systems are often used in long wheelbase cars to reduce turning radius. But using these systems on high speed can lead to lack of stability. That is why the companies use elements of blocking rear wheels' neutral position that have strict precision and possible free play requirements.

7. Prospects of development of active suspension systems in Russia and in the word

Adjusted suspension systems underwent two active phases of introduction into mass production. The first phase is implementation of pneumatic suspension, the second phase – mass implementation of adjusted dampers. Now one can define the third phase – mass implementation of systems of roll stabilization and systems of high intelligent suspension system control.

Russia is traditionally called a country of military technologies. That is why reliability is more important than comfort. In practice car producers have to pay attention also to suspension adjustment to reduce strong vibrations that affect drivers during long marches. Some cars gained special systems but they turned out to be too expensive for civil technologies.This experience let create basic industries for such technologies and develop specialist's experience to work with them. Today the leading project in designing active suspension systems is UMP, the NAMI's project aimed at creating luxurious cars for country leaders.

In general modern intellectual systems let seriously boost quality of car's transport functions by using suspension systems. It improved quality of passenger transportation, cargo's safety and reduced driver's tiredness. Even function of dampers' hard mode implementation reduces rolling speed by 11% [5]. This could be checked by ADAMS model connected with Simulink model, where the vehicle behavior simulated in ADAMS environments, but the control system of damper described in Matlab Simulink program. In such soft ware could be prepared a program which changes the damper resistance from "soft" to "hard" mode according lateral acceleration. The increased resistance usually simulated by additional damper force sent to each damper. The appeared force presented on figure 12 [5].

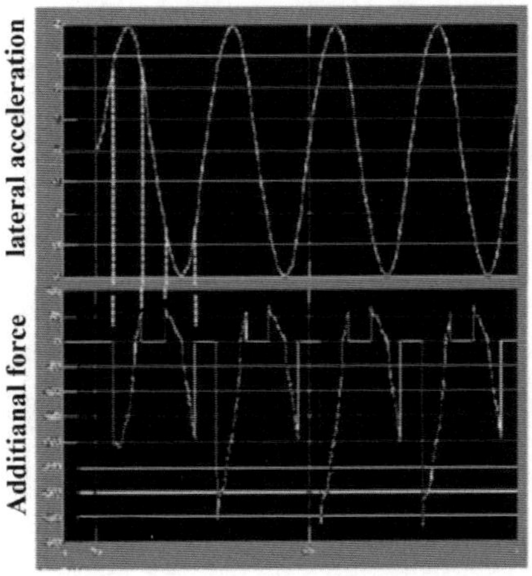

Figure 12. Appeared force description [5].

The result lateral speed decreasing could be seen on figure 13. Presented graphs show the difference betwwenn controlled dampers and passive (not controlled dampers) [5].

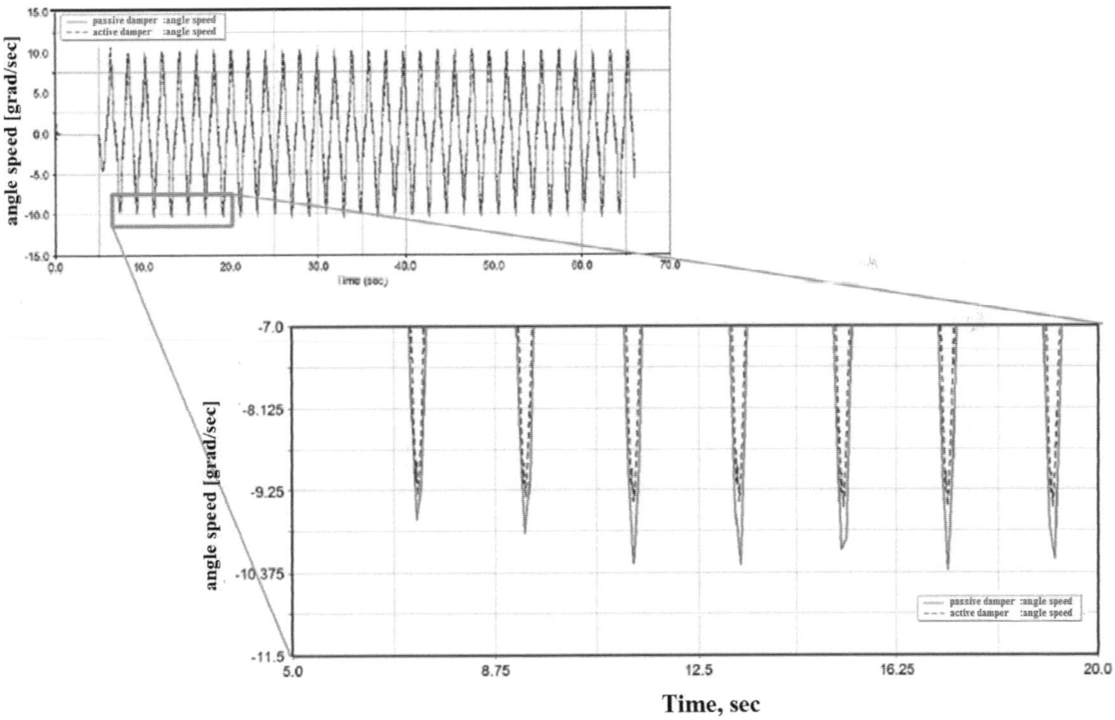

Figure 13. Angle speed decreasing according damper resistance mode [5].

Worldwide the quality of car's behavior is estimated on the basis of a spider graph with ray coming from the center of the diagram and showing the score. The lower the score the closer to the center it stands (and worse it is). The score improves by 1.5-2 points after using intellectual suspension systems on adjusted chassis. Thus the usage of intellectual systems will inevitably continue and in the future their work will let eliminate harmful vibrations affecting passengers.

References

[1] Information on http://www.safholland.bg/sites/usa/en-US/aboutus/photos/Pages/
[2] Nitschke T and Nessel A 2010 The closed air suspension system of the Porsche Panamera *Vehicle Dynamics Expo* (Stuttgart, 22.06.10) p 27
[3] Information on https://studfiles.net/preview/2652747/page:48/
[4] Nefedova T A 2008 Electrorheological effect in suspensions of nanostructured silica *Dissertation of the candidate of chemical sciences: 02.00.04* (Ivanovo) p 136
[5] Information on http://www.tenneco.com/assets/1/7/5498-Kinetic_brochure-9.8.pdf
[6] Bakhmutov S V and Yurlin D V 2017 Simulation of active suspension systems of the vehicle through complex models with external description of control systems *Works of NAMI* **2** (269) pp 6-15

IOP Conf. Series: Materials Science and Engineering **315** (2018) 012030 doi:10.1088/1757-899X/315/1/012030

Comparative analysis of the operation efficiency of the continuous and relay control systems of a multi-axle wheeled vehicle suspension

M M Zhileykin[1], G O Kotiev[2] and M V Nagatsev[3]

Department of Wheeled Vehicles, Bauman Moscow State Technical University, Moscow, Russian Federation

E-mail: [1] jileykin_m@mail.ru, [2] kotievgo@yandex.ru, [3] ngmax@yandex.ru

Abstract. In order to improve the efficiency of the multi-axle wheeled vehicles (MWV) automotive engineers are increasing their cruising speed. One of the promising ways to improve ride comfort of the MWV is the development of the dynamic active suspension systems and control laws for such systems. Here, by the dynamic control systems we mean the systems operating in real time mode and using current (instantaneous) values of the state variables. The aim of the work is to develop the MWV suspension optimal control laws that would reduce vibrations on the driver's seat at kinematic excitation. The authors have developed the optimal control laws for damping the oscillations of the MWV body. The developed laws allow reduction of the vibrations on the driver's seat and increase in the maximum speed of the vehicle. The laws are characterized in that they allow generating the control inputs in real time mode. The authors have demonstrated the efficiency of the proposed control laws by means of mathematical simulation of the MWV driving over unpaved road with kinematic excitation. The proposed optimal control laws can be used in the MWV suspension control systems with magnetorheological shock absorbers or controlled hydropneumatic springs. Further evolution of the research line can be the development of the energy-efficient MWV suspension control systems with continuous control input on the vehicle body.

1. Introduction

Currently, with the use of pneumatic and hydropneumatic suspensions and increase in operational speeds, controlled suspension systems are becoming more and more popular in the global automotive industry.

There are two large groups of controlled suspensions categorized by their control rate: statically controlled and dynamically controlled suspensions.

By the static control (control by the averaged parameters) we mean the process of changing the parameters of the suspension passive elements during the change of the vehicle motion parameters in order to adapt the suspension system of the multiaxle wheeled vehicle (MWV) to these new motion conditions. The motion conditions include the vehicle speed, load, road surface quality, etc. In a steady state, i.e. when the motion conditions are constant, no control occurs. Thus, the objective of the static control is to provide the optimal values of the parameters of the suspension passive elements for each of the steady states of motion.

In this paper, by the dynamic control (control by the instantaneous value of a parameter) we mean the control of the sprung mass oscillations by changing the force applied to the vehicle body from the suspension unit over time. The specific value of this compensating force is defined by the oscillations damping algorithm. This force can be created either by the suspension special active elements using an external power source or by changing the parameters of the suspension passive elements. The systems with dynamic control using external source of energy are known as active suspensions, and the systems with control of the current values of the parameters are called semi-active suspensions. Review of the works in the field of the active suspensions is provided in [1, 2]. Theoretical aspects of the active vibration isolation systems for general cases of the rigid body vibration isolation as well as for special cases like driver's seat suspension engineering have been studied by many researchers: Balagula V. Ya., Gaytsgori M.M., Genkin M.D., Yablonskiy V.V., Eliseev S.A, Inosov S. V., Sinev A. V., Furman F.A. and others [3 – 11]. Automobile active suspensions have been studied in the works of Kol'tsov V.I., Furunzhiev R.I., Ostanin A.N., Sharapov V.D., Sutton H.B., Thompson A.G. and others [12 – 16]. The researchers note that the active suspensions is the most effective choice in terms of ride comfort.

There are two different types of active suspensions. Active suspensions of the first type have no passive springing and damping elements, the body of the vehicle being connected to the wheels by means of non-elastic links capable of changing the force in the suspension according to the control system signal [4, 12]. In the systems of this kind it is preferable to use the disturbance control principle [16].

Active suspension of the second type has passive springing and damping elements generating variable forces in the suspension unit. Active elements of the suspension (for instance, hydraulic actuators) generate forces for body oscillations damping. Active suspensions of this type are of particular interest for practical use due to the following reasons:

1) less energy consumption for active elements actuation due to lower requirements of the control signal processing speed needed to provide the necessary efficiency;

2) lower response speed requirements of the control system;

3) high reliability, firstly, due to simplicity of the control system, secondly, due to ability to sustain vehicle operation in case of the control system failure.

Dynamic control systems can be divided into continuous control systems and discrete (relay) control systems. In the systems of the first type, the control input is defined as additional force to the suspension forces. Work [17] describes an example of this type control system. The relay control can be implemented as a two-level suspension damping control system [18] that provides either high or low damping according to the motion conditions.

Suspension control system (SCS) efficiency and quality depend on many factors: control rate, i.e. the time from parameter registration to the moment of the control input generation; accuracy of the measurement and transfer of the parameters used for the control input generation; accuracy of the control input generation by the actuators. Performance degradation of the SCS parameters in terms of any of the above mentioned factors may lead not only to the lower efficiency of the system but to the full loss of its functional capability.

In real life, there are no control systems with instantaneous response. Signal delay time Δt is determined by the time required for the needed parameters registration, their transfer to the on-board computer, and computation and generation of the control input by the actuators. Increase of the signal delay time can deteriorate the performance of the control system.

It is evident that the accuracy of the control system cannot be higher than the accuracy of the measurements. Control system measurement errors are determined by the following factors: sensor errors; analog-to-digital signal transformation errors; computational errors; feedback errors of the control system actuators.

However, no scientifically grounded criteria has been proposed to assess the performance of a suspension control system so far. As a result, firstly, it is impossible to perform efficiency analysis of new control laws in real-life operational conditions, and, secondly, it is impossible to compare different control laws. The object of this work is to develop criteria for assessment of the efficiency of the suspension control laws at different values of the delay time and input signal measurement errors, another object of this work is to compare efficiency of the continuous and relay MWV suspension control laws by mathematical simulation methods described in [17] and [18].

2. Efficiency criteria for multi-axle wheeled vehicle suspension control system performance

Let's define requirements for the control system efficiency criteria. These criteria must be dimensionless to provide comparison of the control system efficiency in different operation conditions. Besides, efficiency criteria must be based on the overall vibration level measure (GOST 12.1.012-2004), i.e. standard dispersion of the vertical accelerations on the driver's seat.

Control system efficiency criterion taking into account only system elements response delay as a function of the vehicle speed *V* is then

$$K_{\Delta t}(V) = \frac{D_{\Delta t}(V) - D_0(V)}{D_{\Delta t_0}(V) - D_0(V)} 100\% \tag{1}$$

here $D_{\Delta t}(V)$ – standard dispersion of the vertical accelerations on the driver's seat for active or semi-active suspension with total response delay Δt; $D_0(V)$ – standard dispersion of the vertical accelerations on the driver's seat for passive suspension; $D_{\Delta t}(V)$ – standard dispersion of the vertical accelerations on the driver's seat for the active suspension with response delay $\Delta t = 0$ (a perfect system).

Criterion (1) shows relative decrease in control efficiency when the system has non-zero response delay. It is evident that for a perfect system $K_{\Delta t}(V) = 100\%$.

SCS efficiency criterion taking into account only total relative error of measurement as a function of the vehicle speed *V* can be expressed as

$$K_{\Delta A}(V) = \frac{D_{\Delta A}(V) - D_0(V)}{D_{\Delta A_0}(V) - D_0(V)} 100\% \tag{2}$$

here $D_{\Delta A}(V)$ – standard dispersion of the vertical accelerations on the driver's seat for active or semi-active suspension with total measurement error ΔA; $D_{\Delta A_0}(V)$ – standard dispersion of the vertical accelerations on the driver's seat for the active suspension with total measurement error $\Delta A = 0$ (a perfect system).

Criterion (2) shows relative decrease in control efficiency when the system has a non-zero measurement error. It is evident that for a perfect system $K_{\Delta A}(V) = 100\%$.

To compare different SCSs, let us introduce criterion K, describing the difference between the performance of a suspension with a perfect suspension control system (for the case of zero response delay) and the performance of a passive suspension.

$$K(V) = \frac{D_i(V) - D_0(V)}{D_0(V)} 100\% \tag{3}$$

here $D_i(V)$ – standard dispersion of the vertical accelerations on the driver's seat for the case of the active suspension.

Criterion (3) shows the relative decrease in standard dispersion of the vertical accelerations on the driver's seat for the case of the active suspension in comparison with the case of the passive suspension.

3. Response delay effect on the control system efficiency

Works [17] and [18] analyze relay and continuous suspension control systems by the computer simulation of a MWV straight-line motion with speeds from 10 to 70 km/h over an unpaved road. The vehicle motion over the terrain was simulated by the mathematical model developed at the Department of Wheeled Vehicles of the BMSTU [2]. In the MWV mathematical model the speed of the vehicle is not a preset value but is determined by the forces of the interaction between the rolling wheels and the terrain. This provides higher accuracy of the MWV motion simulation. The model is built in MATLAB/Simulink. The model includes a member providing a constant delay of the control signal.

The delay simulates the resultant delay of all control paths of the system. Figure 1 shows the efficiency criterion $K_{\Delta t}(V)$ calculated according to formula (1) for a continuous control system at different vehicle speeds and the signal delay time varying from 0 to 0.3 s.

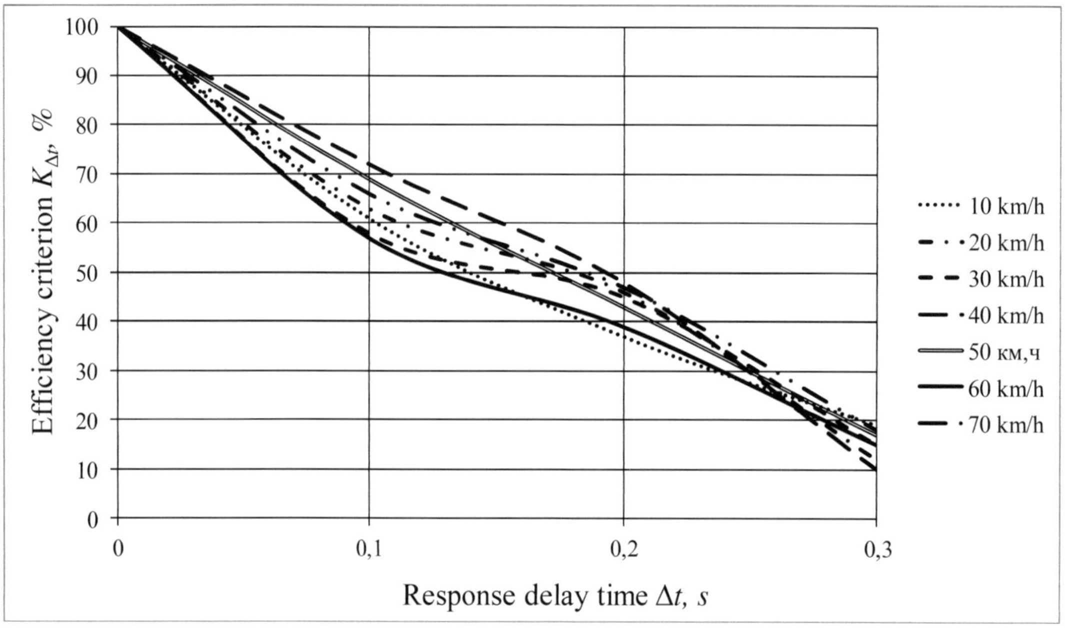

Figure 1. Efficiency criterion $K_{\Delta t}(V)$ for a continuous control system at different vehicle speeds V and different signal delay time Δt.

The dependencies in Figure1 show that the continuous control system is sensitive to the response delay time of its components. The response delay time range from 0.1 to 0.3 s is characteristic for the most control systems. A continuous control system with such a delay time becomes considerably less effective. Therefore, the continuous control systems require higher speeds of response. It is fair to say that the system will remain effective if the response delay time does not exceed 0.1 s.

Similar research has been performed for the relay control system with the response delay time varying from 0 to 0.5 s. The simulation showed that within the said delay time range the relay system kept its efficiency, the criterion $K_{\Delta t}(V)$ value being equal to 100%.

4. Measurement error effect on the suspension control system efficiency

Works [17] and [18] analyze relay and continuous suspension control systems by the computer simulation of a MWV straight-line motion with speeds from 10 to 70 km/h over an unpaved road. Measurement error is simulated as a white noise in the control signal path. The total relative error of measurement is calculated by the formulae

$$A = \frac{\Delta C}{C} = \frac{C_{er} - C}{C}, \qquad D_{er} = (\Delta C)^2, \tag{4}$$

here C_{er} – signal value measured with error; C – true signal value; D_{er} – total standard dispersion of the error.

C_{er} can be calculated by the equation

$$C_{er} = C + x[0;\ D_{er}]. \tag{5}$$

Here x[0; D_{er}] – a white noise with zero mean value and zero standard dispersion $D_{er.}$

Figure 2 shows the efficiency criterion $K_{\Delta A}(V)$ calculated according to formula (2) for a continuous control system at different vehicle speeds and the measurement error varying from 0 to 10 %.

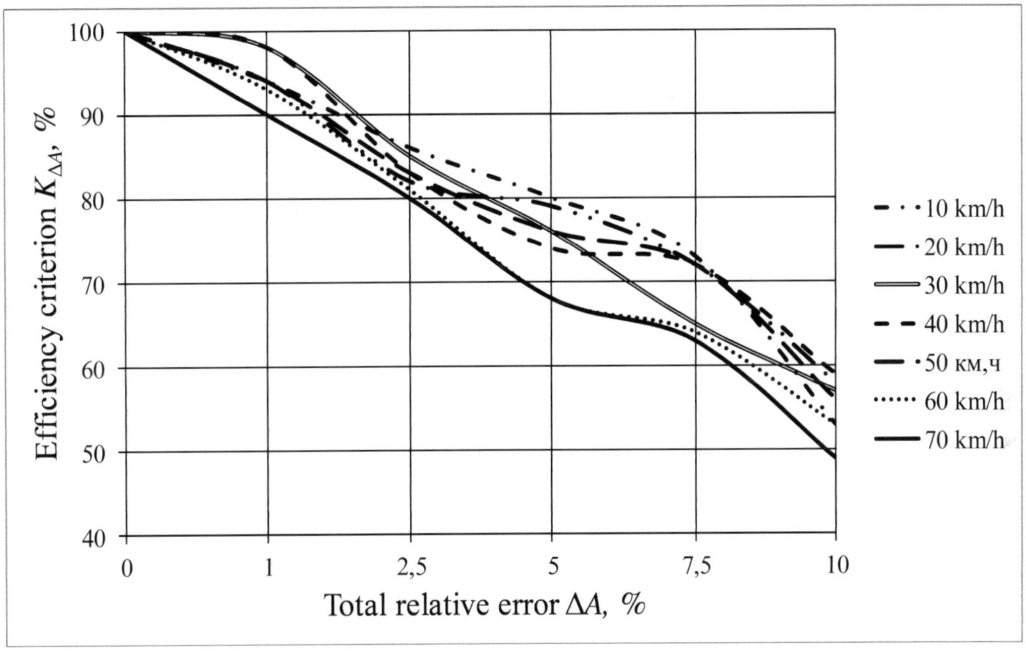

Figure 2. Efficiency criterion $K_{\Delta A}(V)$ for a continuous control system at different vehicle speeds V and different measurement error ΔA.

Figure 2 shows that the measurement errors have significant impact on the efficiency of the suspension continuous control system. Acceptable values of the total relative measurement error of all control paths lie within the range 2.5 – 5 %.

Similar research was carried out for a relay control system with the same range of the total relative error. The simulation showed that within the said range the relay system kept its efficiency, the criterion $K_{\Delta A}(V)$ value being equal to 100%.

5. Comparison of the suspension continuous and relay control system efficiency by the simulation results

Analysis of the simulation of the multi-axle wheeled vehicle motion over unpaved roads allows us to compare the two developed control laws of the suspension: a continuous control law [17] of the active suspension and a relay control law of the suspension with two levels of damping [18]. Figure 3 shows the efficiency criterion *K(V)* according to expression (3) as a function of the vehicle speed.

IOP Conf. Series: Materials Science and Engineering **315** (2018) 012030 doi:10.1088/1757-899X/315/1/012030

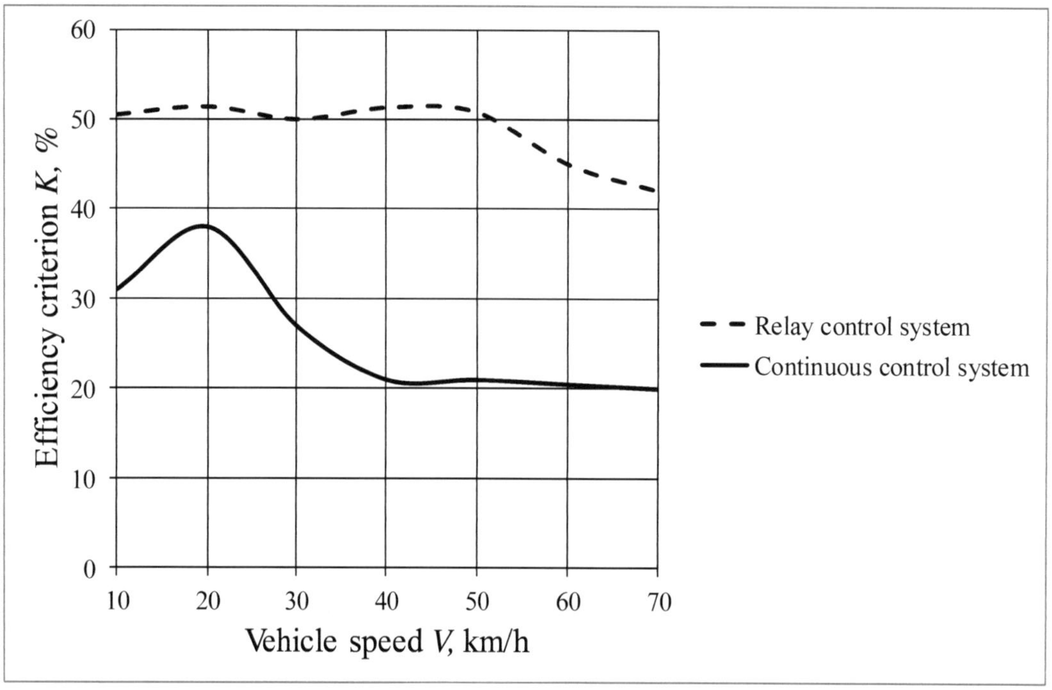

Figure 3. Efficiency criterion *K(V)* as a function of the vehicle speed for continuous and relay control systems over an unpaved road.

Figure 3 shows that the relay control system has practically constant damping level of about 50%. The continuous control system has maximum efficiency for velocities within the range 10 – 30 km/h, then its efficiency goes down and levels off at 20 %. Such a behavior can be explained by the physical limits imposed on the control input.

Table 1 contains the results of the integral comparison of the efficiency of the two control systems (the relay system and the continuous system). Figure 3 shows that for the vehicle speeds in the range 10 – 70 km/h, when the main criterion of a control system efficiency is the degree of decrease in the passenger and cargo vibration levels, the efficiency of the relay control system is higher than the efficiency of the continuous control system. However, when it necessary to provide active damping of the MWV at varying external forces, the continuous control system is much more efficient for a suspension than the relay control system with two-level damping, which is shown in paper [19].

Energy consumption of the active suspension control system is orders of magnitude higher than the one of the two-level damping control system [20], since the latter spends energy only on the actuation of the suspension control elements (for instance, adjustable orifice valves) and the active suspension needs energy for generation of the compensating force in the suspension (for example, by pumping working fluid into the chambers of the hydropneumatic suspension). While the software implementation complexity of the algorithm of the both control systems is equally low, the active suspension is much more complex in design, which results in a considerably higher price and lower reliability in comparison to the relay control suspension. Table 1 shows results of the comparison of the continuous control system of the active suspension with the two-level damping relay control system.

IOP Conf. Series: Materials Science and Engineering **315** (2018) 012030 doi:10.1088/1757-899X/315/1/012030

Table 1. Comparison of the performance efficiency of the suspension control algorithms.

	Compared factor	Continuous control system of the active suspension	Two-level damping relay control system
1	Vibration reduction	+	+
2	Active damping of the MWV body at single inputs	+	–
3	Energy consumption of the control system	–	+
4	Software implementation simplicity	+	+
5	Hardware implementation simplicity	–	+
6	Response delay sensitivity	–	+
7	Measurement error sensitivity	–	+

Therefore, we can define the optimal application fields for the both control systems.

A two-level damping discrete control system should be used in the MWV suspension with relay type actuators (slide valves and pilot valves of the hydropneumatic spring units and hydraulic shock absorbers with valve actuation control system). A continuous control system should be used in the suspension with the primary function of the active damping of the MWV body.

Conclusion

The main results of the work are as follows.

The authors have obtained novel criteria for estimation of the efficiency of a suspension control system with response delay and measurement errors of the input parameters.

The authors have performed a novel analysis of the efficiency of the active continuous and semi-active relay control systems of a suspension for the case when the main control objective is the driver's seat vibration reduction.

The paper has shown that a continuous control system of the active suspension should be used in the case when the main control objective is the active damping of the vehicle body. Discrete control system should operate in driving modes (when there is kinematic excitation of the vehicle body) and provide smooth ride at the lowest energy consumption.

References

[1] Dmitrichenko S S and Zav'yalov Yu A 1983 Procedural framework for the road and field microprofile research for use in tractor dynamic analysis *Inter-university collected research publications* (Moscow) pp 49 – 60

[2] Zheglov L F 2001 Suspension automatic control systems (Moscow)

[3] Balagula V Ya and Gaytsgori M M 1976 Active vibration isolation systems for building machines and road construction machinery Earthmoving equipment research *Russian National Research and Information Centre for Construction Machinery proceedings* 72 pp 44 – 51

[4] Balagula V Ya and Gaytsgori M.M. 1977 Structuring control system of the active suspension for the earthmoving equipment *Russian National Research and Information Centre for Construction Machinery proceedings* **75** pp 39 – 48

[5] Vygovskiy V V 1979 On control synthesis for the active vibration isolation system of a rigid body. Controlled mechanical systems (Irkutsk) pp 41 – 51

[6] Eliseev S V 1979 Theory of the active vibration isolation systems. *Author's abstract of the doctoral thesis* (Irkutsk: Irkutsk State Technical University)

[7] 1974 Theory of the active vibration isolation systems *Collected research papers* **438** (Irkutsk: Irkutsk State Technical University)

[8] Eliseev S V 1982 Dynamics of the controlled mechanical systems (Irkutsk: Irkutsk State Technical University)

[9] Kolovskiy M Z 1976 Automatic control of the vibration isolation systems *Nauka publishing*
[10] Kolovskiy M Z 1977 Optimization of the active vibration isolation systems *Mashinovedenie* **6** pp 42 – 46
[11] Frolov K V and Furman F A 1980 Applied theory of the vibration isolation systems *Mashinostroenie*
[12] Sharapov V D 1980 Active suspensions of the road vehicles *RVVPKU* (Riga)
[13] Furunzhiev P I snd Ostanin A N 1984 Control of the oscillations of the multi-axle vehicles *Mashinostroenie*
[14] Kol'tsov V I 1967 Limits of the suspensions of the terrain vehicles *Ph. D. thesis* (Moscow)
[15] Sutton 1978 Synthesis of active suspension systems *Ph. D. thesis, University of Salford*
[16] Tompson A 1970-71 Design of active suspensions *Proc. Inst. Mech. Eng.* **185** pp 553 – 563
[17] Jileikin M M, Fedotov I V and Mardeeva L R 2013 Development of a continuous control law for a semi-active suspension system with a fuzzy configuration of parameters *Science and education of the Bauman MSTU* **7** Avalable at: http: //technomag. edu. ru/doc/347169.html
[18] Jileikin M M, Fedotov I V and Mardeeva L R 2011 Development of an adaptive relay control law for the two-level damping of the multi-wheel vehicles *Science and education of the Bauman MSTU* **7** Avalable at: http: //technomag. edu. ru/doc/347444.html
[19] Jileikin M M, Fedotov I V and Mardeeva L R 2011 Synthesis of adaptive dynamic continuous system of the vehicle body oscillations damping *Science and education of the Bauman MSTU* **11** Avalable at: http: //technomag. edu. ru/doc/347435.html
[20] Sukhorukov A V 2003 Controling damping elements of the suspension system of a fast-moving tracked vehicle *Ph. D. thesis* (Moscow)

Synthesis of the adaptive continuous system for the multi-axle wheeled vehicle body oscillation damping

M M Zhileykin[1], **G O Kotiev**[2] **and M V Nagatsev**[3]

Department of Wheeled Vehicles, Bauman Moscow State Technical University, Moscow, Russian Federation

E-mail: [1]jileykin_m@mail.ru, [2]kotievgo@yandex.ru, [3]ngmax@yandex.ru

Abstract. In order to meet the growing mobility requirements for the wheeled vehicles on all types of terrain the engineers have to develop a large number of specialized control algorithms for the multi-axle wheeled vehicle (MWV) suspension improving such qualities as ride comfort, handling and stability. The authors have developed an adaptive algorithm of the dynamic damping of the MVW body oscillations. The algorithm provides high ride comfort and high mobility of the vehicle. The article discloses a method for synthesis of an adaptive dynamic continuous algorithm of the MVW body oscillation damping and provides simulation results proving high efficiency of the developed control algorithm.

1. Introduction

In order to improve efficiency of the multi-axle wheeled vehicles (MWV) automotive engineers are increasing their cruising speed [1].

As it is known, the maximum speed of a MWV is limited by the road conditions and the parameters of the vehicle. The limitations imposed by the road conditions can be divided into the following groups: traction limitations, and the so-called direct limitations [2]. Increasing the power-weight ratio of the modern vehicles allows eliminating some limitations from the first group. The second group includes:

- speed limitations determined by the kinematic properties of the powertrain;
- steering limitations connected with the ability of the vehicle to negotiate the actual turns of the track;
- limitations connected with the danger of losing control over the vehicle movement;
- limitations imposed by the vehicle body threshold accelerations which may lead to the crew fatigue and injuries and to the vehicle equipment failure when driving over terrain irregularities.

In view of the above, the problem of the improvement of the MVW suspension to provide higher vehicle speed over irregular terrain remains important today.

One of the promising ways to improve the ride comfort of the MWV is the development of the dynamic active suspension systems and control laws for such systems. By the dynamic control systems here we mean the systems operating in real time mode and using current (instantaneous) values of the state variables. This feature distinguishes them from the static systems using the time-averaged signal. Here, by active control systems we mean the systems using energy supplied from external sources.

Theoretical aspects of the active vibration isolation systems for the general case of the rigid body vibration isolation as well as for special cases like driver's seat suspension engineering have been studied by many Russian researchers: Balagula V. Ya., Gaytsgori M.M., Genkin M.D., Yablonskiy V.V., Eliseev S.A, Inosov S. V., Sinev A. V., Furman F.A. and others [3 – 8].

Automobile active suspensions have been studied in the works of Kol'tsov V.I., Furunzhiev R.I., Ostanin A.N., Sharapov V.D., V.D., V.D. [9, 10, 11], Sutton H.B., Thompson A.G. and others [12, 13].

The researchers note that active suspensions is the most effective choice in terms of ride comfort.

Introduction of the active control systems into the MWV suspension is hindered by the lack of the continuous control laws for the spring and damping elements of the suspension [1].

The aim of the work is to develop the MWV suspension optimal control laws that would decrease vibrations on the driver's seat at kinematic excitation.

2. Synthesis of the dynamic continuous control system for the multi-axle vehicle body oscillation damping

Let us assume that the vehicle body is rigid and the suspension of the wheels is a sliding pillar suspension [1]. The linearized model of the MWV body motion has just three degrees of freedom (see figure 1):

- Translation along the vertical axis Z of the coordinate system fixed to the MWV body;
- Rotation about the principal axis X of inertia in the coordinate system fixed to the vehicle body;
- Rotation about the principal axis Y of inertia in the coordinate system fixed to the vehicle body;
- Rotation about the principal axis Y of inertia in the coordinate system fixed to the vehicle body.

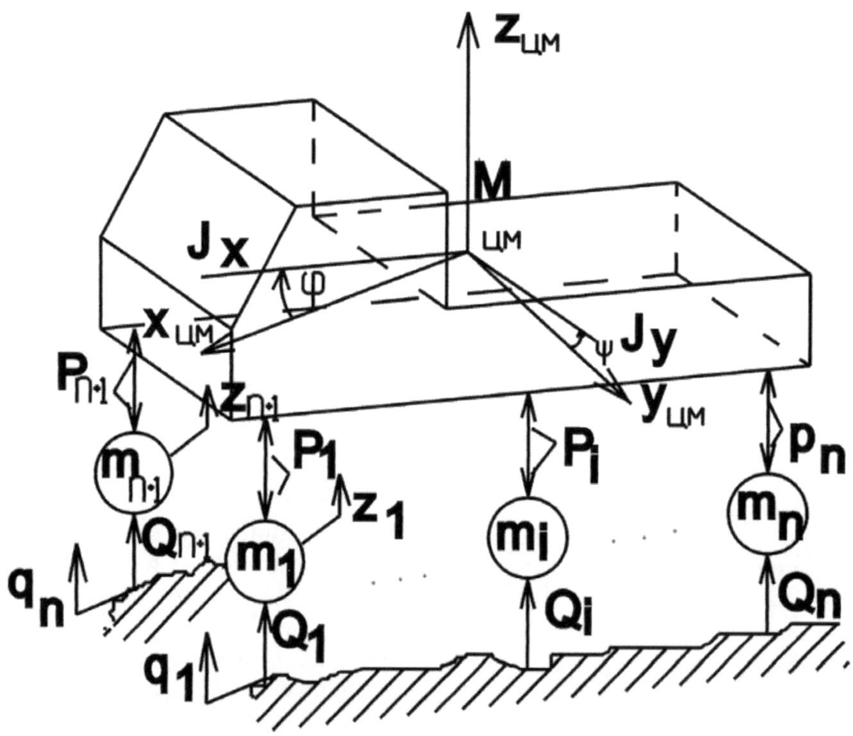

Figure 1. Multi-axle wheeled vehicle modeled as a multiply supported oscillating system.

Simplified equations of the multi-axle vehicle body motion:

$$\begin{cases} \dot{\omega_x} = \frac{1}{J_x} U_1 \\ \dot{\omega_y} = \frac{1}{J_y} U_2 \\ \dot{V_z} = \frac{1}{M_{spm}} U_3 - g \\ \dot{\varphi} = \omega_x \\ \dot{\psi} = \omega_y \\ \dot{z} = V_z \end{cases} \quad (1)$$

here

U_1 – control moment generated by the damping elements of the suspension about X-axis (applied at the vehicle body center of mass);

U_2 – control moment generated by the damping elements of the suspension about Y-axis (applied at the vehicle body center of mass);

U_3 – resultant control force vector generated by the spring and damping elements of the suspension in the vertical direction (applied in the vehicle body center of mass);

ω_x – angular velocity of the vehicle body about x-axis;

ωx – angular velocity of the vehicle body about y-axis;

V_z – linear velocity of the vehicle body along *z*-axis;

φ – angle of the body roll about *x*-axis;

ψ – angle of the body roll about y-axis;

z – translation of the vehicle body center of mass along z-axis;

J_x, J_y – MVW sprung mass inertia moments about axes x and y respectively;

M_{spm} – sprung mass.

Here control inputs U1 and U2 are control torques about the longitudinal axis X and central lateral axis Y of the vehicle respectively. Control input U3 is a control vector whose components are forces parallel to the vertical axis Z. All the control inputs are generated by the controlled spring and damping elements of the multi-axle vehicle suspension.

System (1) can be divided into three independent control channels: two channels for damping the vehicle body angular oscillations and angular stabilization of the vehicle body, and one channel for damping its vertical oscillations.

During synthesis of the optimal systems the developers widely use quadratic integral performance criteria containing the coordinates of the object and of the control input [14, 15].

Initial system (1) of the state equations is a linear system which can be presented as a vector differential equation

$$\frac{dX(t)}{dt} = A(t)X(t) + B(t)u(t) \quad (2)$$

here *X(t)* – n-dimensional state vector;*A(t), B(t) – matrix time functions with dimensions n×n, n×m respectively;*

u(t) - *m*-dimensional control vector.

Due to evident physical reasons, control vector U(t) = [u1(t), …, um(t)]T has limited dimensions, i.e. it belongs to a closed set

$$|u_j(t)| \leq U_{jmax}, j = 1, 2, \dots, m.$$

In order to select the control performance criterion we should define its requirements.

The most universal performance criterion is the criterion based on the optimized integral functional. Use of the integral criteria, the quadratic functionals in particular, allows formulation of the requirements for the transient behavior of the control system by setting large number of their weight coefficients.Practically unrestricted selection of the values of these coefficients provides for the fundamental property of the synthesized system – its asymptotical stability [15]. Iterative directed search of the weight coefficients usually meets the reasonable requirements of the primary transient performance criteria.

The main advantage of this method is that it provides the ability to use results of the optimal control theory and the theory of the analytical design of the optimal controllers for the synthesis of a control system for a complex object. Ever-growing use of the theory of the analytical design of the optimal controllers in the complex industrial object control applications is due to such their advantages as generality, maximum formalization, logical completeness, essential mathematical simplicity.

Thus, for controlled objects (2) we need to solve a problem of the theory of the analytical design of the optimal controllers for the respective integral criterion. Selection of the control performance functional (during the theory of the analytical design of the optimal controllers problem formulation) is performed under the assumption that the control objective for a multi-axle wheeled vehicle suspension is stabilization of the output variables (velocity, acceleration, and displacement of the sprung masses) at a certain level providing lower vibrations and better stability in case of high-amplitude (high power) inputs from the terrain.

At present, in the theoretical and practical optimal control applications, the developers widely use control system performance integral quadratic functionals of the form [16]

$$I = \int_{t_0}^{t_1} [U^T(t)r(t)u(t) + X^T(t)q(t)X(t)]dt. \tag{3}$$

Here, $U(t)$ – m-dimensional control vector; $X(t)$ – n-dimensional state vector; $r(t)>0$, $q(t)>0$ –weight coefficients.

Value of the functional I depends on the initial state $X(t_0)$, time value t_0, and control input $u(t)$ over the interval (t_0,t_1). The objective of the optimal control is to minimize functional (3).

3. Developing optimal control law for damping angular and vertical oscillations of the multi-axle wheeled vehicle

3.1. Damping angular oscillations of the MWV body

Let us assume that x1 = φ – body roll angle; x2 = ω – angular velocity. From equations (1) we can get to the system of equations describing the angular oscillations of the MWV of the form

$$\begin{cases} \dot{x}_1 = x_2 \\ \dot{x}_2 = U^* \\ U^* = \dfrac{U}{J} \end{cases} \tag{4}$$

Here, $J=J_x$ or $J=J_y$.

Here, the control input U is the control torque about the respective central axis of the vehicle. All the control inputs are generated by controlled spring and damping elements of the vehicle suspension.

The unknown is the control law providing minimum value of the functional

$$I(x,U) = \int_0^\infty \left[rU^2 + \sum_{i=1}^{2} q_i x_i^2 \right] dt, \tag{5}$$

here, $r > 0$; $qi > 0$; x – object state vector; U – object control vector.

We'll try to meet the following boundary conditions

$$x_1(0) = x_{10};\ x_2(0) = x_{20};$$
$$x_1(\infty) = 0;\ \ x_2(\infty) = 0. \tag{6}$$

So, we'll be solving the problem without the right-hand boundary since at the dynamic control of the suspension there is no need to impose any limitations on the phase coordinates. This type of control only provides reduction of the phase coordinate values and is not capable of bringing them to zeroes. At this stage of the optimal controller synthesis, we'll not impose any limitations on the phase coordinates and control inputs.

The above-formulated problem is Lagrange problem of the constraint extremum [16]. According to Lagrange problem solution procedure, intermediary functional

$$I(x,U) = \int_0^\infty \left[rU^2 + \sum_{i=1}^{2} q_i x_i^2 + \sum_{i=1}^{2} \lambda_i(t) \left(x_i - \sum_{i=1}^{2} a_{ik}x_k - b_i U_i \right) \right] dt, \tag{7}$$
$$A = \begin{bmatrix} 0 & 1 \\ 0 & 0 \end{bmatrix}, B = \begin{bmatrix} 0 \\ 1/J \end{bmatrix}.$$

Euler-Lagrange equations for functional (7)

$$\frac{\partial \Phi}{dx_i} - \frac{d}{dt}\frac{\partial \Phi}{dx_i} = 0, i = 1, 2,$$
$$\frac{\partial \Phi}{dU_i} - \frac{d}{dt}\frac{\partial \Phi}{dU_i} = 0, i = 1, 2, \tag{8}$$

here $\Phi(x_i, U_i)$ – integrand of functional (7). We should add constraint equations (4) to equations (8). The system of the associated equations will look as follows

$$\begin{cases} \dot{x}_1 = x_2 \\ x_2 = \dfrac{1}{2J^2 r}\lambda_2 \\ \dot{\lambda}_1 = 2q_1 x_1 \\ \dot{\lambda}_2 = 2q_2 x_1 - \lambda_1 \end{cases} \tag{9}$$

The characteristic equation for a linear homogeneous system of differential equations (9)

$$D(\mu) = \begin{bmatrix} -\mu & 1 & 0 & 0 \\ 0 & -\mu & 0 & \dfrac{1}{2\cdot J^2 r} \\ 2\cdot q_1 & 0 & -\mu & 0 \\ 0 & 2\cdot q_2 & -1 & -\mu \end{bmatrix} = 0, \tag{10}$$

or

$$\frac{\mu^4 J^2 r - q_2 \mu^2 + q_1}{J^2 r} = 0$$

Since $r>0$, the roots of the characteristic equation are

$$\mu_{1,2,3,4} = \pm \left[\frac{2}{J^2 r} \left(q_2 \pm \sqrt{q_2^2 - 4J^2 r q_1} \right) \right]^{1/2}. \tag{11}$$

According to boundary conditions (6), eigen values from the right-hand semiplane can be neglected, which brings us to the synthesis of a stable automatic control system. Then, according to the known procedures [16] we arrive to the general solution of system (9)

$$x_1(t) = C_1 exp(\mu_3 t) + C_2 exp(\mu_4 t), \tag{12}$$
$$x_2(t) = C_1 \mu_3 exp(\mu_3 t) + C_2 \mu_4 exp(\mu_4 t), \tag{13}$$

here μ_3, μ_4 – roots of characteristic equation (10), the roots being located in the left-hand semiplane; C_1, C_2 - constants.

Let us solve equations (12) and (13) for $C_1 \exp(\mu_3 t)$ and $C_2 \exp(\mu_4 t)$. We get the following expressions:

$$C_1 exp(\mu_3 t) = \frac{x_1 \mu_4 - x_2}{\mu_4 - \mu_3},$$
$$C_2 exp(\mu_4 t) = \frac{x_2 - \mu_3 x_1}{\mu_4 - \mu_3}.$$

The solution of the second equation of system (4) gives us the expression for the optimal control law of the angular oscillation damping

$$U^* = \mu_4 \mu_3 J \left[-x_1 + \frac{\mu_4 + \mu_3}{\mu_4 \mu_3} x_2 \right] \tag{14}$$

System (4) with control law (14) is stable since the roots μ_3, μ_4 of its characteristic equation are located in the left-hand semiplane. The derived optimal control law can be used for controlling each roll angle of the vehicle body separately.

3.2. Damping vertical oscillations of the MVW body

Let's turn to the development of the control law for damping the vertical oscillations of the vehicle body. Initial differential equation

$$\dot{V}_z = \frac{1}{M_{spm}} U_3 - g \tag{15}$$

Let us introduce new notation. Let us denote $U = \frac{1}{M_{spm}} U_3 - g$. Then equations (15) can be rewritten as

$$\begin{cases} \dot{V}_z = U \\ \dot{z} = V_z \end{cases} \tag{16}$$

Let us denote x1 = z – vertical displacement of the vehicle body center of mass; x2 = Vz – vertical velocity of the vehicle body center of mass. Then, system (16) can be rewritten as

$$\begin{cases} \dot{x}_1 = x_2 \\ \dot{x}_2 = U \end{cases} \tag{17}$$

Control law U_3 is the resultant vector force at the center of mass of the vehicle body. The unknown is the control law providing minimum value of the functional

$$I(x, U) = \int_0^\infty \left[rU^2 + \sum_{i=1}^{2} q_i x_i^2 \right] dt, \tag{18}$$

here, $r > 0$; $q_i > 0$; x – object state vector; U – object control vector.

We'll try to meet the following boundary conditions

$$x_1(0) = x_{10};\ x_2(0) = x_{20};$$
$$x_1(\infty) = 0;\ x_2(\infty) = 0. \tag{19}$$

The above-formulated problem is Lagrange problem of the constraint extremum [16]. According to Lagrange problem solution procedure, intermediary functional

$$I(x,U)=\int_0^{\infty}\left[rU^2+\sum_{i=1}^{2}q_i x_i^2+\sum_{i=1}^{2}\lambda_i(t)\left(x_i-\sum_{i=1}^{2}a_{ik}x_k-b_iU_i\right)\right]dt, \tag{20}$$

$$A=\begin{bmatrix}0&1\\0&0\end{bmatrix}, B=\begin{bmatrix}0\\1/J\end{bmatrix}.$$

Euler-Lagrange equations for functional (20) have the form of (8). We should add constraint equations (17) to equations (8). The system of the associated equations will look as follows

$$\begin{cases}\dot{x}_1=x_2\\ x_2=\dfrac{1}{2r}\lambda_2\\ \dot{\lambda}_1=2q_1x_1\\ \dot{\lambda}_2=2q_2x_1-\lambda_1\end{cases} \tag{21}$$

The characteristic equation for a linear homogeneous system of differential equations (21)

$$D(\mu)=\begin{bmatrix}-\mu & 1 & 0 & 0\\ 0 & -\mu & 0 & \dfrac{1}{2\cdot r}\\ 2\cdot q_1 & 0 & -\mu & 0\\ 0 & 2\cdot q_2 & -1 & -\mu\end{bmatrix}=0, \tag{22}$$

or

$$\frac{\mu^4M^2r-q_2\mu^2+q_1}{r}=0$$

Since r>0, the roots of the characteristic equation are

$$\mu_{1,2,3,4}=\pm\left[\frac{2}{M^2r}\left(q_2\pm\sqrt{q_2^2-4rq_1}\right)\right]^{1/2}. \tag{23}$$

According to boundary conditions (19), as shown above, eigen values from the right-hand semiplane can be neglected, which brings us to the synthesis of a stable automatic control system. Then, according to the known procedures [16] we arrive to the general solution of system (21)

$$x_1(t)=C_1exp(\mu_3t)+C_2exp(\mu_4t), \tag{24}$$

$$x_2(t)=C_1\mu_3exp(\mu_3t)+C_2\mu_4exp(\mu_4t), \tag{25}$$

here μ_3,μ_4 – roots of characteristic equation (22), the roots being located in the left-hand semiplane. Let us solve equations (24) and (25) for $C_1\exp(\mu_3t)$ and $C_2\exp(\mu_4t)$. This gives the following expressions

$$C_1exp(\mu_3t)=\frac{x_1\mu_4-x_2}{\mu_4-\mu_3},$$
$$C_2exp(\mu_4t)=\frac{x_2-\mu_3x_1}{\mu_4-\mu_3}.$$

Solution of the second equation of system (17) gives us the expression for the optimal control law of the vertical oscillation damping

$$U = \mu_4\mu_3\left[-x_1 + \frac{\mu_4+\mu_3}{\mu_4\mu_3}x_2\right] \quad (26)$$

System (17) with control law (26) is stable since the roots μ_3, μ_4 of its characteristic equation are located in the left-hand semiplane. The final expression of the control law has the following form

$$U_3 = M_{spm}(U + g) = M_{spm}\left[\mu_4\mu_3\left\{-x_1 + \frac{\mu_4+\mu_3}{\mu_4\mu_3}x_2\right\} + g\right]. \quad (27)$$

4. Simulation results

The developed optimal control laws were tested by simulation of the straight-line steady motion of a multi-axle 8x8 wheeled vehicle with gross weight 60 ton over an unpaved road with different speeds. The vehicle motion over the terrain was simulated by the mathematical model developed at the Department of Wheeled Vehicles of the BMSTU [17]. In the MWV mathematical model the speed of the vehicle is not a preset value but is determined by the forces of the interaction between the rolling wheels and the terrain. This provides high accuracy of the MWV motion simulation.

Figures 2 – 4 show vibration levels on the driver's seat in octave bands of frequencies for active and passive suspensions over an unpaved road for speeds 10 – 70 km/h.

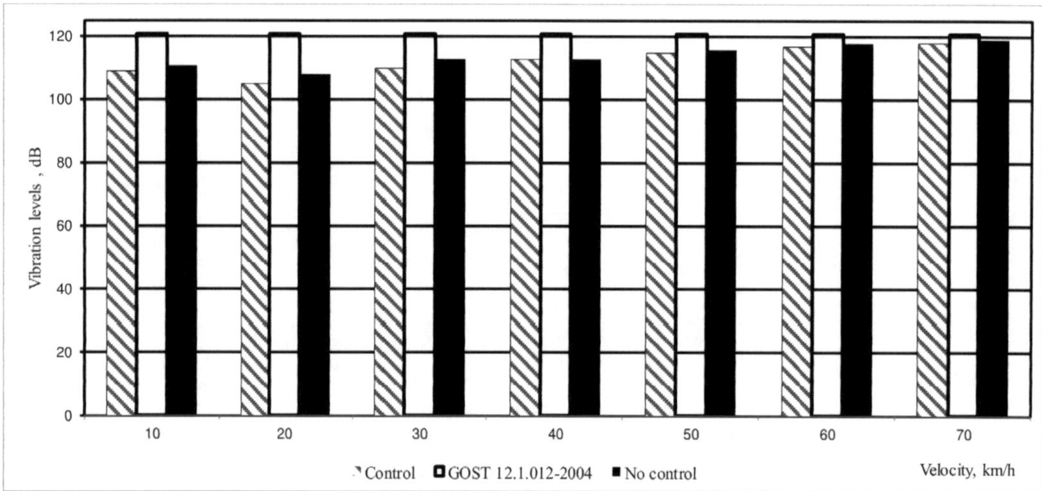

Figure 2. Vibration levels on the driver's seat (octave I).

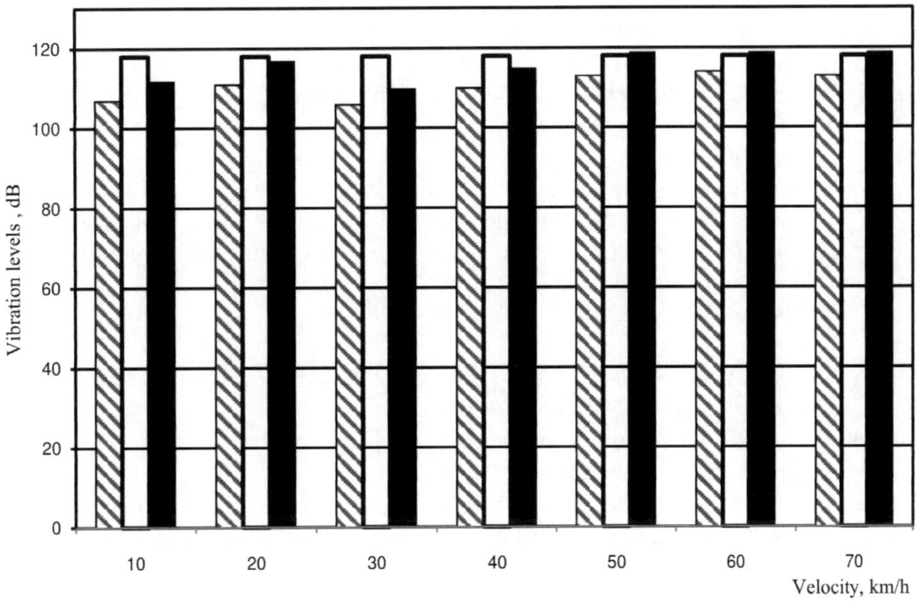

Figure 3. Vibration levels on the driver's seat (octave II).

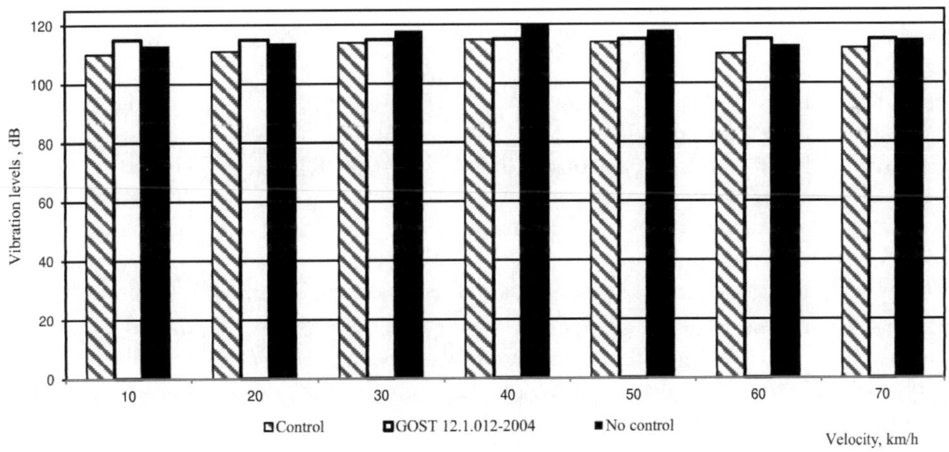

Figure 4. Vibration levels on the driver's seat (octave III).

Analysis of the charts in figures 2 – 4 brings us to the following conclusions.

- The developed optimal control laws for damping the MVW vehicle body are effective for damping oscillations of the sprung mass at all frequencies from the operational range (0 – 5 Hz).
- Over the unpaved road for the active suspension vibration levels at different octaves on the driver's seat go down by 2 – 6 dB according to the vehicle speed.
- The simulation results show that for a vehicle moving over the unpaved road with the active suspension controlled according to the proposed control laws, maximum speed can be increased from 20 km/h to 70 km/h by ride comfort criterion (according to the standards of GOST 12.1.012-2004).

Conclusion

The authors have developed the optimal control laws for damping the oscillations of the MWV body, which allows reduction of the vibration level on the driver's seat and increase in the maximum speed of the vehicle, the laws are characterized in that they provide generating the control inputs in real time mode. The authors have demonstrated the efficiency of the proposed control laws by means of mathematical simulation of the MWV driving over an unpaved road with kinematic excitation.

The proposed optimal control laws can be used in the MWV suspension control systems with magnetorheological shock absorbers or controlled hydropneumatic springs. Further evolution of the research line can be the development of the energy-efficient MWV suspension control systems with continuous control input on the vehicle body.

References

[1] Belousov B N and Popov S D 2014 Heavy-duty wheeled vehicles: Design, Theory, Calculations *SAE International* ISBN 978-0-7680-7723-0
[2] Sukhorukov A V 2003 Controling the damping elements of the suspension system of a fast-mooving tracked vehicle (Moscow, *Ph. D. thesis*)
[3] Balagula V Ya and Gaytsgori M M 1977 Structuring control system of the active suspension for the earthmoving equipment *Russian National Research and Information Centre for Construction Machinery proceedings* **75** pp 39 – 48
[4] Vygovskiy V V 1979 On control synthesis for the active vibration isolation system of a rigid body *Controlled mechanical systems* (Irkutsk) pp 41 – 51
[5] Eliseev S V 1979 Theory of the active vibration isolation systems *Author's abstract of the doctoral thesis* (Irkutsk: Irkutsk State Technical University)
[6] Eliseev S V 1982 Dynamics of the controlled mechanical systems (Irkutsk: Irkutsk State *Technical* University)
[7] Kolovskiy M Z 1977 Optimization of the active vibration isolation systems *Mashinovedenie* **6** pp 42 – 46
[8] Frolov K V and Furman F A 1980 Applied theory of the vibration isolation systems *Mashinostroenie* (Moscow) 276 p.
[9] Sharapov V D 1980 Active suspensions of the road vehicles RVVPKU (Riga)

[10] Furunzhiev P I Ostanin A N 1984 Control of the oscillations of the multi-axle vehicles *Mashinostroenie* (Moscow)
[11] Kol'tsov V I 1967 Limits of the suspensions of the terrain vehicles Ph. D. thesis (Moscow)
[12] Sutton 1978 Synthesis of active suspension systems *Ph. D. thesis University of Salford*
[13] Tompson A 1970-71 Design of active suspensions *Proc. Inst. Mech. Eng.* **185** pp 553 – 563
[14] Andryushchenko V A 1990 Theory of the automatic control systems (Leningrad, *Leningrad State University*)
[15] Aliev F A, Larin V B and Naumenko K I et al 1978 Optimization of the linear time-invariant control systems *Naukova Dumka* (Kiev)
[16] Ivanov V A and Faldin F V 1981 Theory of the optimal automatic control systems *Nauka* (Moscow)
[17] Afanas'yev B A, Belousov B N and Zheglov L F et al. 2008 All-wheel drive vehicle engineering **3** (Moscow: BMSTU)

IOP Conf. Series: Materials Science and Engineering **315** (2018) 012032 doi:10.1088/1757-899X/315/1/012032

The impact of electric vehicles on the outlook of future energy system

A Zhuk and E Buzoverov

Joint Institute for High Temperatures of the Russian Academy of Sciences (JIHT RAS), Moscow, Russia

E-mail: teoconsult@inbox.ru

Abstract: Active promotion of electric vehicles (EVs) and technology of fast EV charging in the medium term may cause significant peak loads on the energy system, what necessitates making strategic decisions related to the development of generating capacities, distribution networks with EV charging infrastructure, and priorities in the development of battery electric vehicles and vehicles with electrochemical generators. The paper analyses one of the most significant aspects of joint development of electric transport system and energy system in the conditions of substantial growth of energy consumption by EVs. The assessments of per-unit-costs of operation and depreciation of EV power unit were made, taking into consideration the expenses of electric power supply. The calculations show that the choice of electricity buffering method for EV fast charging depends on the character of electricity infrastructure in the region where the electric transport is operating. In the conditions of high density of electricity network and a large number of EVs, the stationary storage facilities or the technology of distributed energy storage in EV batteries – vehicle-to-grid (V2G) technology may be used for buffering. In the conditions of low density and low capacity of electricity networks, the most economical solution could be usage of EVs with traction power units based on the combination of air–aluminum electrochemical generator and a buffer battery of small capacity.

1. Introduction

Current trends in motor transport evolution indicate the inevitability of transition to electric vehicles (EVs). It is forecasted that by the year 2030 their number will reach 150 million units (10% of the total number of motor vehicles), and by the year 2060 – 1.2 billion units (60%). Significant growth of EV fleet may have substantial impact on the power supply system, especially taking into consideration the development of quick charging technologies which imply usage of charging terminals with capacity up to 350 kW. In case of estimated increase of EV fleet, their electricity consumption can reach 12% in overall power balance by the year 2030, and 32% by the year 2050 [1]. It should be noted that the increase of EV number may lead not only to growth of overall energy consumption, but also to significant increase of peak electricity consumption, for example in the morning or evening hours.

One of the possible solutions to the problem of growth of peak electricity consumption may be participation of EV fleet in the regulation of power grid regimes based on the use of informational possibilities of smart grids. Another, even more radical, option of smoothing the peaks of consumption loads is the V2G concept, enabling usage of EV batteries for regulation of power grid regimes.

An alternative solution to the problem of covering peak loads on the power system caused by EVs is to install stationary electrochemical storage facilities.

Another option of reducing peak loads is to use EVs with electrochemical generators on hydrogen or aluminum [2-4]. The intermediate energy carrier can be produced outside large cities, thus avoiding extra loads on the municipal grids. When using aluminum or hydrogen, EV charging procedure is shortened to 5 -10 min., i.e. it becomes comparable with the duration of fueling of a conventional vehicle. In this case no significant changes of the grid infrastructure are required.

The purpose of this work is to assess of economic efficiency of the listed above options for reducing peak loads on the power system taking into consideration the differences in the procedures of recharging battery electric vehicles (BEVs) and vehicles with traction power units based on electrochemical generators.

2. Problem formulation

This work contains the comparison of expenses for electric energy (energy carrier) and depreciation of the power unit of electric vehicle. It is assumed that other expenses connected with ownership of electric vehicles belonging to one class are approximately the same.

Two types of electric vehicles are being considered:

- electric vehicles with lithium-ion batteries;
- electric vehicles with combined power units based on air-aluminum electrochemical generators and lithium-ion batteries.

It is assumed that BEVs will employ the fast charging option at public charging stations of capacity up to 350 kW. The peak loads on electric grid created by the charging stations are compensated by storage facilities also connected to the grid and located in the same cluster with the particular fast charging station. Two sub-options of energy storage are analyzed. In one case, the electric energy for rapid charging of electric vehicles is accumulated by stationary storage facilities which capacities are specially reserved for these purposes. In the other case, the capacities of EV batteries are employed which are not used at the current time, but are connected to the charging terminal by a bidirectional inverter. The batteries of such group of EVs are provided by their owners to the power system as components of the distributed energy storage system.

The electric vehicle on aluminum fuel is using only the electrochemical generator as the power unit, and the battery is used only as the source of power for acceleration and for recovery of braking energy. Replacement of anodes and electrolyte is performed at special recharging points.

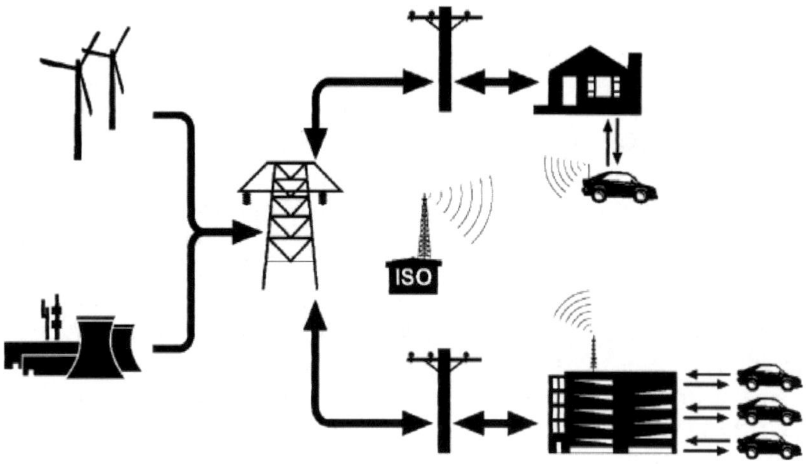

Figure 1. Scheme of V2G technology implementation.

3. Methods

3.1. Expenses related to operation of a battery electric vehicle with buffering of electric energy through V2G

A special feature of this scheme is the usage of EV batteries for two purposes: for EV traction and for buffering of electric energy (figure 1) required for charging of other EVs.

In case of using the charging stations for BEVs with the possibility of using smart grid and V2G for leveling peak loads, the cost of electricity will be the sum of the expenses for purchasing electricity considering the costs of use of storage facilities, and depreciation of the EV battery:

$$C_{100\kappa m} = \left(\frac{C_{el}^{year}}{L^{year}} + \frac{C_{V2G}^{year}}{L^{year}} + \frac{C_{bat.EV}^{year}}{L^{year}}\right) \cdot 100, \quad (1)$$

where C_{el}^{year} - yearly cost of electricity for EV charging, \$/year; C_{V2G}^{year} - expenses for V2G-regulation, \$/year; $C_{bat.EV}^{year}$ - EV battery wear when used for EV traction, \$/year; L^{year} - EV yearly kilometrage, km/year; 100 – coefficient for normalizing the expenses per 100 km.

$$C_{el}^{year} = \frac{W_{el}^{year} \cdot C_{el}}{\eta}, \quad (2)$$

where C_{el} - weighted average price of electricity, \$/kWh, W_{el}^{year} - yearly consumption of electricity for EV running, kWh/year, η - energy efficiency of the storage facility.

Yearly electricity consumption is determined by the specific electricity consumption and kilometrage:

$$W_{el}^{year} = \frac{L_{year} \cdot v}{100}, \quad (3)$$

where L_{year} - EV yearly kilometrage, km, v - EV specific electricity consumption, kWh/100 km, 100 – coefficient for normalizing the expenses per 100 km.

Expenses for V2G regulation are determined by specific cost of EV batteries, their lifetime number of cycles and depth of discharge. In order to calculate the expenses for V2G regulation, additional expenses for V2G system are estimated which are connected with battery deterioration during leveling of peak consumption C_{bat}^{year}, \$/year, expenses for EV charging equipment and other infrastructure $C_{\inf r}^{year}$, \$/year.

$$C_{V2G}^{year} = C_{bat}^{year} + C_{\inf r}^{year}, \quad (4)$$

Yearly expenses associated with EV battery deterioration resulting from participation in V2G C_{bat}^{year}, are calculated as:

$$C_{bat}^{year} = \frac{C_{EV}^{bat}}{N_{cycle} \cdot DoD} \cdot W_{el}^{year}, \quad (5)$$

where C_{EV}^{bat} - specific cost of battery, \$/kWh; N_{cycle} - battery lifetime number of cycles, DoD - specific depth of discharge of the battery.

Expenses for V2G infrastructure per one serviced electric vehicle are determined by: its useful capacity P_{V2G}, kW, and service life, $T_{\inf r}^{V2G}$. Assuming that one EV takes part in not more than one V2G regulation cycle per day, specific expenses for V2G infrastructure are:

$$C_{\inf r}^{year} = \frac{C_{\inf r}^{V2G}}{T_{\inf r}^{V2G} \cdot P_{V2G} \cdot 365 \cdot \lambda_{V2G}} W_{el}^{year}, \tag{6}$$

where $C_{\inf r}^{V2G}$ - cost of V2G infrastructure per one EV, \$, P_{V2G} - useful capacity, kW, $T_{\inf r}^{V2G}$ - infrastructure service life, years, 365 – number of regulation cycles in a year assuming that one EV takes part in one V2G regulation cycle per day, days/year, λ_{V2G} - the ratio of output energy per cycle of V2G- regulation to battery power, kWh /(kW·cycle).

Expenses associated with EV battery deterioration resulting from EV travel are calculated as follows:

$$C_{bat.EV}^{year} = \frac{C_{EV}^{bat} \cdot W^{bat}}{T_{bat}}, \tag{7}$$

where C_{EV}^{bat} - specific cost of EV battery, \$/kWh, W^{bat} - EV battery capacity, kWh, T_{bat} – battery service life, years.

$$T_{bat} = \frac{N_{cycle}}{365}, \tag{8}$$

where N_{cycle} - battery lifetime number of cycles, 365 – number of battery operating cycles per year, cycles/year.

3.2. Expenses for BEV kilometrage with buffering of electric energy in stationary storage facilities

In this scheme, buffering of electric energy is implemented by stationary storage facilities located in the energy system (figure 2), necessary for recharging of other electric vehicles.

When stationary storage facilities in the system are used for leveling peak loads caused by fast charging of electric vehicles, the cost of electricity will be the sum of expenses for purchase of electricity, expenses for usage of storage facilities, and depreciation cost of the battery normalized to the run of 100 km. The calculation procedure is generally similar to V2G, but instead of the expenses for EV battery depreciation, the expenses for stationary storage facilities depreciation are taken into account.

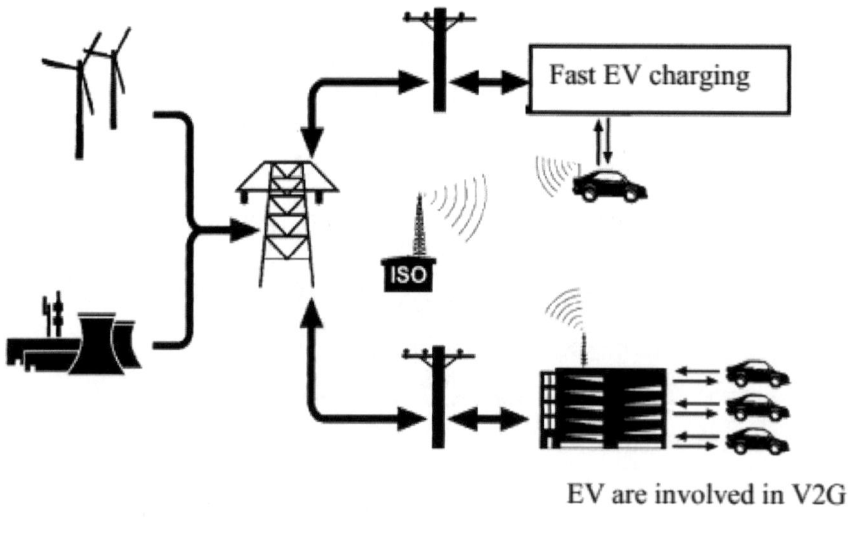

Figure 2. Scheme of EV charging using stationary storage facilities.

3.3. Expenses for running of an electric vehicle with air-aluminum electrochemical generator

Cruising range of the electric vehicle on aluminum fuel [2, 3, 5] is not less than 350 km without recharging of the electrochemical generator. This is possible due to separation of the energy source and the power source – figure 3. Substantial quantity of stored energy is provided by the air–aluminum electrochemical generator (AA ECG), while the peak power necessary during speed-up and the recuperation of braking energy is provided by the rechargeable battery.

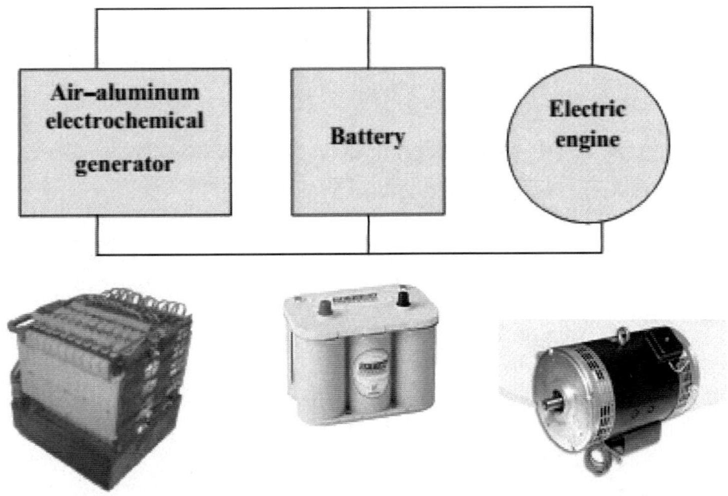

Figure 3. Combined traction power unit based on air–aluminum electrochemical generator [2].

Traction power units of such type, recharging procedure and associated equipment are described in detail in the monograph [2].

AA ECG, rechargeable battery and inverter system comprise the integrated power unit (IPU) of an electric vehicle. AA ECG anodes and electrolyte are replaced as needed. The procedure of replacement of AA ECG is comparatively simple, and the replacement station is not much more complicated than a tire shop.

AA ECG anodes are made of aluminum with alloying additives – indium and tin. Cost of the anode alloy may be reduced if the closed fuel cycle is arranged. The exhausted electrolyte containing products of electrochemical reaction – Al hydroxide – is sent to alumina–aluminum enterprise for restoration of original aluminum. If the closed cycle of energy carrier is organized, the overall cost of AA ECG recharging is low, allowing the EVs on aluminum fuel to compete successfully with battery EVs and hydrogen EVs. More detailed economic calculations and description of the closed fuel cycle are given in [3, 5, 6].

Expenses associated with consumption of aluminum, electricity, IPU depreciation are calculated by the formula:

$$C_{100\kappa m.AA} = (\frac{C_{fuel.AA}^{year}}{L^{year}} + \frac{+C_{AA}^{year}}{L^{year}}) \cdot 100 , \quad (9)$$

where $C_{fuel.AA}^{year}$ - yearly cost of aluminum for the EV, \$/year; C_{AA}^{year} - normalized expenses for EV IPU, \$/year, 100 – coefficient for normalizing the expenses per 100 km.

$C_{fuel.AA}^{year}$ is calculated basing on the cost of anode alloy being the consumable component in AA ECG C_{al}, \$/kWh, and yearly energy expenditures:

$$C_{fuel}^{year} = C_{al} \cdot W_{el}^{year} , \quad (10)$$

Annual electricity consumption is calculated in accordance with (3).

$$C_{AA}^{year} = \frac{C_{bat} + C_{AA}}{T_{AA}}, \tag{11}$$

where C_{bat} - cost of rechargeable battery being the component of IPU, \$, C_{AA} - cost of AA ECG, \$, T_{AA} - IPU service life which is determined by the service life of rechargeable battery, years.

Cost of rechargeable battery being the component of IPU:

$$C_{bat} = C_{EV}^{bat} \lambda_{EV} P^{EV}, \tag{12}$$

where P^{EV} - capacity of EV power unit, kW, λ_{EV} - ratio of useful energy during one cycle of battery discharge to EV maximum power, kWh/kW.

Cost of AA ECG being the component of IPU:

$$C_{AA} = C_{AA}^{kWh} \cdot W^{AA}, \tag{13}$$

where C_{AA}^{kWh} - specific cost of AA ECG, \$/kWh, W^{AA} - energy intensity of AA ECG, kWh.

$$W^{AA} = \frac{L_{day} \cdot v}{100}, \tag{14}$$

where L_{day} - EV maximum run until replacement of AA ECG anodes, km, v - EV specific electricity consumption, kWh/100 km, 100 – coefficient for normalizing the expenses per 100 km.

3.4. Adjustment of cost indicators for inflation

The above calculations may be adjusted for the time-related devaluation of money by applying *CRF* (capital return factor) - %/year, represented by the formula [8]:

$$CRF = \frac{d}{1-(1+d)^{-T}}, \tag{15}$$

where d – weighted average cost of capital, percent per year; T – service life of the device, years.

4. Assumptions

Assumptions for calculations are given in table 1.

Table 1. Assumptions for calculations.

Parameter	Designation	UOM	Value	Reasoning
General data				
Discount factor	d		10%	[8]
Cost of electricity	C_{el}	\$/kWh	0,1	[10]
Energy efficiency of the storage system	η		73%	[11]
Battery depth of discharge	DoD		0.8	
Data related to V2G system				
Specific cost of EV battery participating in V2G regulation. Current / prospective value	$C_{bat.EV}$	\$/kWh	485 / 128	[1]

Parameter	Designation	UOM	Value	Reasoning
Battery resource	N_{cycle}	cycles	4 000	[12]
Battery power	P^{EV}	kW	70	[12]
Ratio of battery energy to battery power when participating in V2G regulation	λ_{V2G}	kWh / (kW·cycle)	1.5	[12]
Cost of electric grid infrastructure for V2G regulation. Capacity of EV invertor: 15 kW.	$C_{\inf r}^{V2G}$	$/ EV	1 900	[11]
Infrastructure service life	$T_{\inf r}^{V2G}$	years	30	[3]
Stationary storage facility				
Lifetime number of cycles	N_{cycle}	cycles	4 000	[14]
Battery cost	$C_{bat.ESS}$	$/kWh	1 200	[14]
Cost of infrastructure	$C_{\inf r.ESS}$	$/kW	125	[14]
Operating costs	$C_{oper.ESS}^{year}$	$/kW·year	10	[14]
Ratio of power and capacity of the storage facility	λ_{ESS}	kWh /(kW·cycle)	0.5	[12]
Infrastructure useful life	$T_{\inf r}^{ESS}$	years	30	
Electric vehicle				
Energy consumption for running of the electric vehicle	v	kW·h/100 km	20	[2]
Maximum run before recharging	L_{day}	km	350	
Ratio of power and capacity of EV battery on the run	λ_{EV}	kW / kWh	0.25	[12]
Air–aluminum electrochemical generator				
Specific cost of the generator	C_{AA}	$/kWh	77	[6]
Specific cost of anode plates	C_{al}	$/kWh	1.0	[6]

5. Results

The dependence of expenses on energy and depreciation of EV power unit on kilometrage is shown in figures 4 and 5.

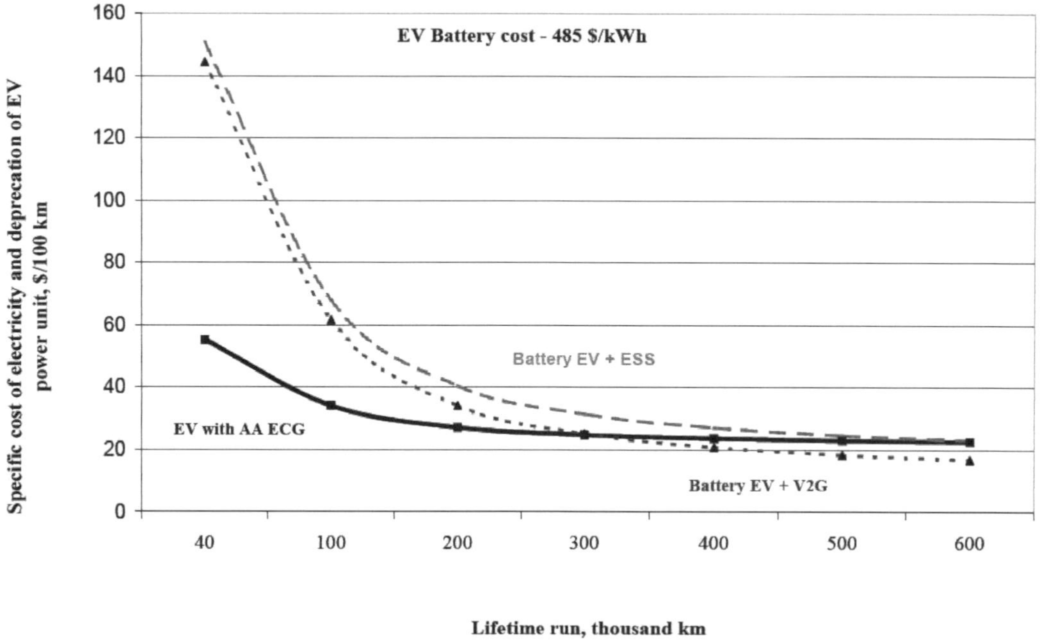

Figure 4. Dependence of expenses on energy and depreciation of EV power unit on kilometrage. Cost of EV battery – 485 $/kWh.

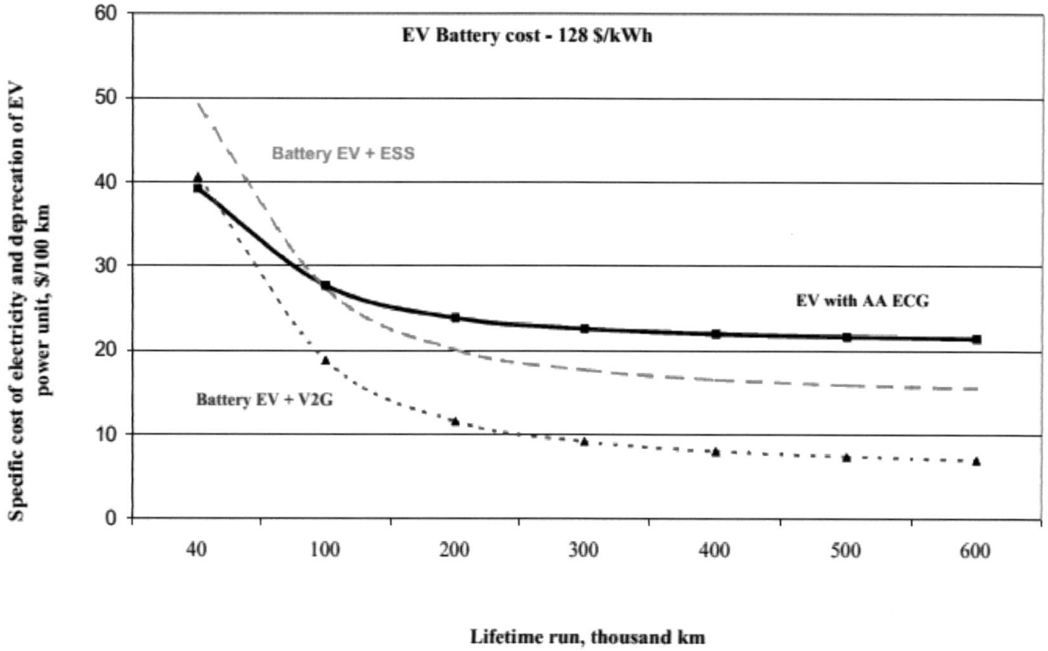

Figure 5. Dependence of expenses on energy and depreciation of EV power unit on kilometrage. Cost of EV battery – 128 $/kWh (minimum value forecasted for the year 2025).

6. Conclusion

1. The most economical option for providing fast charging of electric transport is the buffering of electricity in the batteries of electric vehicles. This option is feasible when sufficiently large number of electric vehicles is present within a limited area. However, permanent connection of idle electric vehicles to the grid must be provided in this case, which is not always possible.

2. Usage of high-capacity stationary storage facilities allows avoiding permanent connection of EVs to the charging terminal. High-capacity stationary storage facilities support the operation of fast charging terminals. At the same time they solve the problem of integration and stabilization of energy streams coming from distributed energy sources, in particular, from energy generators using renewable energy sources.

3. The competitiveness of electric vehicles employing electrochemical generators will largely depend on the dynamics of decrease of batteries cost. Aluminum as energy carrier may be produced at a sufficient distance from large cities, using cheap electricity produced by high-capacity regional power plants. The energy carrier is transported to the places of consumption, and the products of electrochemical reaction are returned to the production cycle for metal recovery. In this case the refueling infrastructure is similar to the existing system of petrol refueling stations, but simpler, cheaper and more eco-friendly.

Acknowledgments

The study was supported through a grant of the Russian Science Foundation (project No. 14-50-00124).

References

[1] Average Price of Electricity to Ultimate Customers by End-Use Sector [Online] Available: https://www.eia.gov/electricity/monthly/epm_table_grapher.php?t=epmt_5_6_a [Accessed: 10-Oct-2017]

[2] Zhuk A Z, Klejmenov B V, Fortov V E and Shejndlin A E 2012 *Jelektromobil' na aljuminievom toplive* (Moscow: Nauka)

[3] Turton H and Moura F 2008 Vehicle-to-grid systems for sustainable development: An integrated energy analysis *Technological Forecasting & Social Change* **75** 1091–108

[4] Peterson S and Michalek J 2013 Cost-effectiveness of plug-in hybrid electric vehicle battery capacity and charging infrastructure investment for reducing US gasoline consumption *Energy Policy* **52** 429-38

[5] Yang S and Knickle H 2002 Design and analysis of aluminum/air battery system for electric vehicle *Journal of Power Sources* **112** 162-73

[6] Ilyukhina A, Kleymenov B and Zhuk A 2017 Development and study of aluminum-air electrochemical generator and its main components *Journal of Power Sources* **342** 741-9

[7] Ereev S Y and Patel M K 2012 Standardized cost estimation for new technologies (SCENT) – methodology and tools *Journal of Business Chemistry* **9** 31-48

[8] Sandia National Lab, 2011. Energy storage systems cost update. A study for DOE Storage systems program [Online]. Available: http://prod.sandia.gov/techlib/access-control.cgi/2011/112730.pdf [Accessed: 02-Oct-2017]

[9] National energy technology laboratory, 2010. Investment Decisions for Baseload Power Plants [Online]. Available: http://www.netl.doe.gov/File%20Library/Research/Coal/ewr/water/InvestmtDecsnsBsldPP4.pdf [Accessed: 12-Oct-2017]

[10] IEA. Electric Power Monthly with Data for May 2017 [Online]. Available: https://www.eia.gov/electricity/Monthly/pdf/epm.pdf [Accessed: 08-Oct-2017]

[11] Global EV Outlook 2017 [Online]. Available: http://www.iea.org/publications/freepublications/publication/global-ev-outlook-2017.html [Accessed: 12-Semp-2017]

[12] Hein R, Kleindorfer P and Spinler S 2012 Valuation of electric vehicle batteries in vehicle-to-grid and battery-to-grid systems *Technological Forecasting & Social Change* **79** 1654–71

[13] Forrest K E, Tarroja B, Zhang L, Shaffer B and Samuelsen S 2016 Charging a renewable future: The impact of electric vehicle charging intelligence on energy storage requirements to meet renewable portfolio standards *Journal of Power sources* **336** 63-74
[14] Zhuk A Z, Buzoverov E A and Sheindlin A E 2015 Distributed energy storage systems on the basis of electric vehicle fleets *Therm. Eng.* **62** 1–6